Umwelt
78,- €
1/19

Practice of Optimisation Theory in Geotechnical Engineering

Zhen-Yu Yin · Yin-Fu Jin

Practice of Optimisation Theory in Geotechnical Engineering

同济大学出版社
TONGJI UNIVERSITY PRESS

Springer

Zhen-Yu Yin
Department of Civil and Environmental Engineering
Hong Kong Polytechnic University
Hong Kong, China

Yin-Fu Jin
Department of Civil and Environmental Engineering
Hong Kong Polytechnic University
Hong Kong, China

This book was funded by B & R Book Program.

ISBN 978-981-13-3407-8 ISBN 978-981-13-3408-5 (eBook)
https://doi.org/10.1007/978-981-13-3408-5

Jointly published with Tongji University Press, Shanghai, China
The print edition is not for sale in Mainland China. Customers from Mainland China please order the print book from: Tongji University Press.

Library of Congress Control Number: 2018968095

This Springer imprint is published by the registered company Springer Nature Singapore Pte Ltd.
The registered company address is: 152 Beach Road, #21-01/04 Gateway East, Singapore 189721, Singapore

To our families, teachers and students.

Foreword

Soil, whether it be this natural material used for a foundation or an artificial material used for roads and dams, behaves in such highly complex ways given its polyphasic nature that it is difficult to always model its behaviour with accuracy and dependency. In the geotechnical field, serious accidents occur from shortcomings of initial design because the engineer has not fully appreciated the vital relationship between building strategies and soil behaviour. Therefore, at stake from the beginning of an engineering project is the selection of the appropriate constitutive soil model capable of analysing globally the geotechnical construction system.

A great number of models can be found in the literature. But not enough research has been done to analyse the pertinence of these models in relation to the projects they undertake (foundations, tunnels, retaining walls…). Which model is best may be an interesting or perhaps naive theoretical question but as a practical issue, this question is rarely treated with seriousness. To do so, one must take into account different factors such as the difficulties due to parameter determination, the importance of errors induced by the model itself, and the consequences for the calculation of a project as a result of using a determined model.

A simplified model may facilitate identifying parameters, but it could also lead to making faulty assumptions in numerical simulations for the reason that the hypotheses emitted in the first place are overly simplistic. A more complex model, we know, contains, in general, a greater number of parameters but in the end, this may push the user towards more objective determination procedures which simply do not have to depend upon the user's judgment. Among these procedures can be found the different optimisation methods that reveal which principal traits of soil behaviour are necessary to be reproduced by a particular model.

This book proposes a basis for selecting a model through the use of genetic algorithms. The authors have studied the effectiveness of different algorithms and have conceived a new algorithm that suggests the best solution for any given problem. Numerous examples are presented to allow the reader to become familiarised with different identification procedures and their subsequent application to practical examples. Different examples based on approaches more or less complex (elastic-perfectly plastic, non linear hardening, concept of critical state, double yield surface, consideration of inherent or loading induced anisotropy, time-dependent behaviour of fine materials) are discussed. The parameter optimisation procedure has been applied to each of these approaches with the avail of a database containing oedometric and drained and undrained triaxial tests.

These methods have equally served to estimate the nature of the tests, thereby providing the most adequate information for parameter identification and the number of tests needed to reach a sufficient level of precision for determining the optimised value of the parameters.

Another feature of this book is its analysis of existing constitutive models and each of their applicability to solve day-to-day geotechnical problems. In pointing out the advantages and limitations of the currently most widely used models, it provides a helpful guide to engineers. We all know that improper use of any constitutive model can lead to a faulty analysis of a geotechnical problem. Indeed, the paucity of geotechnical research concerning an adequate utilisation of these models has caused such problems in the past that the profession has within the last decades introduced the notion of risk analysis to remedy the most severe of problems generated by faulty design procedures. The aim of this book is to provide the objective elements for choosing a model that will reduce the geotechnical risks related to the conception of design.

The two authors have brought to this work their considerable competence in the field of numerical modelling. They have provided with clarity the source codes for optimisation algorithms and constitutive models which will enable readers to use this work for their own applications.

Finally, as a close witness of Zhen-Yu Yin's impressive academic career since the time he came to study in France at the *Ecole Centrale de Nantes*, I am pleased that he has in turn transmitted his insight and experience to a younger generation of researchers. The present book written with Yin-Fu Jin is an admirable example of a successful teacher–student collaboration.

I am confident that this book, which many readers will recognise as the fruit of a certain school of thought, will benefit engineers and scientists alike. It is my hope that all students interested in this topic will find here a stimulating introduction to the mechanics of geomaterials. The text of this Foreword has been translated from French to English by Pearl-Angelika Lee.

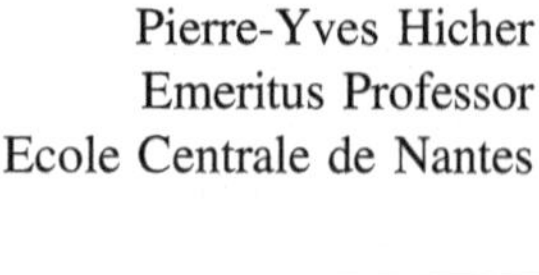

Nantes, France
January 2019

Pierre-Yves Hicher
Emeritus Professor
Ecole Centrale de Nantes

Preface

Large-scale infrastructure constructions are increasingly carried out around the world involving many engineering issues. However, for a long time, the engineering practice is out of line with the existing theory, and the advanced geomechanical theory could not be well applied to the engineering design, construction, and operation and maintenance, which leaves a safety hazard for the actual project and creates a resource pole.

The book is written back to 2013, when Prof. Hong-Wei Huang of Tongji University, China, and Prof. Pierre-Yves Hicher of Ecole Centrale de Nantes (ECN), France, were conducting an HSFC-ANR joint project: "The selection of soil constitutive model based on risk analysis". Dr. Zhen-Yu Yin happened to be one of the main investigators of the project mainly responsible for the parameter identification of constitutive models. Dr. Yin-Fu Jin, as a joint doctoral student of Shanghai Jiao Tong University (SJTU) and Ecole Centrale de Nantes, just went to ECN in September 2013 for exchange. Under the guidance of Dr. Zhen-Yu Yin and Prof. Shui-Long Shen of SJTU, Dr. Yin-Fu Jin first carried out a review of the identification of parameters based on optimization theory and then gradually went deep into the constitutive model selection, from the beginning of the laboratory test, slowly transition to the field construction until the draft today.

In recent decades, various advanced constitutive models for soils have sprung up. The identification of parameters and selection of constitutive models have become one of the hot spots and difficulties in geotechnical engineering, even the engineers are at a loss. Therefore, the purpose of this book is to deal with how to intelligently identify models and their parameters, and how to apply them to engineering practice. To this end, the book first systematically introduces the basic principles and application fields of various optimization algorithms, compares the advantages and disadvantages of several typical optimization algorithms in parameter identification, and then presents how to improve the performance of optimization-based parameter identification taking the genetic algorithm and differential evolution algorithm as examples. Several typical geotechnical problems are selected to illustrate the practice of optimization theory, such as (1) the development of empirical correlations for key mechanical parameters of soils;

(2) the parameter identification of natural soft structured clay and granular materials with proposing the least number of objective tests and the type of objective tests; (3) the model selection among various advanced models and critical-state-related formula; (4) the improvement of multi-objective differential evolution algorithm with its application in updating of predictions during the engineering construction.

With the rapid development of computer technique, the optimization theory has made important progress, which has greatly promoted the development of intelligent geotechnics. This book is a rare monograph, which systematically describes the optimization theory in geotechnical practice. This book brings together the authors' original results in optimization theory, soil testing, constitutive models, numerical analysis, etc. Starting from the actual case of model selection and parameter identification, readers can quickly and accurately establish the concept of how to apply optimization theory to solve engineering problems. Moreover, this book uses a combination of theoretical analysis and case discussion to facilitate the reader's understanding and application. In addition, this book provides MATLAB source codes for a variety of optimization algorithms and ABAQUS UMAT for a variety of constitutive models, which can be directly used to analyze or exercise for readers.

In addition, model selection and parameter identification in geotechnical engineering is a challenging task. It not only requires an outstanding algorithm for searching the global maximum value, but also challenges a variety of complex engineering problems. The optimization-based model selection and parameter identification introduced in this book can provide technical support and useful reference for the application of advanced constitutive theories in practice.

In view of the limited theoretical and technical level of the authors, there are inevitable flaws in the book. Readers and peers are expected to criticize and correct.

Hong Kong, China Zhen-Yu Yin

Yin-Fu Jin

Acknowledgements

In the process of writing this book, I have received much sincere guidance and help from experts and colleagues. Here, I would like to express my sincere gratitude to Prof. Pierre-Yves Hicher; Ms. Angelika (Jia-Yuan) Lee; Dr. Yvon Riou; Dr. Christophe Dano from Ecole Centrale de Nantes, France; Prof. Hong-Wei Huang and Prof. Dong-Mei Zhang from Tongji University, China; and Prof. Shui-Long Shen from Shanghai Jiao Tong University.

The publication of this book has been supported by the National Natural Science Foundation of China (51579179), and the French National Research Agency Program (ANR-RISMOGEO).

About This Book

This book systematically introduces the application of optimization algorithms in geotechnical engineering and summarizes the advantages and disadvantages of different optimization methods in identifying parameters and other aspects. Five typical optimization algorithms (genetic algorithm GA, simulated annealing algorithm SA, particle swarm optimization PSO, differential evolution algorithm DE, and artificial bee colony algorithm ABC) are first compared in the identification of Mohr–Coulomb parameters from the pressuremeter test and excavation. The improvement and verification of real-coded genetic algorithm and differential evolution algorithm are detailed. Taking the secondary compression coefficient of remolded clay as an example, the application of optimization method in evolutionary polynomial regression technique is introduced. A method for determining the parameters of creep combined with destructuration of natural soft clay based on optimization method is proposed. A procedure for identifying parameters of model for granular materials considering grain breakage effect is also proposed. The type, quantity, and strain level of objective tests in parameter identification are discussed using optimization method, and the selection of the critical-state-related formula for critical-state-based sand models is also discussed. Taking the excavation as an example, the improvement of multi-objective evolutionary difference algorithm and its application in updating of prediction are detailed. Finally, an optimization-based parameter identification platform is developed and presented. This book selects the simple and easy-to-understand optimization algorithm, so that readers can master the optimization method to analyze and solve the problem in a short time. In addition, this book provides MATLAB codes for various optimization algorithms and ABAQUS UMAT source codes for constitutive models so that readers can directly analyze and practice.

This book can be used as a postgraduate textbook for civil engineering, hydraulic engineering, transportation, railway, engineering geology, and other majors in colleges and universities, and as an elective course for senior undergraduates. It can also be used as a reference for relevant professional scientific researchers and engineers.

Contents

About the Authors

Dr. Zhen-Yu Yin was graduated from Zhejiang University, China, in 1997 for B.Sc. and then worked as Engineer at Zhejiang Jiahua Architecture Design Institute, China, for 5 years. He obtained his M.Sc. and Ph.D. in geotechnical engineering at Ecole Centrale de Nantes, France, in 2003 and 2006, respectively. After Ph.D., he worked as Research and Teaching Fellow mainly at Helsinki University of Technology (Finland), University of Strathclyde (UK), University of Massachusetts (Umass-Amherst, USA), Shanghai Jiao Tong University (China), Tongji University (China), and Ecole Centrale de Nantes (France) and now is Associate Professor of geotechnical engineering at The Hong Kong Polytechnic University.

His research interests include laboratory testing for soil properties and behaviors; constitutive modeling from micro to macro; multi-scale and multi-physics modeling for geotechnical engineering; soil–structure interaction (SSI) and macro-element modeling; numerical analyses in geotechnical engineering; and artificial intelligence in geotechnical engineering. Since 2008, he has published over 100 articles in peer-reviewed international journals.

Dr. Yin-Fu Jin received the bachelor degree in civil engineering from Northwest A&F University in 2011 and the Ph.D. from Ecole Centrale de Nantes in 2016. Then, he did his postdoctoral research at the University of Macau on 2017–2018, and moved to The Hong Kong Polytechnic University as Postdoctoral Research Fellow up to now. He is mainly engaged in the research of soil mechanics, geotechnical engineering, numerical analysis, and engineering application of artificial intelligence. Since 2013, he has published more than 20 papers in peer-reviewed international journals.

Abbreviations

a	Constant of fines content effect in silty sand (SIMSAND+fr)
a	Target inclination of yield surface related to volumetric strain (ASCM)
A_d	Constant of magnitude of stress–dilatancy (0.5–1.5)
b	Constant controlling the amount of grain breakage (SIMSAND+Br)
b	Target inclination of yield surface related to deviatoric plastic strain
C_{aei}	Intrinsic secondary compression index (remolded clay)
$\mathbf{D}$	Stiffness matrix of material
E	Young's modulus
e, e_0	Void ratio and initial void ratio
E_0	Referential Young's modulus (dimensionless)
e_{c0}	Initial critical state void ratio (SIMSAND); virgin initial critical state void ratio before breakage
e_{cuf}	Fractal initial critical state void ratio due to breakage
e_d	General shear strain
E_h, E_v	Horizontal Young's modulus and vertical Young's modulus
$e_{hc,c0}$	Initial critical state void ratio of pure fine soils (f_c = 0%)
$e_{hf,c0}$	Initial critical state void ratio of pure coarse soils (f_c = 100%)
e_{max}	Maximum void ratio
E_u	Undrained Young's modulus
f_{th}	Threshold fines content from coarse to fine grain skeleton (20–35%)
f	Fines content
G	Shear modulus
G_0	Referencial shear modulus

G_{vh}	Shear modulus
I_1, I_2, I_3	The first, second, and third invariants of the stress tensor
I'_1, I'_2, I'_3	The first, second, and third invariants of the strain tensor
J_1, J_2, J_3	The first, second, and third invariants of the deviatoric stress tensor
J'_1, J'_2, J'_3	The first, second, and third invariants of the deviatoric strain tensor
K	Bulk modulus
K_0	The coefficient of earth pressure at rest
k_p	Plastic modulus-related constant in SIMSAND; plastic modulus-related parameter in ASCM
K_w	Bulk modulus of water
M	Constraint modulus in elasticity; slope of critical state line in p'-q plane
m	Constant of fines content effect in sandy silt
M_c	Slope of critical state line in triaxial compression in p'-q plane
n	Porosity of soil; elastic constant controlling nonlinear stiffness
n_d	Phase transformation angle-related constant ($\approx$1)
n_p	Peak friction angle-related constant ($\approx$1)
p'	Mean effective stress
p_{at}	Atmosphere pressure
p_{b0}	Initial bonding adhesive stress
p_{c0}	Initial size of yield surface; initial size of yield surface of grain breakage (SIMSAND+Br)
p_{excess}	Excess pore pressure
p_{steady}	Steady pore pressure
q	Deviatoric stress
R_d	Ratio of mean diameter of sand to silt D_{50}/d_{50}
R_α	Stress relaxation coefficient
s_{ij}	Deviatoric stress tensor
u_x, u_y, u_z	Displacements
α_{k0}	Initial inclination of yield surface
β	Rate-dependency coefficient
χ_0	Initial bonding ratio
δ_{ij}	Kronecker symbol
ε_1, ε_2, ε_3	Principle strains
ε_a, ε_r	Axial strain and radial strain
ε_{ij}	Strain tensor
ε_m	Mean strain
ε_v	Volumetric strain
γ_{xy}, γ_{yx}, γ_{yz}, γ_{zy}, γ_{zx}, γ_{xz}	Engineering shear strains
φ	Friction angle

κ	Swelling index of the isotropic compression test (in e-lnp' plane)
κ_i	Intrinsic swelling index (of remolded soil, in e-lnp' plane)
λ	Lame constant in elasticity; compression index (in e-lnp' plane); constant controlling the nonlinearity of CSL in SIMSAND
λ'	Compression index under the plane of log$e$-logp'
λ_i	Intrinsic compression index (of remolded soil, in e-lnp' plane)
v_{u}	Undrained Poisson's ratio
v'_{vh}	Horizontal Poisson's ratio
v'_{vv}	Vertical Poisson's ratio
θ	Lode angle
ρ	Constant controlling the movement of CSL
σ_{a}, σ_{r}	Axial stress and radial stress
σ_{ij}	Stress tensor
$\sigma_{\mathrm{m}}(p)$	Mean stress
σ_{n}, σ_{h}	Vertical and horizontal stresses
σ_{p0}	Preconsolidation pressure
σ_w	Pore water pressure
σ_x, σ_y, σ_z	Normal stresses
σ_1, σ_2, σ_3	First, second, and third principle stresses
τ	Reference time (oedometer test τ = 24 h) (ANICREEP)
τ_{xy}, τ_{yx}, τ_{yz}, τ_{zy}, τ_{zx}, τ_{xz}	Shear stresses
υ	Poisson's ratio
ω	Absolute rotation rate of the yield surface
ω_{d}	Rotation rate of the yield surface related to the deviatoric plastic strain
ξ	Constant controlling the nonlinearity of CSL (SIMSAND); absolute rate of bond degradation
ξ_{b}	Degradation rate of the interparticle cohesive bonding
ξ_{d}	Constant controlling the deviatoric strain-related bond degradation rate
ψ	Dilatancy angle

List of Figures

List of Tables

Chapter 1
Need of Optimization Theory in Geotechnical Engineering

1.1 Engineering Requirements Overview

Due to the complex mechanical properties of the soil, the variability is large, and the risk level of geotechnical engineering is high. The major safety accidents occurring are mainly geotechnical engineering. Compared with structural engineering, the proportion of accidents caused by design in geotechnical engineering is much higher than that of structural engineering. For example, in structural engineering, accidents caused by design errors account for about 3% of total accidents [1]; in deep foundation pits, accidents caused by design errors account for up to 50% of total accidents [2]. In order to reduce geotechnical engineering accidents and reduce the risk level of geotechnical engineering, we should first consider how to conduct risk analysis and control from the design level to minimize the probability of occurrence of risks.

The constitutive model of soil is the basic mechanical model for simulating the stress–strain relationship of soil, and it is also the key to geotechnical analysis and design. At present, the researchers have proposed hundreds of different soil constitutive models (see Shen [3], Li [4], Zheng et al. [5], Yao et al. [6], Huang et al. [7], etc.). However, due to the complexity of soil, each model has its own limitations, and no model can describe the properties of all types of soils. Potts [8] pointed out that some of the most commonly used constitutive models may also have obvious unreasonable predictions when analyzing conventional engineering, resulting in "traps" in numerical simulation analysis. For example, the Mohr–Coulomb (MC) model is one of the most commonly used constitutive models in geotechnical engineering; however, if MC is used for excavation analysis, the ground always produces upward displacement, which is exactly the opposite of the actual situation [9, 10]; for another example, the modified Cam-Clay (MCC) model is commonly used in simulating clay behaviors; when using the MCC model to predict the long-term settlement of the tunnel, the obtained settlement is often too small, making the analysis results dangerous for tunnel construction. A lack of understanding of the applicability and limitations of constitutive models can lead to serious safety incidents. One of important reasons

Z. Yin and Y. Jin, *Practice of Optimisation Theory in Geotechnical Engineering*,
https://doi.org/10.1007/978-981-13-3408-5_1

for the instability of excavation of Nicoll Highway in Singapore in 2003 was the adoption of an inappropriate constitutive model [11].

Choosing different constitutive models will result in different numerical simulation results, which may lead to different engineering decisions, affecting the safety, economy, and risk level of geotechnical engineering. In geotechnical engineering analysis, the influence of constitutive model on decision making and its related consequences should be fully considered. The existing researches have focused on how to propose a more accurate soil constitutive model, but there are still little systematic studies on the applicability evaluation, selection, and application of existing models. In practical applications, the selection of constitutive models is often determined according to user preferences and past experience, which is quite subjective. Neglecting the selection of constitutive model has become one of the important sources of risk for geotechnical engineering accidents.

The selection of constitutive model should first consider the model accuracy. Model accuracy can be measured by the magnitude of the model error which can be obtained by systematically comparing the measured data with the model prediction data. In the process of model error, the input parameters of used model also have certain uncertainties. Gilbert and Tang [12] first proposed a method for identifying the mean value of model errors considering the uncertainty of model parameters and applied the method to coastal engineering; Juang et al. [13] proposed a model recognition technique for soil liquefaction prediction. Zhang et al. [14] also proposed a model error recognition method based on Bayesian theory framework and used this method to calculate the accuracy of slope stability. Thus, significant advances have been made in model identification techniques that take into account the uncertainty of model input parameters.

When using theoretical models for prediction, the accuracy of prediction is not only related to the accuracy of the model itself, but also to the accuracy of the parameters that the model needs to input. From Hooke's law (i.e., the elastic model, although not a true constitutive model of soil, but to some extent can be used as a constitutive model of soil for geotechnical calculations), to the Mohr–Coulomb model (an perfect elastoplastic constitutive model of soil), to models that can describe the nonlinear characteristics of the soil (such as the hardening soil model in PLAXIS), to models based on the critical state concept (such as the modified Cam-Clay model), and then to models based on the concept of bounding surface (see Prof. Dafalias' work [15, 16]), even to micromechanical-based models (e.g., Chang and Hicher [17]; Yin and Chang [18]) and models based on hypoplasticity (see Prof. Wu Wei's work [19–21]), the development of constitutive model is becoming more and more advanced, and the ability to describe the characteristics of soil is becoming stronger and stronger. The relationship between model parameters and increasingly complex constitutive formulas is shown in Fig. 1.1. When the model parameters increase, the number of parameters without physical meaning increases, which makes the difficulty of determining parameters more and more, as shown in Fig. 1.2. The increase of parameters and the increase of difficulty will restrict the application of advanced constitutive models in geotechnical engineering, which is not conducive to better solving geotechnical problems. For advanced models, the simple conventional lab-

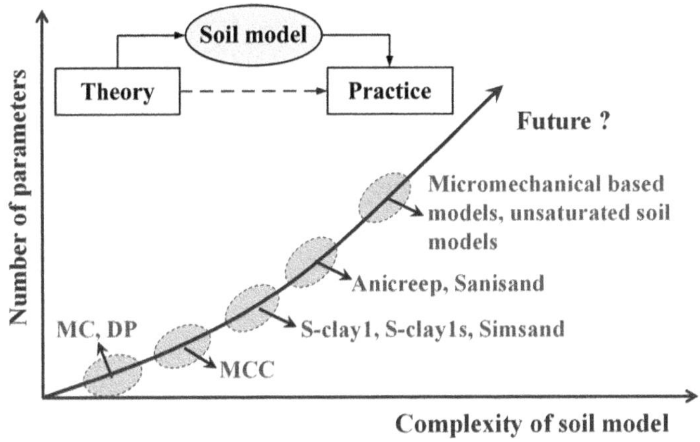

Fig. 1.1 Relationship between the complexity of the constitutive model and the number of parameters

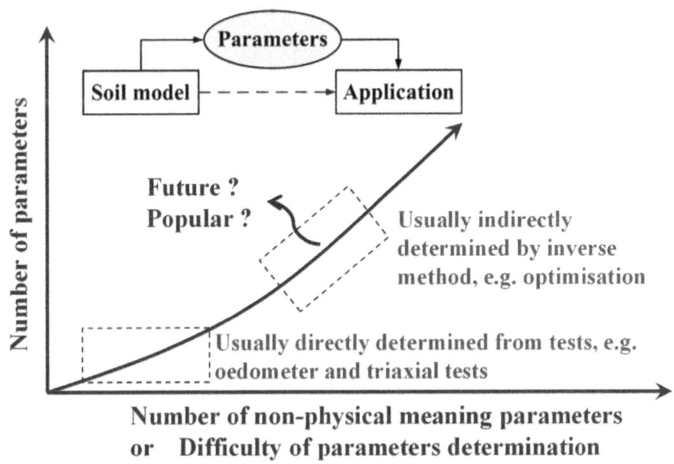

Fig. 1.2 Relationship between the number of model parameters and the number of parameters without physical meaning or the difficulty of parameters

oratory tests do not meet the need for parameter determination, so that some model parameters can only be determined by trial and error (Taiebat and Dafalias [22]). But the problem for determining parameters is that it is difficult to find a parameter that is satisfactory for all experiments, unless you can traverse all possibilities within the range of parameter (such as Monte Carlo method or other random methods). However, those determination methods result in waste of resources and increase the computational costs.

In recent decades, many researchers have studied how to obtain more precise parameters (Zentar et al., Rangeard et al., Dano et al., Yin and Hicher, and Papon et al. [23–27]). The shortcoming is that these studies mostly focus on relatively simple constitutive models, while the identification methods for other advanced soil model parameters remain to be studied. Therefore, finding a way to effectively identify the parameters is very helpful for promoting the application of soil constitutive models in geotechnical engineering, especially the advanced models, which in turn will accelerate the development of geotechnical constitutive theory. And the application of optimization methods can also help to improve the development of intelligent construction technology.

The adaptive selection and parameter identification method of constitutive model is one of the key problems faced by geotechnical engineering design, construction, operation, and maintenance. The selection method and parameter identification technology of the constitutive model will help to control the risk level of geotechnical engineering to minimize the accidents. Under the background of a large number of projects all the world, the study of this issue will help to improve the safety level of geotechnical engineering.

1.2 Overview of Parameter-Based Back Analysis Methods Based on Optimization

Hicher and Shao [28] distinguished three approaches, namely analytical methods, empirical correlations, and optimization methods, to determine soil parameters based on experimental data. Among these approaches, the inverse analysis by optimization has been successfully used in the geotechnical area [29–32] because it produces a relatively objective determination of the parameters for an adopted soil model, even of those that have no direct physical meaning (e.g., for the Mohr–Coulomb model, the Young modulus E model is simply an average secant modulus which stretched to describe the Hookean elasticity; the friction angle ϕ' reflect the angle of internal friction that is attained when failure just occurs in response to a shearing stress; the cohesion c indicates the interaction force among soil particles), and this approach can be applied to any testing procedure and to any constitutive model. For an inverse formulation of the parameter identification, the variables are the model parameters. A way to find their values is to simulate several sets of laboratory or field tests and to minimize the differences between experimental and numerical values of stresses, strains, and other typical data (e.g., void ratio, excess pore pressure, …). This type of problem is usually solved by using optimization techniques which can be divided into two categories, (1) deterministic techniques and (2) stochastic techniques, as shown in Fig. 1.3. However, the advantages and disadvantages of these optimization techniques are rarely systemically summarized and compared for the same geotechnical problem. Therefore, a review and comparative study are necessary for a good understanding of the differences between the various techniques, which may help select the appropriate optimization method to solve geotechnical engineering problems.

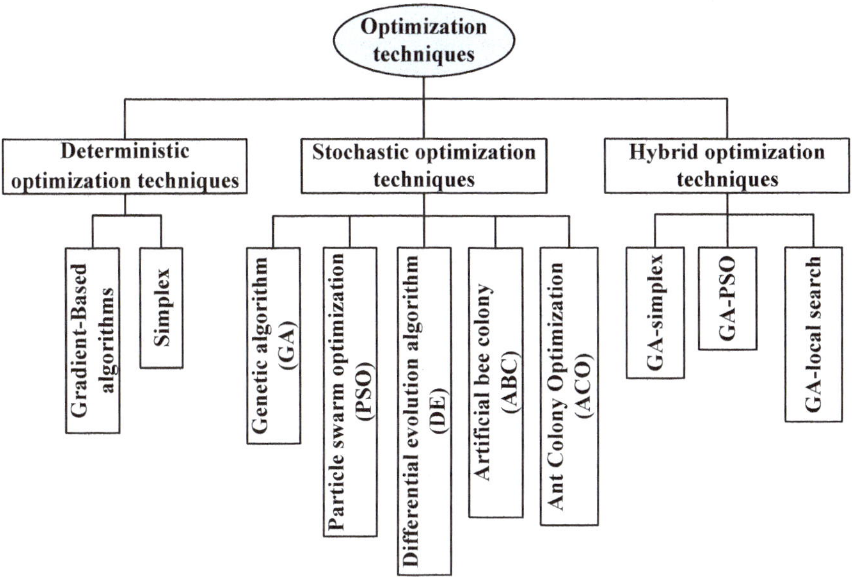

Fig. 1.3 Optimization algorithm classification

1.3 Summary

It is necessary to analyze and compare the differences between different algorithms on the same geotechnical engineering parameter identification problem, improve the disadvantages of existing algorithms based on the analysis and comparison results, and use the improved algorithm for advanced constitutive models or complex geotechnical engineering problems. Based on parameter identification, the most suitable model can be found among the existing constitutive models to deal with geotechnical engineering problems.

References

1. Faber MH (2007) Assessing and managing risks due to natural hazards. In: ISGSR2007 first international symposium on geotechnical safety and risk
2. Bian Y (2006) Selection of supporting system of deep excavation in soft soil area based on risk analysis. Tongji University
3. Shen J-J (1989) Development of constitutive modelling of geological materials (1985–1988). Rock Soil Mech 10(2):3–13
4. Li G-X (2006) Characteristics and development of Tsinghua elasto-plastic model for soil. Chin J Geotech Eng 28(1):1–10

5. Zheng Y-R, Duan J-L, Chen Y-Y (2000) Theory of yield surface and stress-strain relation in generalized plastic mechanics. Rock Soil Mech 21(3):305–308
6. Yao Y-P, Zhang B-Y, Zhu J-G (2012) Behaviors, constitutive models and numerical simulation of soils. China Civ Eng J 45(3):127–150
7. Huang M-S, Yao Y-P, Yin Z-Y, Liu E-L, Lie H-Y (2016) An overview on elementary mechanical behaviors, constitutive modeling and failure criterion of soils. China Civ Eng J 49(7):9–35
8. Potts DM (2003) 42nd Rankine lecture: numerical analysis: a virtual dream or practical reality? Géotechnique 53(6):535–573
9. Schweiger HF (1998) Results from two geotechnical benchmark problems. Springer, Vienna
10. Lim A, Ou C-Y, Hsieh P-G (2010) Evaluation of clay constitutive models for analysis of deep excavation under undrained conditions. J Geoeng 5(1):9–20
11. Simpson B, Nicholson D, Banfi M, Grose B, Davies R (2009) Collapse of the Nicoll Highway excavation. Singapore, Thomas Telford
12. Gilbert RB, Tang WH (1995) Model uncertainty in offshore geotechnical reliability
13. Juang CH, Yang SH, Yuan H, Khor EH (2004) Characterization of the uncertainty of the Robertson and Wride model for liquefaction potential evaluation. Soil Dyn Earthq Eng 24(24):771–780
14. Zhang J, Zhang LM, Tang WH (2009) Bayesian framework for characterizing geotechnical model uncertainty. J Geotech Geoenviron Eng 135(7):932–940
15. Dafalias YF (1986) Bounding surface plasticity. I: mathematical foundation and hypoplasticity. J Eng Mech 112(9):966–987
16. Dafalias YF, Herrmann LR (1986) Bounding surface plasticity. II: application to isotropic cohesive soils. J Eng Mech 112(12):1263–1291
17. Chang C, Hicher P-Y (2005) An elasto-plastic model for granular materials with microstructural consideration. Int J Solids Struct 42(14):4258–4277
18. Yin ZY, Chang CS (2009) Microstructural modelling of stress-dependent behaviour of clay. Int J Solids Struct 46(6):1373–1388
19. Wu W, Bauer E (1994) A simple hypoplastic constitutive model for sand. Int J Numer Anal Methods Geomech 18(12):833–862
20. Wu W, Bauer E, Kolymbas D (1996) Hypoplastic constitutive model with critical state for granular materials. Mech Mater 23(1):45–69
21. Wu W, Kolymbas D (2000) Hypoplasticity then and now. Constitutive modelling of granular materials, pp 57–105
22. Taiebat M, Dafalias YF (2008) SANISAND: Simple anisotropic sand plasticity model. Int J Numer Anal Methods Geomech 32(8):915–948
23. Dano C, Hicher PY, Rangeard D, Marchina P (2007) Interpretation of dilatometer tests in a heavy oil reservoir. Int J Numer Anal Methods Geomech 31(10):1197–1215
24. Papon A, Riou Y, Dano C, Hicher PY (2012) Single-and multi-objective genetic algorithm optimization for identifying soil parameters. Int J Numer Anal Methods Geomech 36(5):597–618
25. Rangeard D, Hicher PY, Zentar R (2003) Determining soil permeability from pressuremeter tests. Int J Numer Anal Methods Geomech 27(1):1–24
26. Yin ZY, Hicher PY (2008) Identifying parameters controlling soil delayed behaviour from laboratory and in situ pressuremeter testing. Int J Numer Anal Methods Geomech 32(12):1515–1535
27. Zentar R, Hicher P, Moulin G (2001) Identification of soil parameters by inverse analysis. Comput Geotech 28(2):129–144
28. Hicher P-Y, Shao J-F (2002) Modèles de comportement des sols et des roches: Lois incrémentales viscoplasticité, endommagememt: Hermès Science
29. Gioda G, Maier G (1980) Direct search solution of an inverse problem in elastoplasticity: identification of cohesion, friction angle and in situ stress by pressure tunnel tests. Int J Numer Methods Eng 15(12):1823–1848

30. Wood DM, Mackenzie N, Chan A (1992) Selection of parameters for numerical predictions. In: Predictive soil mechanics: proceedings of the wroth memorial symposium, Oxford, UK Thomas Telford, London, pp 496–512
31. Simpson AR, Priest SD (1993) The application of genetic algorithms to optimisation problems in geotechnics. Comput Geotech 15(1):1–19
32. Pal S, Wije Wathugala G, Kundu S (1996) Calibration of a constitutive model using genetic algorithms. Comput Geotech 19(4):325–348

Chapter 2
Optimization-Based Parameter Identification Theory

2.1 Introduction

In this chapter, the optimization techniques for identifying parameters in geotechnical engineering are reviewed. The identification methodology with its three main parts, i.e., error function, search strategy, and identification procedure, is first introduced and summarized. Then, the existing optimization methods are reviewed and classified into three categories with an introduction to their basic principles and applications in geotechnical engineering.

2.2 Methodology of Parameter Identification

The mathematical procedure of optimization consists essentially of two parts: (a) the formulation of an error function measuring the difference between model responses (or analytical responses) and experimental results and (b) the selection of an optimization strategy to enable the search for the minimum of this error function.

2.2.1 Formulation of an Error Function (Fitness Function)

For the optimization problem of identifying parameters of constitutive models based on experimental or observed data, the parameters of the constitutive model play the role of the variables to be optimized. Theoretically, more reliable model parameters can be obtained if many qualitatively different experimental tests from the database for the optimization. In order to carry out an inverse analysis, a function that can evaluate the error between the experimental and numerical results must be defined, and then be minimized.

Z. Yin and Y. Jin, *Practice of Optimisation Theory in Geotechnical Engineering*,
https://doi.org/10.1007/978-981-13-3408-5_2

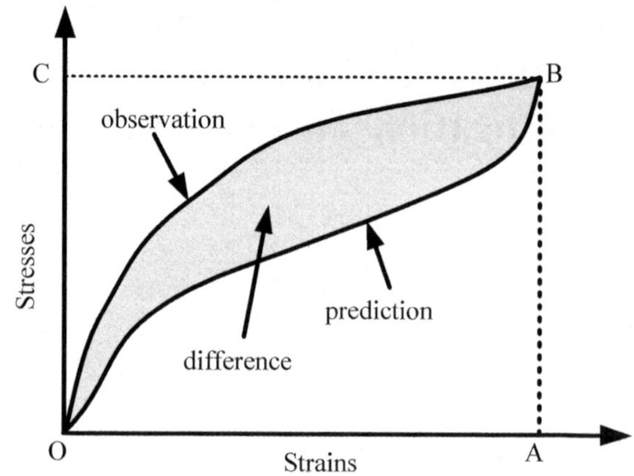

Fig. 2.1 Definition of an error function

For each test involved in the optimization, the difference between the experimental result and the numerical prediction is measured by a norm value, referred to as an individual norm, which forms an error function Error(x), as shown in Fig. 2.1,

$$\text{Error}(x) \rightarrow \min \tag{2.1}$$

where x is a vector containing the parameters to be optimized. Bound constraints are introduced on these variables,

$$x_l \leq x \leq x_u \tag{2.2}$$

where x_l and x_u are, respectively, the lower and upper bounds of x.

As the first step in the formulation of an error function, an expression for the individual norm (e.g., the deviatoric stress q) has to be established. In general, the individual norm is based on Euclidean measures between discrete points, composed of the experimental and the numerical results. The simplest error function can take the following expression:

$$\text{Error}(x) = \frac{1}{N}\left(\sum_{i=1}^{N}\left|U_{\text{exp}}^{i} - U_{\text{num}}^{i}\right|\right) \tag{2.3}$$

where N is the number of values; U_{exp}^{i} is the value of the measurement point i; U_{num}^{i} is the value of the calculation at point i.

Another formulation of the error function was introduced:

$$\text{Error}(x) = \frac{1}{N}\left(\sum_{i=1}^{N}\left(U_{\text{exp}}^{i} - U_{\text{num}}^{i}\right)^{k}\right)^{\frac{1}{k}} \tag{2.4}$$

where k is a non-null positive value with $k = 1$ for the sum of error at every point and $k = 2$ for the least square function.

However, Eq. (2.3) present some disadvantages when they are used for measuring the fitness between simulated and objective curves. For example, if the triaxial tests are selected as the objectives, poor performance of the simulation can result at small strain level if the same fitness is required at different strain levels, because the value of the deviatoric stress is smaller at a small strain level than at a high strain level. Additionally, the number of measured points in different objective curves could also affect the fitness.

In order to make the error independent of the type of test and the number of measurement points, an advanced error function can be adopted with two modifications of 100% and adding weight to each calculation point (Levasseur et al. [1]). The average difference between the measured and the simulated results is expressed in the form of the least square method,

$$\text{Error}(x) = \sqrt{\frac{1}{N}\sum_{i=1}^{N} w_i\left(\frac{U_{\text{exp}}^{i} - U_{\text{num}}^{i}}{U_{\text{exp}}^{i}}\right)^{2}} \times 100 \tag{2.5}$$

where w_i is weight for the calculation at point i.

The scale effects on the fitness between the experimental and the simulated results can be eliminated by this normalized formula. For many types of curves with variables of different order of magnitude involved in the objective, the error magnitude for different curves can be normalized to the same order by this formula. Additionally, the objective error calculated by this function is a dimensionless variable; thus, any difference in error can be avoided for different objectives with different variables, e.g., the void ratio and deviatoric stress in triaxial test.

The next step is to formulate a final norm, a total error function, based on the individual norms computed using the above methods for each experimental test involved in the optimization. Two different final norms have been used and either can be employed for the total error function. The maximum norm and the combined norm are defined as follows:

$$F_{\max} = \max_{1 \le i \le m}\left(\text{Error}^{i}\right) \text{ and } F_{\text{comb}} = m \cdot F_{\max} + \sum_{i=1}^{m}\left(\text{Error}^{i}\right) \tag{2.6}$$

where m is the number of experimental tests involved in the optimization and Error^{i} is the individual norm for Test number i.

Generally, deformation and stress are two extremely important indicators to represent the mechanical behavior of soils. For identifying soil parameters, the error function should involve these two important indicators. Therefore, the generalized individual error function can be expressed as follows

$$\min[\text{Error}(x)] = \min[\text{Error(stress)},\ \text{Error(deformation)}] \quad (2.7)$$

For mono-objective problems, the total error function is expressed as:

$$\text{Error\_total}(x) = \sum_{i=1}^{\text{Num}} (l_i \cdot \text{Error}(x)_i) \quad (2.8)$$

where *Num* is the number of individual errors; $\text{Error}(x)_i$ is the value of the individual error corresponding to the objective i. l_i is the weight factor with $\sum(l_i) = 1$. Finally, the set of parameters with the minimum error value can be selected as the optimal set.

For multi-objective problems, the final error can be expressed as follows,

$$\min[\text{Error}(x)] = \min\begin{bmatrix} \text{Error(stress)} \\ \text{Error(deformation)} \\ \ldots \end{bmatrix} \quad (2.9)$$

Several sets of parameters on the Pareto frontier can finally be found. The optimal parameters can be determined according to the actual requirements of the user.

2.2.2 *Selection of the Search Strategy*

After formulating the error function, the selection of a search strategy is the key step concerning whether the optimized solution can be found or not. The solution to an optimization problem is a vector x_0 which, for any $x_l \leq x \leq x_u$, satisfies the following condition, which is a global minimum:

$$F(x_0) \leq F(x) \quad (2.10)$$

However, most search strategies can guarantee finding a local solution. For obtaining a more accurate solution, a highly efficient optimization method with the ability to search for a global minimum should be adopted. Different optimizers applied in geotechnical engineering are introduced in Sect. 3.

2.2.3 *Procedure of Parameter Identification*

Whether the search strategy used in the optimization is simple or complex, a procedure with a clear structure is necessary and important for the successful identification of parameters. The function of the procedure is to conduct the error function and search strategies together. Therefore, the procedure should be presented before conducting the optimization. Calvello and Finno [2] gave a three-step procedure for a

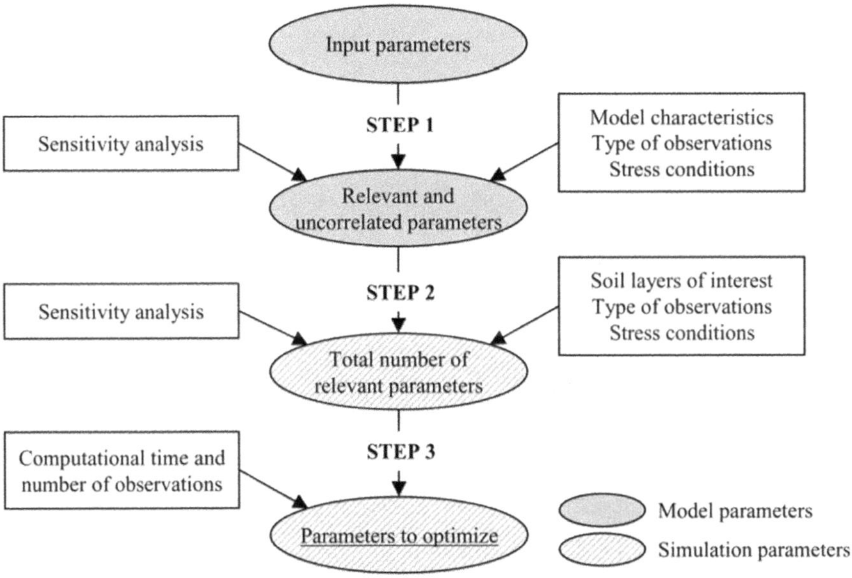

Fig. 2.2 Identification of soil parameters to optimize by inverse analysis

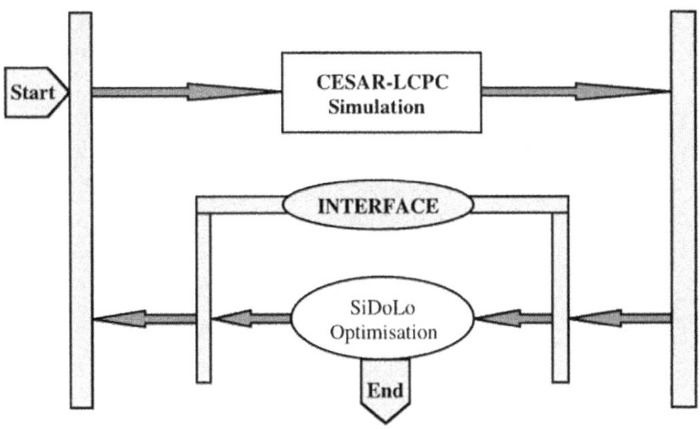

Fig. 2.3 Numerical process to identify soil parameters

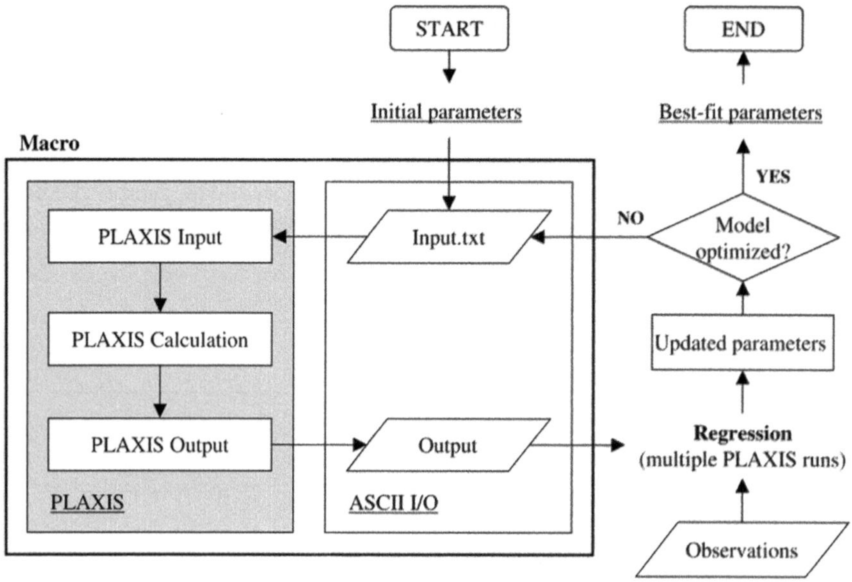

Fig. 2.4 Inverse analysis with UCODE and PLAXIS

general identification of soil parameters, as shown in Fig. 2.2; Zentar and Hicher [3] presented a simplified procedure to combine the finite element code CESAR-LCPC and the SiDoLo optimization tool to identify modified Cam-Clay (MCC) parameters from pressuremeter tests, as shown in Fig. 2.3; Finno and Calvello [4] presented a relatively complex procedure to combine the computer code UCODE and the software tool PLAXIS for identifying hardening soil (HS) model parameters from excavation, as shown in Fig. 2.4; Obrzud et al. [5] presented a procedure employing a two-level neural network tool to conduct the parameters identification, as shown in Fig. 2.5; Zhang et al. [6] presented a procedure involving the MUSEFEM finite element code and particle swarm optimization for identifying the soil parameters of an unsaturated model from pressuremeter tests, as shown in Fig. 2.6; Zhao et al. [7] presented an optimization procedure involving a differential evolution algorithm and ABAQUS software for identifying MCC parameters from an excavation, as shown in Fig. 2.7.

The procedures presented above, and others which are not presented here, are summarized in Fig. 2.8. Most identification procedures are based on two different codes: the FEM code (e.g., PLAXIS [2], FLAC [8], and ABAQUS [7]) or single Gauss point integration of a constitutive model [9, 10] and Ye et al. [11] for the simulation, and the search method code for finding the optimal solution.

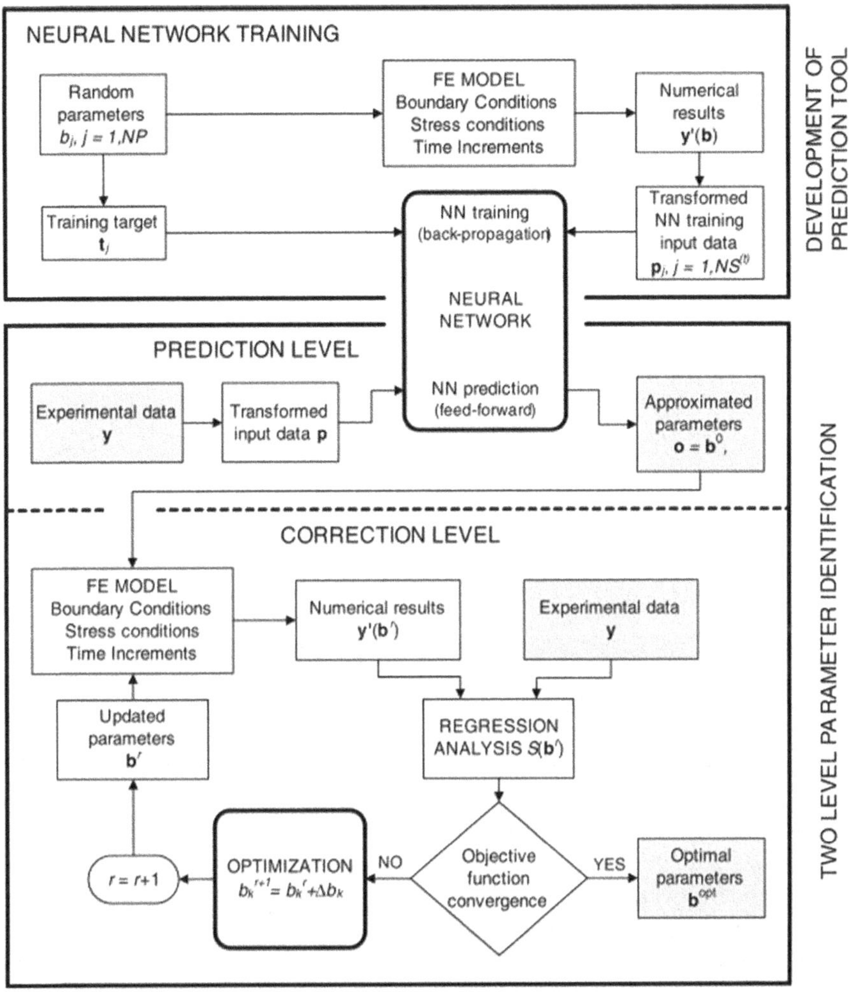

Fig. 2.5 Scheme of the two-level parameter identification using neural network tool

For the initialization step shown in Fig. 2.8, there are two main methods used for sampling initialization: uniform and random. For uniform sampling, a method introduced by Sobol [12] is usually adopted. The SOBOL method is a deterministic algorithm that imitates the behavior of a random sequence. The aim is to obtain a uniform sampling of the design space. It has been reported to be suitable for problems with up to 20 variables [13], and is therefore used in optimizing geotechnical engineering problems. For random sampling, a particular method named Latin

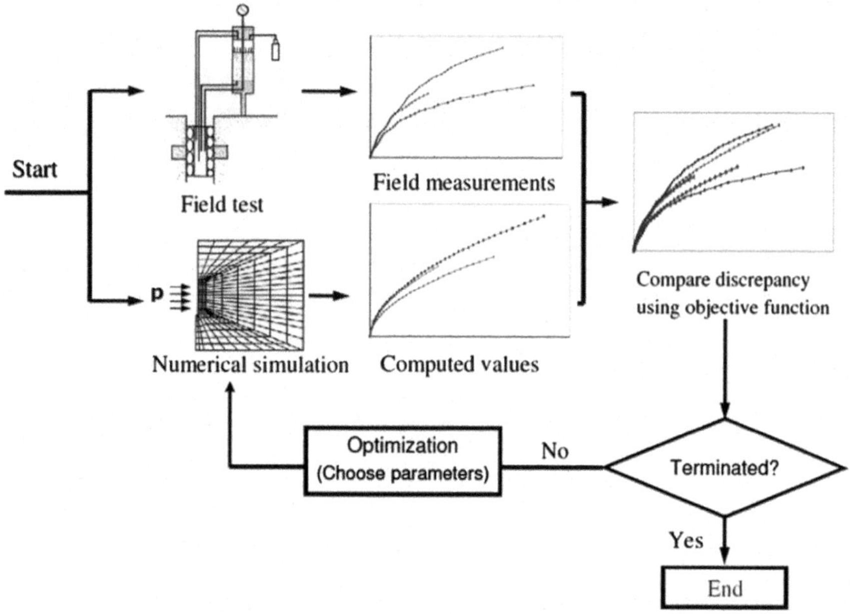

Fig. 2.6 Flowchart for identifying soil parameters using particle swarm optimization from pressuremeter tests

Hypercube Sampling (ULH), proposed by McKay et al. [14] is usually adopted. ULH is an advanced random (Monte Carlo) sampling. Compared to the commonly used random (Monte Carlo) method, ULH is better at mapping the marginal probability distributions (i.e., the statistical distribution of each single variable), especially in cases where there is a small number of generated designs.

For objective tests, laboratory tests or field tests can be adopted in the optimization for calibrating model parameters. These test results are usually displayed in the form of a displacement–stress curve, which implies the softening or hardening, the contraction or dilation of soil. In other words, the results of selected tests can provide information to optimize the model parameters, which is the basis of parameter identification with an optimization method. For laboratory tests, the isotropic or anisotropic compression and conventional triaxial tests are usually recommended for use within the industry [9, 10]. For field tests, the pressuremeter test [1, 2, 13, 15, 16], pile [17], excavation [7, 18–21], and tunneling [8, 22–24] are usually employed.

Either one of the error functions introduced above can be adopted to calculate the fitness value to the results of the optimization method.

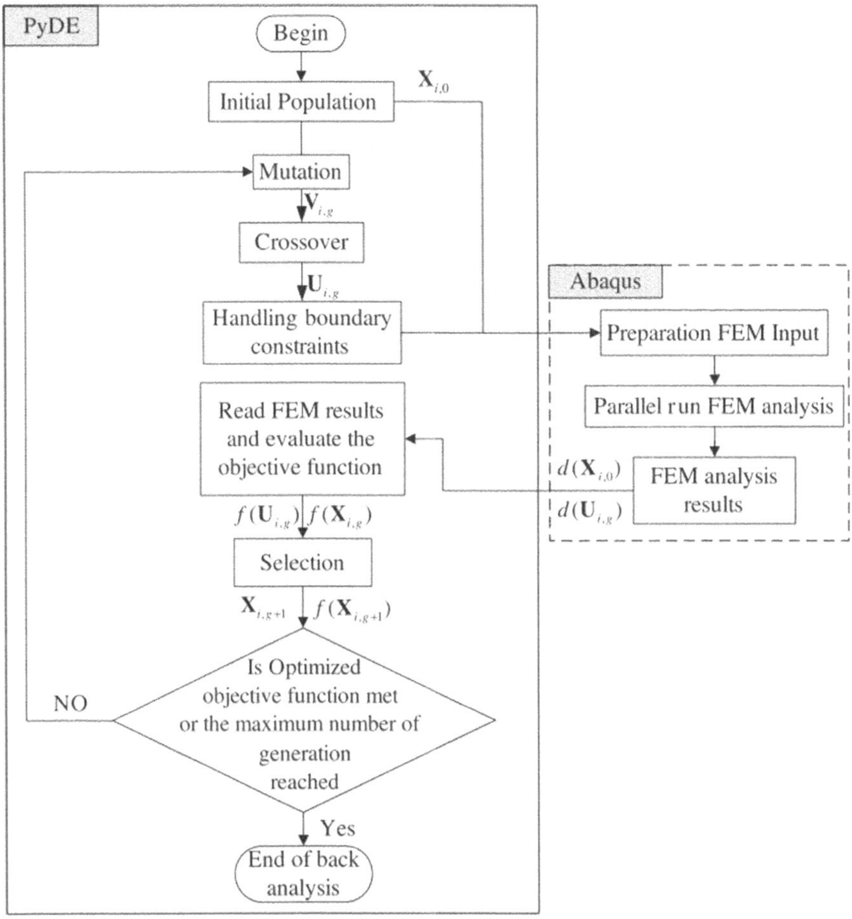

Fig. 2.7 Back analysis with an interaction between the differential evolution algorithm and ABAQUS

For the optimization algorithms, the deterministic techniques (e.g., gradient-based algorithms and simplex) or stochastic techniques (e.g., genetic algorithms, particle swarm optimization, and differential evolution algorithms) can be employed to minimize the error. The optimization process does not stop until the convergence criterion is attained.

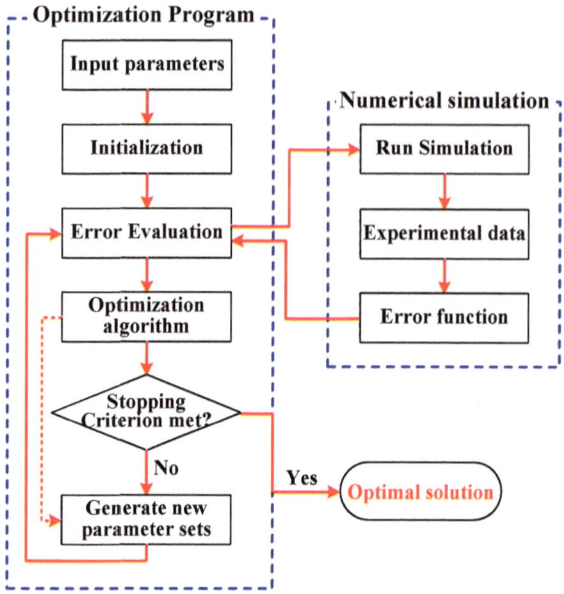

Fig. 2.8 General identification procedure

2.3 Review of Optimization Techniques

In this section, several optimization techniques widely used in geotechnical engineering are reviewed and their basic principles are introduced.

2.3.1 Deterministic Optimization Techniques

2.3.1.1 Gradient-Based Algorithms

The gradient method is probably one of the oldest optimization algorithms, going back to 1847 with the initial work of Cauchy. The gradient method is an algorithm for examining the directions defined by the gradient of a function at the current point. Based on the basic principle, different gradient-based methods have been developed, such as the steepest descent method, the conjugate gradient method, the Levenberg-Marquardt method [25, 26], the Newton method and several Quasi-Newton methods, the Davidon-Fletcher-Powell (DFP), and the Broyden-Fletcher- Goldfarb-Shanno (BFGS) methods.

A rapid convergence is the primary advantage of a gradient-based method. Clearly, the effective use of gradient information can significantly enhance the speed of convergence compared to a method that does not compute gradients. However, gradient-based methods have some limitations, being strongly dependent on user skills (e.g., the basic knowledge of typical values of parameters and the ability to selecting ranges of parameters), due to the need to choose the initial trial solutions. Also, they can easily fall into local minimums, mainly when the procedure is applied to multi-objective functions, as it is the case for material parameter identification with a nonlinear soil model. The requirement of derivative calculations makes these methods non-trivial to implement. Another potential weakness of the gradient-based methods is that they are relatively sensitive to difficulties such as noisy objective function spaces, inaccurate gradients, categorical variables, and topology optimization.

The gradient-based methods have been used for solving different geotechnical engineering problems, such as identifying mechanical soil parameters [3, 4, 15] or soil permeability coefficient [27], optimizing the tunneling-induced ground movement [28], and analyzing the excavation-induced wall deflection [29]. However, due to their limitations stemming from lack of enough information, the gradient-based methods cannot be satisfactorily applied to complex nonlinear optimization problems.

2.3.1.2 Nelder–Mead Simplex

The simplex algorithm is a nonlinear optimization algorithm developed by Nelder and Mead [30] for minimizing an objective function in a poly-dimensional space, which adopts a direct search strategy. The method uses the concept of a simplex, which is a polytope of $N + 1$ vertices in N dimensions, in order to find a locally optimal solution to a problem with N variables when the objective function varies monotonically.

The Nelder–Mead simplex can change in five different ways during iteration in two dimensions, as shown in Figs. 2.9 and 2.10. For example, the number of selected variables is N. Then $N + 1$ set of parameters (or $N + 1$ individuals) should be generated. Then, the error of all individuals can be calculated. Thus, the ascending order of all individuals is obtained. The worst point of the simplex at iteration k (point X_3 in the figure) is selected to be reflected. The $\bar{X}$ point is the mean of parameter sets $X_1 \sim X_{\mathrm{N}}$ and is taken as the reflection center. After the reflection, the X_{r} and its error $f(X_{\mathrm{r}})$ are obtained. Then, the $f(X_{\mathrm{r}})$ is compared to previous errors $f(X_1 \sim X_{N+1})$. Based on the results of comparison, there are three possibilities to update the worst point (expansion, outside contraction, and inside contraction). If the updated point is better than the worst point, then the worst point is replaced by the updated point. Otherwise, apart from the X_1, the individuals $X_2 \sim X_{N+1}$ will be updated by using the shrink; the errors of all updated individuals are then calculated. Finally, the convergence criterion is checked; if yes, the individual with the minimum error is considered as the optimal parameter set; if no, continue to next iteration.

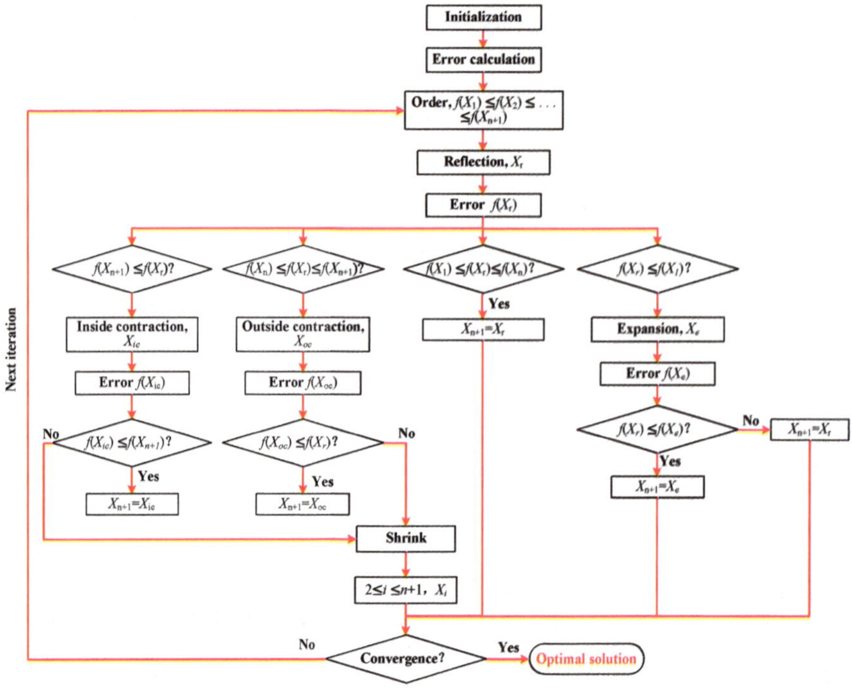

Fig. 2.9 Flowchart of Nelder–Mead simplex algorithm

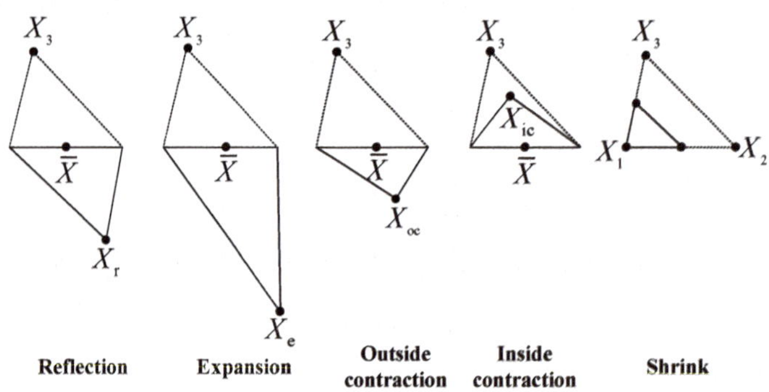

Fig. 2.10 Structure of Nelder–Mead simplex algorithm

The Nelder–Mead simplex can lead to the best solution using a limited number of calculations. In that sense, it can be fast and efficient. However, most direct search strategies, such as the gradient-based and simplex methods described above, are only capable of searching for a local minimum. Generally, it is difficult to verify whether the local minimum is the global one in the multi-dimensional parameter space. A possible solution to this problem is to start the search from different initial positions and, if the local minimum remains the same, then this is most probably also the global minimum.

The pseudo code of the simplex is given below.

Initialization: *Evaluate the function value at each vertex point and order the n + 1 vertices to satisfy* $f(X_1) \le f(X_2) \le \ldots \le f(X_{n+1})$.
Reflection: *Compute the reflection point* X_r *as follows,* $X_r = \bar{X} + \alpha \cdot (\bar{X} - X_{n+1})$ *and evaluate* $f(X_r)$. *If* $f(X_1) \le f(X_r) \le f(X_n)$, *replace* X_{n+1} *with* X_r.
Expansion: *If* $f(X_r) \le f(X_1)$ *then compute the expansion point* X_e *from* $X_e = \bar{X} + \beta \cdot (X_r - \bar{X})$ *and evaluate* $f(X_e)$. *If* $f(X_r) \le f(X_e)$, *replace* X_{n+1} *with* X_e; *otherwise replace* X_{n+1} *with* X_r
Outside Contraction: *If* $f(X_n) \le f(X_r) \le f(X_{n+1})$, *compute the outside contraction point* $X_{oc} = \bar{X} + \gamma \cdot (X_r - \bar{X})$ *and evaluate* $f(X_{oc})$. *If* $f(X_{oc}) \le f(X_r)$, *replace* X_{n+1} *with* X_{oc}; *otherwise go to Shrink.*
Inside Contraction: *If* $f(X_r) \ge f(X_{n+1})$, *compute the inside contraction point* X_{ic} *from* $X_{ic} = \bar{X} + \gamma \cdot (X_r - \bar{X})$ *and evaluate* $f(X_{ic})$. *If* $f(X_{ic}) < f(X_{n+1})$, *replace* X_{n+1} *with* X_{ic}; *otherwise, go to Shrink.*
Shrink: *for* $2 \le i \le n + 1$, *define* $X_i = X_1 + \delta(X_i - X_1)$ *where* $\alpha = 1$, $\beta = 1 + 2/n$, $\gamma = 0.75 - 1/2n$ *and* $\delta = 1 - 1/n$; *n is number of variables.*
The condition of convergence is max(max(abs(X(2: n + 1:)-X(1: n,:))))≤tol. The convergence criterion is $tol = 10^{-4}$.

However, most direct search strategies, such as the gra(*(dient-based and simplex methods described above, are only capable of searching for a local minimum. Generally, it is difficult to verify whether the local minimum is the global one in the multi-dimensional parameter space. A possible solution to this problem is to start the search from different initial positions and, if the local minimum remains the same, then this is most probably also the global minimum.

Nevertheless, there are still many applications of simplex due to its excellent convergence speed, such as identifying the cohesion and friction angle of an elastic—plastic model and the initial stresses using a flexible polyhedron (modified simplex) strategy [31]; estimating hydraulic properties from field data [32]; identifying parameters of a hardening soil model based on pressuremeter tests [13], and identifying both creep and destructuration-related parameters for soft clays [11].

2.3.2 *Stochastic Optimization Techniques*

2.3.2.1 Genetic Algorithms (GA)

The genetic algorithm (GA) originally developed by Holland [33] is a simulation mechanism of Darwinian natural selection and a genetics computational model of the biological evolutionary process. It is also a process to search for the optimal solution by simulating the natural evolution.

The procedure of a genetic algorithm is presented in Fig. 2.11. Once the genetic representation and the fitness function are defined, the GA proceeds by initializing

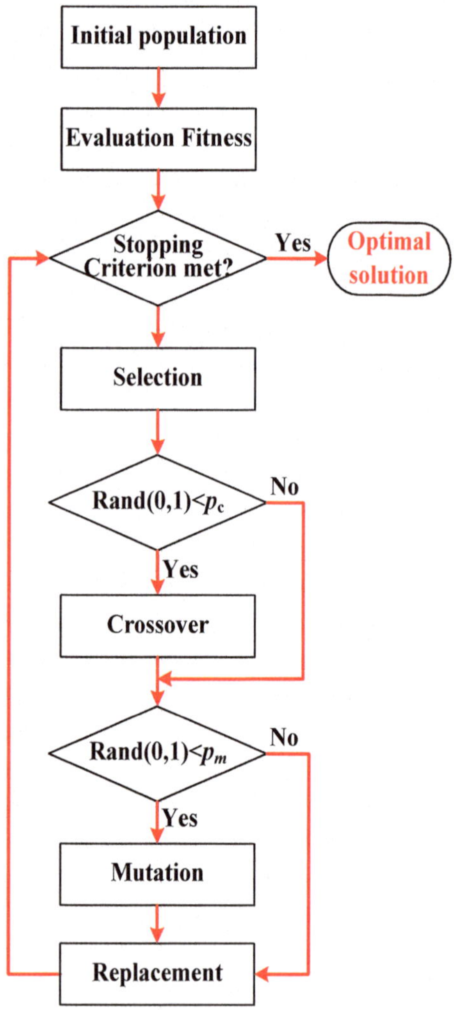

Fig. 2.11 General flowchart of GA

a population of solutions and then improving it through repetitive applications of the selection, crossover, inversion, and mutation operators. Genetic algorithms work with a population of solutions so that they can provide a set of satisfactory solutions. They also do not use any gradient information and they are based on stochastic principles. Moreover, the optimal solution for GAs does not depend on the initial trial solutions compared to gradient-based method. Therefore, they are considered more robust than the gradient methods.

In geotechnical engineering, the GAs have been widely employed to solve various problems such as parameter identification of constitutive models [1, 9, 10, 13, 18, 34–36], prediction of soil hydraulic parameters [37–39], identification of critical slip surfaces in slope stability analysis [40–44], prediction of vertical settlement [45], optimization of pile group design [17, 46], reliability analysis [47], and prediction of soil–water characteristic curves for unsaturated soils [48].

2.3.2.2 Particle Swarm Optimizations (PSO)

Particle swarm optimization (PSO) is a population-based stochastic global optimization algorithm which was first suggested by Kennedy and Eberhart [49] in an attempt to simulate the graceful choreography of flocks of birds, as part of a socio-cognitive study on the notion of "collective intelligence" in biological populations.

Figure 2.12 shows the procedure of PSO to find the best solution. A number of simple entities, the "particles", are randomly placed in the search space of a given problem or a given function, and each entity evaluates the objective function at a particular location. Then, each particle determines its movement through the search space by combining some aspect of the history of its own actual and best (best-fitness) locations with those of one or more members of the swarm, with some random perturbations, as shown in Fig. 2.13. The new position $\boldsymbol{x}_i(t)$ of the ith particle can be defined as,

$$\boldsymbol{x}_i(t) = \boldsymbol{x}_i(t-1) + \boldsymbol{V}_i(t) \times 1,\ i = 1, 2, \ldots, N_p \tag{2.11}$$

where N_p is the total number of particles. The new velocity $\boldsymbol{V}_i(t)$ of the ith particle is calculated by

$$\boldsymbol{V}_i(t) = w(t)\boldsymbol{V}_i(t-1) + c_1 r_1(p\mathrm{Best}_i - \boldsymbol{x}_i(t-1)) + c_2 r_2(g\mathrm{Best} - \boldsymbol{x}_i(t-1)) \tag{2.12}$$

where r_1 and r_2 are random numbers between 0 and 1; $pBest_i$ is the local best position (the best among all previous positions at time $(t-1)$); $gBest$ is the global best position (the best particle position among all known particle positions within the whole swarm); $w(t)$ is the inertia weight used to control the impact of the previous particle velocities on the current velocity and it is usually taken as slightly less than 1 [50]. The learning factors c_1 (cognitive weight) and c_2 (social weight) are positive

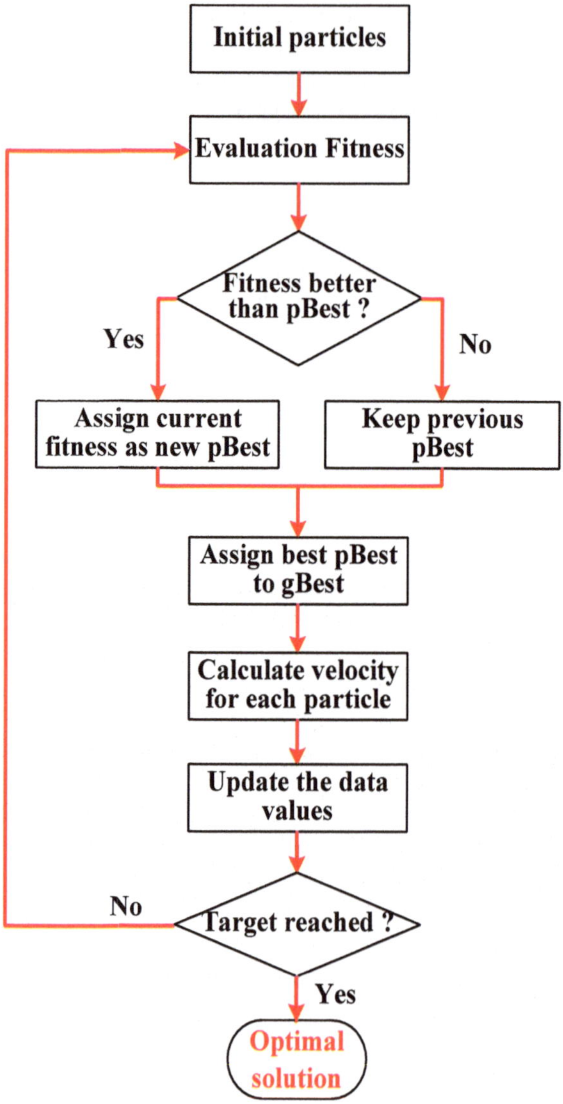

Fig. 2.12 General flowchart of PSO

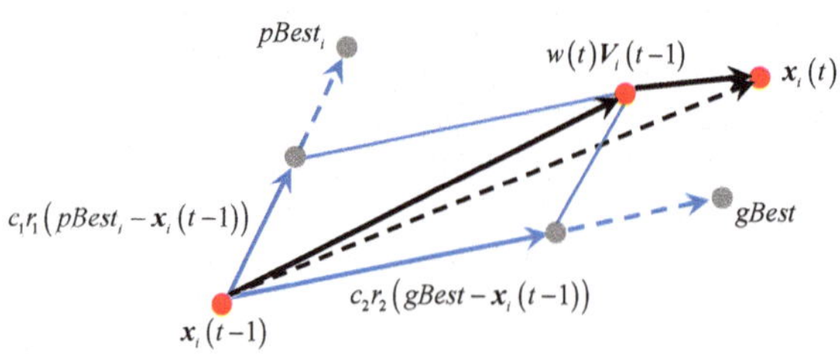

Fig. 2.13 Schematic diagram of the method of the PSO

constants which determine how much the particle is directed toward the good position and are adapted to the individual task to be solved by manual variations; in practice, they are set as equal to 2 [50]. The local best position for the ith particle is updated if

$$F(\boldsymbol{x}_i) < F(pBest_i) \tag{2.13}$$

and the global best position is updated if

$$F(\boldsymbol{x}_i) < F(gBest) \tag{2.14}$$

The next iteration takes place after all particles have been moved. Eventually, the swarm as a whole, like a flock of birds collectively foraging for food, is likely to move close to an optimum of the fitness function.

In contrast to GA with its binary encoding, the design variables can take any value in PSO, based on their actual position in the design space and the calculated velocity vector. PSO has also no evolution operator such as crossover or mutation, which makes it ideal for asynchronous parallel implementation. However, both algorithms have similar functions for finding the best solution. The comparison between GA and PSO has already been investigated in terms of computational effectiveness and efficiency by several researchers [51, 52]. Each algorithm has its unique advantages for solving different types of problems.

PSO has been shown to provide valuable results in various inverse geotechnical problems, such as parameter identification of constitutive models [16, 50, 53–59], identification of hydraulic parameters for unsaturated soils [60], parameter identification in soil-structure interaction [61], location of the critical non-circular failure surface in slope stability analysis [62], and parameter estimation of diffusion transport of contaminants [63].

2.3.2.3 Simulated Annealing (SA)

Simulated annealing (SA) is a random-search technique which uses an analogy between the way a metal cools and freezes into a minimum energy crystalline structure (the annealing process) and the search for a minimum in a more general system [64]. Due to the inherent statistical nature of simulated annealing, local minima can be obtained in principle more easily than in gradient-based methods [65].

Figure 2.14 shows a general structure of the simulated annealing algorithm. A simple form of local search (a descent algorithm) starts with an initial solution. A neighbor of this solution is then generated by some suitable mechanism and the change in cost is calculated. If a reduction in cost is found, the current solution is replaced by the generated neighbor; otherwise, the current solution is retained. The process is repeated until no further improvement can be found in the neighborhood of the current solution and so the descent algorithm terminates at a local minimum.

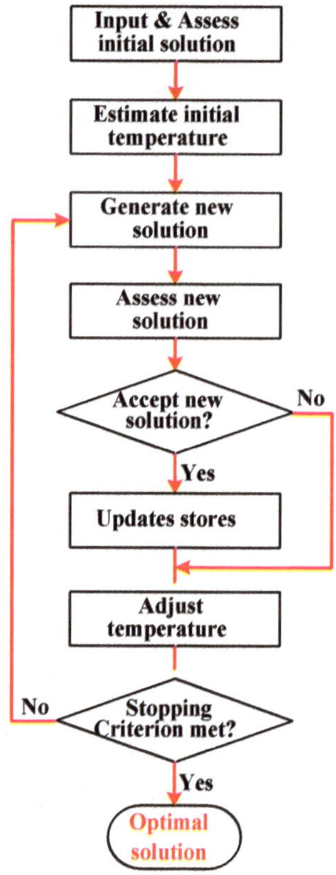

Fig. 2.14 General structure of SA algorithm

Further details about the strengths and weaknesses of simulated annealing, and a comparison with other methods, can be found in Busetti [64]. There are few applications of SA in the geotechnical field; one example was to determine a safety factor for slope stability [66].

2.3.2.4 Differential Evolution Algorithm (DE)

The differential evolution (DE) algorithm proposed by Price and Storn [67, 68] is a simple, but yet powerful population-based stochastic search technique, which is an efficient and effective global optimizer in the continuous search domain. Like other population-based optimization algorithms, DE involves also two phases: initialization and evolution. In the initialization phase, the DE population is generated randomly if nothing is known about the problem. In the evolution phase, individuals from the population undergo repeatedly mutation, crossover, and selection processes until the termination criterion is met. DE uses a mutation operation as a search mechanism and a selection operation based on the differences in randomly sampled pairs of solutions in the population:

$$\begin{aligned} &\boldsymbol{x}_i = \boldsymbol{x}_{r3} + F^k(\boldsymbol{x}_{r2} - \boldsymbol{x}_{r1}), \\ &r1, r2, r3 \in \left\{1, 2, \ldots, N_p\right\}, k \in \{1, 2\}, F^k \in [0, 1] \end{aligned} \tag{2.15}$$

where F^k is a scaling factor which is closely related to the convergence speed; k indicates the number of difference vectors taking part in the mutation operation; $\boldsymbol{x}_{r1}$, $\boldsymbol{x}_{r2}$ and $\boldsymbol{x}_{r3}$ are selected individuals from the population which are different from the running individual $\boldsymbol{x}_i$.

As evolution proceeds, the population of DE may move through different regions in the search space, within which certain strategies associated with specific parameter settings may be more effective than others (Qin et al. [69]). However, the performance of the conventional DE algorithm depends highly on the chosen trial vector generation strategy and the associated parameter values. DE does not guarantee the convergence to the global optimum. It is easily trapped into a local optimum resulting in a low optimizing precision or even a failure (Jia et al. [70]).

In geotechnical engineering, DE has been applied to different optimization problems, such as parameter identification of constitutive models [7], and back analysis of tunneling [8].

2.3.2.5 Artificial Bee Colony (ABC)

The artificial bee colony (ABC) algorithm is an optimization algorithm simulating the intelligent foraging behavior of honeybee swarms and was originally developed by Karaboga [71]. It is a very simple, robust, and population-based stochastic optimization algorithm.

Following Karaboga [71], the flow chart of the algorithm is shown in Fig. 2.15. The ABC algorithm describes the foraging behavior, learning, memorizing, and information sharing characteristics of honeybees. A basic model of the foraging behavior of honeybee swarms consists of two essential components and defines two leading modes of behavior. The artificial bee colony consists of three groups of bees: employed bees, onlookers, and scout bees. Further details of the ABC algorithm can be found in Karaboga and Basturk [72].

However, as in other evolutionary algorithms, the ABC algorithm faces also challenging problems. For example, the convergence speed of an ABC algorithm is typically slower than those of representative population-based algorithms (e.g., DE and PSO) when handling unimodal problems [73]. In addition, an ABC algorithm can easily be trapped in a local optimum when solving complex multi-modal problems.

The ABC algorithms have been applied to different problems in geotechnical engineering, such as to locate the critical slip surface of a slope [74, 75]; to predict the uplift capacity of suction caissons [76]; and to analyze the reliability of geotechnical engineering solutions [77, 78].

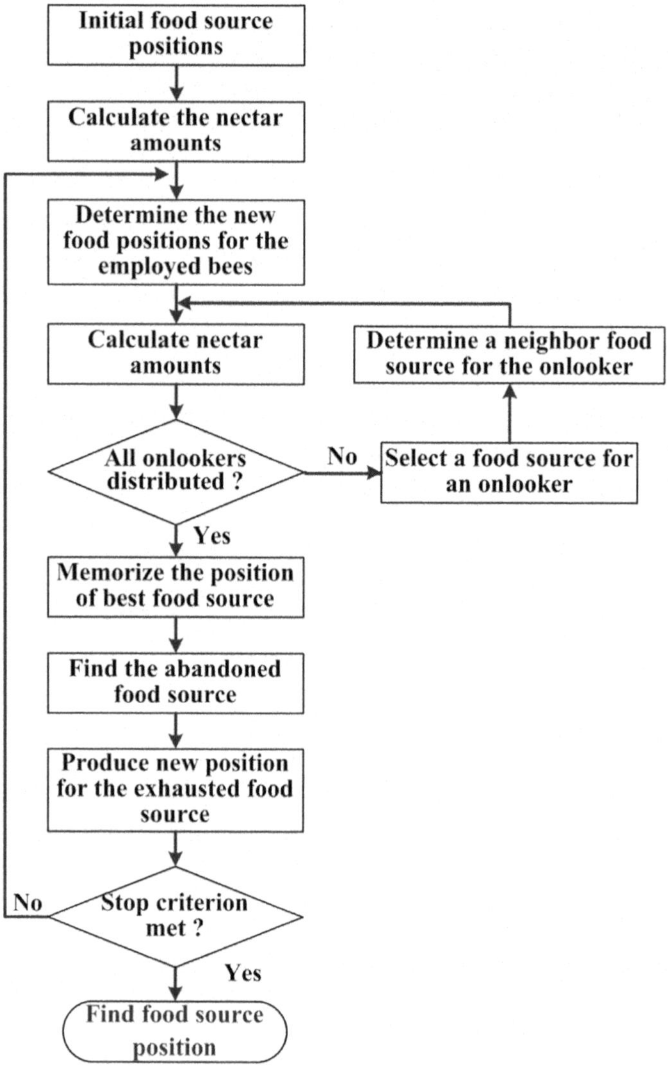

Fig. 2.15 Flowchart of the ABC algorithm

2.3.2.6 Others

Apart from the previously mentioned optimization methods, there are several other algorithms which were initially developed for modeling the behavior of social insects, such as ants and bees, for the purpose of solving various optimization problems. Among them, ant colony optimization (ACO) is a metaheuristic method for solving difficult combinatorial optimization problems. It was introduced by M. Dorigo et al. [79] as a novel nature-inspired metaheuristic method for the solution of hard combinatorial optimization (CO) problems. The inspiration behind ACO is the pheromone trail laying, based on the behavior of real ants, which use pheromones as a communication medium. More information about ACO can be found in Dorigo [80]. In the geotechnical field, applications of ACO include estimating unsaturated soil hydraulic parameters [81] and determining the critical failure surface for slope stability analysis [82].

In a similar way to ACO, the Honey Bee Mating algorithm (HBMO) [83], the bacterial foraging optimization algorithm (BFOA) [84] and the Krill herd (KH) algorithm [85] are also considered as typical bio-based approaches to optimization. However, they have not yet been used in geotechnical engineering.

2.3.3 Hybrid Optimization Techniques

Conventional optimization methods suffer usually from local optimality problem and slow convergence speed, which limits their applicability to a small range of problems, even sometimes causing a failure of the optimization procedure. In order to enhance the performance of these traditional algorithms, an efficient method is to combine the advantages of each approach using a hybrid strategy (e.g., Tsai et al. [86]). The hybrid strategies can generally be divided into three groups: (1) hybrids of different operators (e.g., different crossovers in GAs [87]); (2) hybrids of the local search (e.g., DE with chaotic [70]); and (3) hybrids of different optimization techniques (e.g., GA and DE [8]). Due to their high performance, the hybrid optimization algorithms have become more and more popular in geotechnical engineering and have been applied to many optimization problems, such as the optimization of pile groups [46], the identification of geomechanical parameters [88], the slope reliability analysis [89], and the prediction of the uplift capacity of suction caissons [76]. In addition to these cases, other geotechnical problems are likely to be solved in the future by using hybrid optimization techniques.

2.4 Summary

This chapter reviews the optimization techniques in the field of parameter identification. Firstly, the methods of parameter identification, the error function of various measurement experiments and numerical simulation results, and the basic principles and optimization processes of various optimization algorithms are introduced.

The wide application of optimization technology in the geotechnical field includes the following three aspects: (1) deterministic algorithm (gradient-based optimization algorithm and Nelder–Mead simplex algorithm); (2) stochastic algorithms (GA, PSO, SA, DE, ABC, and others); (3) hybrid optimization algorithm. This chapter systematically reviews the above algorithms and briefly introduces its basic principles and its applications in geotechnical fields. Table 2.1 summarizes the advantages and dis-

Table 2.1 Comparison of several optimization algorithms

Algorithm	Advantages	Disadvantages	Approach to improve
Gradient-based algorithms	Fast convergence	The initial value depends heavily; the linear independence of the first N search directions in each iteration is not guaranteed, which may lead to the solution failure	Change the initial value setting and the search step
Simplex	It is suitable for solving linear programming problems, and can also solve general nonlinear problems; it is convenient to implement and suitable for medium and small scale problems	Poor global search ability; easy to fall into local extremes; serious dependence on initial values	Change the initial value setting and search step; Adjust the position of the reflection point; control the expansion coefficient and contraction coefficient
GA	Self-organizing, adaptive and intelligent; The idea is simple and easy to understand, the steps are standardized, and it is easy to implement the application; parallel computing can be realized	Poor local search ability; premature convergence	Improved genetic operators; genetic algorithms based on multiple groups; optimization of control parameters; combined with other intelligent algorithms
SA	No need to find partial derivatives and large matrix equations; Easy to join constraints; simple programming	The actual operation efficiency is not high	Increase the number of constraints; reduce the perturbation range of the model
PSO	The concept is simple and easy to implement; fewer parameters need to be adjusted	Easy to fall into local optimal solution; premature phenomenon	Improve the value of parameter weights; increase the diversity of PSO algorithms; combined with other intelligent algorithms

(continued)

Table 2.1 (continued)

Algorithm	Advantages	Disadvantages	Approach to improve
DE	The global optimization evolution algorithm based on real coding is unique in that it introduces the differential mutation mode for iterative search, which has strong global search ability and convergence rate	Conventional differential evolution algorithms may fall into local optimum, and there is stagnation, which makes it easy for the algorithm to converge	Avoid falling into local optimum; enhance its continuous space optimization ability; merge with other algorithms
AB	Distributed computing; easy to integrate with other optimization algorithms; easy to implement; low hardware and software requirements for computers	Search time is too long; easy to fall into local optimal solution	Avoid falling into local optimum; enhance its continuous space optimization ability; merge with other algorithms
ABC	Distributed computing; easy to integrate with other optimization algorithms; easy to implement; low hardware and software requirements for computers	Search time is too long; easy to fall into local optimal solution	Avoid falling into local optimum; enhance its continuous space optimization ability; merge with other algorithms
Hybrid optimization algorithms	Advantages of the integrated algorithm, enhance search capabilities, and accelerate convergence	Need to find excellent algorithms for mixing	Introducing advanced search operators

advantages, scope of application, and improvement methods of some of the above algorithms. It can be seen that the intelligent optimization algorithm has greater advantages than the traditional method, but each method has its advantages and disadvantages and scope of application, and the optimal method should be selected according to the actual situation. There are also many scholars who combine various optimization methods to give play to their respective advantages, or to improve the original optimization algorithms, and have achieved good results.

Other details can be found in Yin et al. [90].

References

1. Levasseur S, Malécot Y, Boulon M, Flavigny E (2008) Soil parameter identification using a genetic algorithm. Int J Numer Anal Methods Geomech 32(2):189–213
2. Calvello M, Finno RJ (2004) Selecting parameters to optimize in model calibration by inverse analysis. Comput Geotech 31(5):410–424
3. Zentar R, Hicher P, Moulin G (2001) Identification of soil parameters by inverse analysis. Comput Geotech 28(2):129–144
4. Finno RJ, Calvello M (2005) Supported excavations: observational method and inverse modeling. J Geotech Geoenviron Eng 131(7):826–836
5. Obrzud RF, Vulliet L, Truty A (2009) A combined neural network/gradient-based approach for the identification of constitutive model parameters using self-boring pressuremeter tests. Int J Numer Anal Methods Geomech 33(6):817–849
6. Zhang L, Zuo Z, Ye G, Jeng D, Wang J (2013) Probabilistic parameter estimation and predictive uncertainty based on field measurements for unsaturated soil slope. Comput Geotech 48:72–81
7. Zhao B, Zhang L, Jeng D, Wang J, Chen J (2015) Inverse analysis of deep excavation using differential evolution algorithm. Int J Numer Anal Methods Geomech 39(2):115–134
8. Vardakos S, Gutierrez M, Xia C (2012) Parameter identification in numerical modeling of tunneling using the Differential Evolution Genetic Algorithm (DEGA). Tunn Undergr Space Technol 28:109–123
9. Jin Y-F, Yin Z-Y, Shen S-L, Hicher P-Y (2016) Selection of sand models and identification of parameters using an enhanced genetic algorithm. Int J Numer Anal Methods Geomech 40(8):1219–1240
10. Jin Y-F, Yin Z-Y, Shen S-L, Hicher P-Y (2016) Investigation into MOGA for identifying parameters of a critical-state-based sand model and parameters correlation by factor analysis. Acta Geotech 11(5):1131–1145
11. Ye L, Jin Y-F, Shen S-L, Sun P-P, Zhou C (2016) An efficient parameter identification procedure for soft sensitive clays. J Zhejiang University SCIENCE A 17(1):76–88
12. Sobol IM (1967) On the distribution of points in a cube and the approximate evaluation of integrals. USSR Comput Math Math Phys 7(4):86–112
13. Papon A, Riou Y, Dano C, Hicher PY (2012) Single-and multi-objective genetic algorithm optimization for identifying soil parameters. Int J Numer Anal Methods Geomech 36(5):597–618
14. McKay MD, Beckman RJ, Conover WJ (1979) Comparison of three methods for selecting values of input variables in the analysis of output from a computer code. Technometrics 21(2):239–245
15. Yin ZY, Hicher PY (2008) Identifying parameters controlling soil delayed behaviour from laboratory and in situ pressuremeter testing. Int J Numer Anal Methods Geomech 32(12):1515–1535
16. Zhang Y, Gallipoli D, Augarde C (2013) Parameter identification for elasto-plastic modelling of unsaturated soils from pressuremeter tests by parallel modified particle swarm optimization. Comput Geotech 48:293–303

17. Ardalan H, Eslami A, Nariman-Zadeh N (2009) Piles shaft capacity from CPT and CPTu data by polynomial neural networks and genetic algorithms. Comput Geotech 36(4):616–625
18. Levasseur S, Malecot Y, Boulon M, Flavigny E (2010) Statistical inverse analysis based on genetic algorithm and principal component analysis: Applications to excavation problems and pressuremeter tests. Int J Numer Anal Methods Geomech 34(5):471–491
19. Rechea C, Levasseur S, Finno R (2008) Inverse analysis techniques for parameter identification in simulation of excavation support systems. Comput Geotech 35(3):331–345
20. Hashash YMA, Levasseur S, Osouli A, Finno R, Malecot Y (2010) Comparison of two inverse analysis techniques for learning deep excavation response. Comput Geotech 37(3):323–333
21. Huang Z, Zhang L, Cheng S, Zhang J, Xia X (2014) Back-Analysis and Parameter Identification for deep excavation based on pareto multiobjective optimization. J Aerosp Eng 28(6):A4014007
22. Gens A, Ledesma A, Alonso E (1996) Estimation of parameters in geotechnical backanalysis—II. Application to a tunnel excavation problem. Comput Geotech 18(1):29–46
23. Gong W, Luo Z, Juang CH, Huang H, Zhang J, Wang L (2014) Optimization of site exploration program for improved prediction of tunneling-induced ground settlement in clays. Comput Geotech 56:69–79
24. Yu H, Li S, Duan H, Liu Y (2011) A procedure of parameter inversion for a nonlinear constitutive model of soils with shield tunneling. Comput Math Appl 61(8):2005–2009
25. Moré JJ (1978) The Levenberg-Marquardt algorithm: implementation and theory. Numerical analysis. Springer, p 105–16
26. Marquardt DW (1963) An algorithm for least-squares estimation of nonlinear parameters. J Soc Ind Appl Math 11(2):431–441
27. Rangeard D, Hicher PY, Zentar R (2003) Determining soil permeability from pressuremeter tests. Int J Numer Anal Methods Geomech 27(1):1–24
28. Chi S-Y, Chern J-C, Lin C-C (2001) Optimized back-analysis for tunneling-induced ground movement using equivalent ground loss model. Tunn Undergr Space Technol 16(3):159–165
29. Tang Y-G, Kung GT-C (2009) Application of nonlinear optimization technique to back analyses of deep excavation. Comput Geotech 36(1):276–290
30. Nelder JA, Mead R (1965) A simplex method for function minimization. Comput J 7(4):308–313
31. Gioda G, Maier G (1980) Direct search solution of an inverse problem in elastoplasticity: identification of cohesion, friction angle and in situ stress by pressure tunnel tests. Int J Numer Methods Eng 15(12):1823–1848
32. Ritter A, Hupet F, Muñoz-Carpena R, Lambot S, Vanclooster M (2003) Using inverse methods for estimating soil hydraulic properties from field data as an alternative to direct methods. Agric Water Manag 59(2):77–96
33. Holland J (1992) Adaptation in natural and artificial systems. MIT Press, Cambridge, MA
34. Jin Y-F, Wu Z-X, Yin Z-Y, Shen JS (2017) Estimation of critical state-related formula in advanced constitutive modeling of granular material. Acta Geotech 12(6):1329–1351
35. Jin Y-F, Yin Z-Y, Shen S-L, Zhang D-M (2017) A new hybrid real-coded genetic algorithm and its application to parameters identification of soils. Inverse Prob Sci Eng 25(9):1343–1366
36. Jin Y-F, Yin Z-Y, Wu Z-X, Zhou W-H (2018) Identifying parameters of easily crushable sand and application to offshore pile driving. Ocean Eng 154:416–429
37. Mahbod M, Zand-Parsa S (2010) Prediction of soil hydraulic parameters by inverse method using genetic algorithm optimization under field conditions. Arch Agron Soil Sci 56(1):13–28
38. Schneider S, Jacques D, Mallants D (2013) Inverse modelling with a genetic algorithm to derive hydraulic properties of a multi-layered forest soil. Soil Res 51(5):372–389
39. Ines AV, Droogers P (2002) Inverse modelling in estimating soil hydraulic functions: a Genetic Algorithm approach. Hydrol Earth Syst Sci Dis 6(1):49–66
40. Zolfaghari AR, Heath AC, McCombie PF (2005) Simple genetic algorithm search for critical non-circular failure surface in slope stability analysis. Comput Geotech 32(3):139–152
41. Xue J-F, Gavin K (2007) Simultaneous determination of critical slip surface and reliability index for slopes. J Geotech Geoenviron Eng 133(7):878–886

42. McCombie P, Wilkinson P (2002) The use of the simple genetic algorithm in finding the critical factor of safety in slope stability analysis. Comput Geotech 29(8):699–714
43. Goh AT (1999) Genetic algorithm search for critical slip surface in multiple-wedge stability analysis. Can Geotech J 36(2):382–391
44. Sun J, Li J, Liu Q (2008) Search for critical slip surface in slope stability analysis by spline-based GA method. J Geotech Geoenviron Eng 134(2):252–256
45. Park HI, Park B, Kim YT, Hwang DJ (2009) Settlement prediction in a vertical drainage-installed soft clay deposit using the Genetic Algorithm (GA) back-analysis. Mar Georesour Geotechnol 27(1):17–33
46. Chan C, Zhang L, Ng JT (2009) Optimization of pile groups using hybrid genetic algorithms. J Geotech Geoenviron Eng 135(4):497–505
47. Cui L, Sheng D (2005) Genetic algorithms in probabilistic finite element analysis of geotechnical problems. Comput Geotech 32(8):555–563
48. Pedroso DM, Williams DJ (2011) Automatic calibration of soil–water characteristic curves using genetic algorithms. Comput Geotech 38(3):330–340
49. Kennedy J (2011) Particle swarm optimization. Encyclopedia of machine learning. Springer, pp 760–766
50. Knabe T, Datcheva M, Lahmer T, Cotecchia F, Schanz T (2013) Identification of constitutive parameters of soil using an optimization strategy and statistical analysis. Comput Geotech 49:143–57
51. Eberhart RC, Shi Y (1998) Comparison between genetic algorithms and particle swarm optimization. Evolutionary Programming VII. Springer, pp 611–616
52. Elbeltagi E, Hegazy T, Grierson D (2005) Comparison among five evolutionary-based optimization algorithms. Adv Eng Inform 19(1):43–53
53. Mulia A (2012) Identification of Soil constitutive soil model parameters using multi-objective particle swarming optimization
54. Nguyen-Tuan L, Lahmer T, Datcheva M, Stoimenova E, Schanz T (2016) A novel parameter identification approach for buffer elements involving complex coupled thermo-hydro-mechanical analyses. Comput Geotech 76:23–32
55. Sadoghi Yazdi J, Kalantary F, Sadoghi Yazdi H (2011) Calibration of soil model parameters using particle swarm optimization. Int J Geomech 12(3):229–238
56. Zhang Y, Gallipoli D, Augarde CE (2009) Simulation-based calibration of geotechnical parameters using parallel hybrid moving boundary particle swarm optimization. Comput Geotech 36(4):604–615
57. Meier J, Schaedler W, Borgatti L, Corsini A, Schanz T (2008) Inverse parameter identification technique using PSO algorithm applied to geotechnical modeling. J Artif Evol Appl 2008:1–14
58. Schanz T, Zimmerer MM, Datcheva M, Meier J (2006) Identification of constitutive parameters for numerical models via inverse approach. Felsbau 24(2):11–21
59. Knabe T, Schweiger HF, Schanz T (2012) Calibration of constitutive parameters by inverse analysis for a geotechnical boundary problem. Can Geotech J 49(2):170–183
60. Zhang Y, Augarde C, Gallipoli D (2008) Identification of hydraulic parameters for unsaturated soils using particle swarm optimization. In: Proceedings 1st European conference on unsaturated soils, 2008, p 765–71
61. Fontan M, Ndiaye A, Breysse D, Bos F, Fernandez C (2011) Soil–structure interaction: Parameters identification using particle swarm optimization. Comput Struct 89(17):1602–1614
62. Cheng Y, Li L, S-c Chi, Wei W (2007) Particle swarm optimization algorithm for the location of the critical non-circular failure surface in two-dimensional slope stability analysis. Comput Geotech 34(2):92–103
63. Bharat TV, Sivapullaiah PV, Allam MM (2009) Swarm intelligence-based solver for parameter estimation of laboratory through-diffusion transport of contaminants. Comput Geotech 36(6):984–992
64. Busetti F (2003) Simulated annealing overview. JP Morgan, Italy
65. Xiang Y, Gubian S, Suomela B, Hoeng J (2013) Generalized simulated annealing for global optimization: the GenSA Package. R Journal 5(1):13–28

66. Li S, Liu Y, He X, Liu Y (2003) Global search algorithm of minimum safety factor for slope stability analysis based on annealing simulation. Chin J Rock Mech Eng 22(2):236–240
67. Price K, Storn R (1997) Differential evolution: a simple evolution strategy for fast optimization. Dr Dobb's j 22(4):18–24
68. Storn R, Price K (1997) Differential evolution–a simple and efficient heuristic for global optimization over continuous spaces. J Global Optim 11(4):341–359
69. Qin AK, Huang VL, Suganthan PN (2009) Differential evolution algorithm with strategy adaptation for global numerical optimization. IEEE Trans Evol Comput 13(2):398–417
70. Jia D, Zheng G, Khan MK (2011) An effective memetic differential evolution algorithm based on chaotic local search. Inf Sci 181(15):3175–3187
71. Karaboga D (2005) An idea based on honey bee swarm for numerical optimization. Technical report-tr06, Erciyes university, engineering faculty, computer engineering department, 2005
72. Karaboga D, Basturk B (2007) A powerful and efficient algorithm for numerical function optimization: Artificial Bee Colony (ABC) algorithm. J Global Optim 39(3):459–471
73. Karaboga D, Akay B (2009) A comparative study of artificial bee colony algorithm. Appl Math Comput 214(1):108–132
74. Kang F, Li J, Ma Z (2013) An artificial bee colony algorithm for locating the critical slip surface in slope stability analysis. Eng Optim 45(2):207–223
75. Kang F, Li J, Ma Z, Li H (2011) Artificial bee colony algorithm with local search for numerical optimization. J Software 6(3):490–497
76. Cheng M-Y, Cao M-T, Tran D-H (2014) A hybrid fuzzy inference model based on RBFNN and artificial bee colony for predicting the uplift capacity of suction caissons. Autom Constr 41:60–69
77. Kang F, Li J (2015) Artificial bee colony algorithm optimized support vector regression for system reliability analysis of slopes. J Comput Civil Eng 30(3):04015040, 1–13
78. Zhao H, Zhao M, Zhu C (2015) Reliability-based optimization of geotechnical engineering using the artificial bee colony algorithm. KSCE J Civil Eng, pp 1–9
79. Dorigo M, Maniezzo V, Colorni A (1996) Ant system: optimization by a colony of cooperating agents. IEEE Trans Syst Man Cyber Part B: Cyber 26(1):29–41
80. Dorigo M, Birattari M, Stützle T (2006) Ant colony optimization. In: Computational Intelligence Magazine, IEEE 1(4): 28–39
81. Abbaspour K, Schulin R, Van Genuchten MT (2001) Estimating unsaturated soil hydraulic parameters using ant colony optimization. Adv Water Res 24(8):827–841
82. Kahatadeniya KS, Nanakorn P, Neaupane KM (2009) Determination of the critical failure surface for slope stability analysis using ant colony optimization. Eng Geol 108(1):133–141
83. Afshar A, Haddad OB, Mariño MA, Adams B (2007) Honey-bee mating optimization (HBMO) algorithm for optimal reservoir operation. J Franklin Inst 344(5):452–462
84. Ali E, Abd-Elazim S (2011) Bacteria foraging optimization algorithm based load frequency controller for interconnected power system. Int J Electr Power Energy Syst 33(3):633–638
85. Gandomi AH, Alavi AH (2012) Krill herd: a new bio-inspired optimization algorithm. Commun Nonlinear Sci Numer Simul 17(12):4831–4845
86. Tsai J-T, Liu T-K, Chou J-H (2004) Hybrid Taguchi-genetic algorithm for global numerical optimization. IEEE Trans Evol Comput 8(4):365–377
87. Herrera F, Lozano M, Sánchez AM (2005) Hybrid crossover operators for real-coded genetic algorithms: an experimental study. Soft Comput 9(4):280–298
88. Jiang A, Wang S, Tang S (2011) Feedback analysis of tunnel construction using a hybrid arithmetic based on support vector machine and particle swarm optimisation. Autom Constr 20(4):482–489
89. Khajehzadeh M, Taha MR, El-Shafie A (2011) Reliability analysis of earth slopes using hybrid chaotic particle swarm optimization. J Central South Univ Technol 18(5):1626–1637
90. Yin Z-Y, Jin Y-F, Shen JS, Hicher P-Y (2018) Optimization techniques for identifying soil parameters in geotechnical engineering: comparative study and enhancement. Int J Numer Anal Methods Geomech 42(1):70–94

Chapter 3
Comparative Study of Typical Optimization Methods

3.1 Introduction

Generally, each optimization technique has its advantages and drawbacks, which means that not all optimization problems can be effectively solved by a given optimization method. For a given optimization problem, it is necessary to evaluate the optimizing performance of different methods and then select the most appropriate method to conduct the optimization. In order to evaluate their search ability and their convergence speed in identifying material parameters, these different techniques need to be applied to the same optimization problem. Therefore, the five mostly common optimization methods, including genetic algorithms (GA), particle swarm optimization (PSO), simulated annealing (SA), the differential evolution (DE) algorithm, and the artificial bee colony (ABC) algorithm, were selected for comparison. Two typical cases, i.e., the identification of parameters from pressuremeter tests and from excavation measurements, were adopted for the optimization process.

3.2 Case 1: Pressuremeter Test

3.2.1 Pressuremeter Test Simulated by Mohr–Coulomb Model

In order to conduct the comparative study, a parameter identification using the Mohr–Coulomb (MC) model from a pressuremeter test (PMT) result was selected. For a fair comparison and absolute evaluation of the computing techniques, a synthetic objective pressuremeter test was generated in the FEM code ABAQUS with a given set of the MC parameters. The pressuremeter test simulated by ABAQUS was a displacement-controlled test, and the small deformation condition was assumed (see Zentar et al. [1] and Papon et al. [2]). For generating the stress–displacement

Z. Yin and Y. Jin, *Practice of Optimisation Theory in Geotechnical Engineering*,
https://doi.org/10.1007/978-981-13-3408-5_3

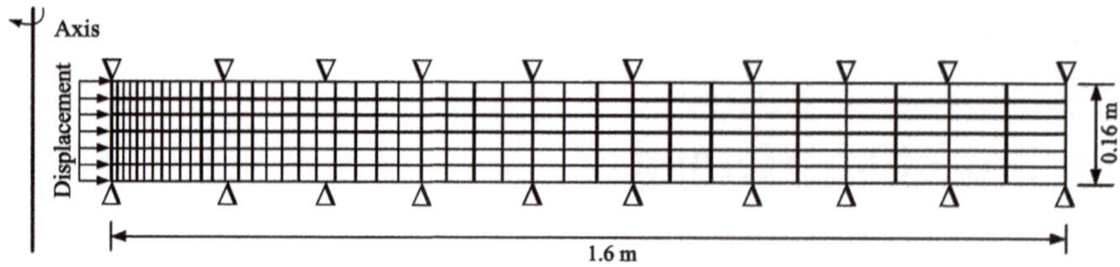

Fig. 3.1 Geometry model of PMT in ABAQUS

Fig. 3.2 Result of synthetic objective test

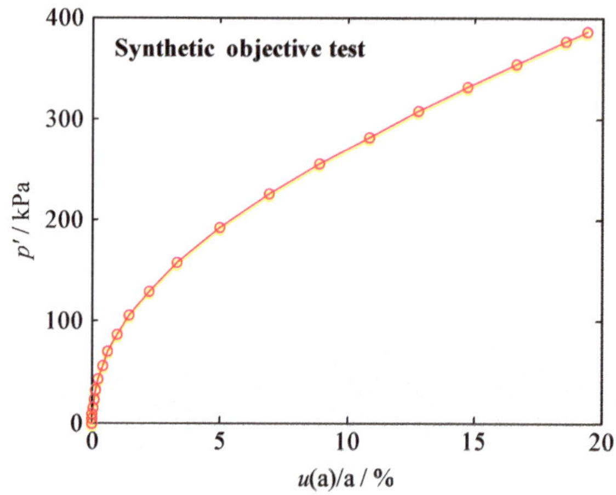

curve of the PMT, a 2D finite element model with an axisymmetric condition was created, as shown in Fig. 3.1. The upper and bottom sides are only fixed for vertical displacement, and the right side is fixed for horizontal displacement. The loading can then be generated by applying horizontal displacement at the left side. Therefore, the horizontal displacement has a biggest value at the left side and gradually decreases to zero up to the right side. A total of 240 4-node reduced integration elements (CAX4R) were used to simulate the soil. For reproducing the in situ conditions, the initial state of stress was defined by the K_0 condition. The initial vertical and horizontal stresses were, respectively, 31 and 22 kPa, consistent with field tests [2]. The same displacement as in a typical field test was applied, and at each step, the same displacement increment was applied.

Using a typical set of MC parameters (elastic modulus $E = 30$ MPa, Poisson's ratio $\nu = 0.30$, friction angle $\phi = 35°$, cohesion $c = 5$ kPa, and dilatancy angle $\psi = 5°$) which corresponds to a dense sand with a little moisture content, a synthetic result for a pressuremeter test was generated and the results are shown in Fig. 3.2, where the p'_{PMT} is the pressure applied on the left side, a is the radius of the pressuremeter probe, and $u(a)$ is the displacement of the left side. Note that the objective test can be generated by ABAQUS with any set of MC parameters as long as the parameters are reasonable in describing a general soil behavior.

3.2.2 Sensitivity Analysis

To evaluate the relative importance of each model parameter on the pressuremeter test result, a sensitivity analysis should be performed prior to the optimization procedure [3, 4]. The composite scaled sensitivity (CSS_j) analysis proposed by Hill [5] was adopted to conduct the sensitivity analysis. The composite scaled sensitivity analysis indicates the amount of information provided by the ith observation for the estimation of the jth parameter and is defined as:

$$CSS_j^i = \sqrt{\left(\frac{1}{N}\sum_{i=1}^{N}\left(\left(\frac{\partial y_i}{\partial x_j}\right)\cdot x_j\sqrt{\omega_i}\right)^2\right)} \quad (3.1)$$

where $\partial y_i / \partial x_j$ can be expressed as:

$$\frac{\partial y_i}{\partial x_j} = \frac{y_i(x + \Delta x_j) - y_i(x - \Delta x_j)}{2\Delta x_j} \quad (3.2)$$

where y_i is the ith simulated value; x_j is the jth estimated parameter; $\partial y_i / \partial x_j$ is the sensitivity of the ith simulated value with respect to the jth parameter (see Fig. 3.3); N is the number of observations; ω_i is the weighting factor, which is related to the ith observation and can be evaluated based on the statistics (i.e., variance, or standard deviation, or coefficient of variation of the error of the observation). See Calvello and Finno [3] for more details concerning ω for the laboratory data. Note that the sensitivity of the same parameter will change with the change of the parameter range. Therefore, in order to more accurately reflect the sensitivity of a parameter, the value of CSS_j should be calculated multiple times on different parameter values. In case of more than three times, i.e., $N > 3$.

The composite scaled sensitivities indicate the total amount of information provided by the observations for the estimation of parameter j and measure the relative

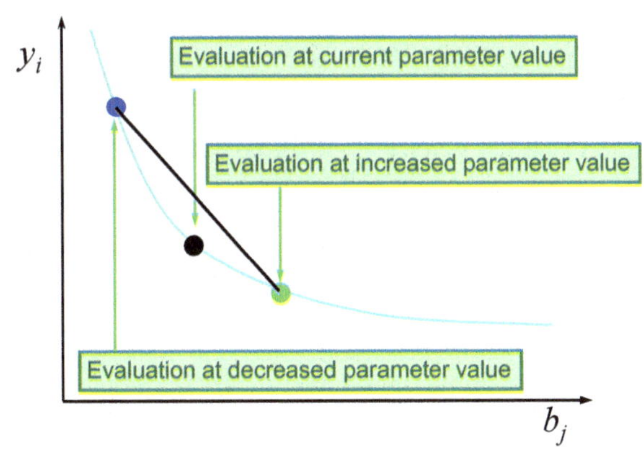

Fig. 3.3 Illustration of parameter sensitivity calculation

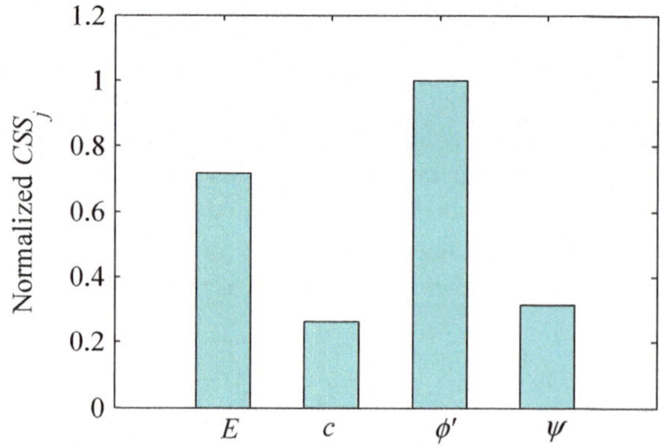

Fig. 3.4 Composite scaled sensitivity (CSS_j) of MC model parameters on PMT

importance of the input parameters being simultaneously estimated. Low values of CSS_{j} indicate high uncertainty in the parameter estimate, and the parameter can be considered to be poorly identified using the observations.

Apart from the Poisson's ratio ν, the rest of the MC parameters were selected to evaluate the sensitivity on the PMT. Figure 3.4 shows the normalized composite scaled sensitivity (CSS_j) of the selected parameters. It can be seen that the friction angle, ϕ, and Young's modulus, E, have significant effects on the simulation of the PMT, which indicates that more attention should be paid to these two parameters. The dilatancy angle, ψ, and the cohesion, c, have relatively minor effects on the simulated PMT, which demonstrates that a large step size for these parameters can be given in the optimization procedure. Note that the composite scaled sensitivity (CSS_j) analysis can be applied to any constitutive model to analyze the sensitivity of each parameter on a given geotechnical problem.

Thus, based on the sensitivity results shown in Fig. 3.4, the intervals of the selected parameters which are larger than those corresponding to their typical values are given in Table 3.1.

Table 3.1 Search domain for MC parameters in the optimization

Case	PMT (excavation)				
MC parameters	ν	E	ϕ	ψ	c
Lower bound	–(0.1)	10000 (10000)	20 (20)	0 (–)	0 (0)
Upper bound	–(0.4)	50000 (50000)	50 (50)	20 (–)	20 (20)
Step size	–(0.02)	1000 (1000)	0.1 (0.1)	0.5 (–)	0.5 (1.0)

Remark $\nu = 0.25$ is fixed for PMT; values in ( ) are for excavation

3.2.3 Optimization Results and Discussion

The error function shown in Eq. (4.5) was used to measure the error between the objective test results and the simulations. For all the selected methods, the numbers of initial individuals and the maximum evaluations were set at 30 and 3000, respectively. All the initial individuals (or populations) were generated by SOBOL. For the GA, the probabilities of crossover were set at 0.8, and the probability of mutation was set at 0.05; for PSO, the learning factors c_1 and c_2 were set at 2.0; for DE, the mutation factor was set at 0.8 and the probability of crossover was set at 1.

Following the procedure shown in Fig. 2.8, the optimizations using the selected methods were conducted. For each optimization, the calculation does not stop until reaching the maximum number of evaluations set previously even though the convergence has been reached in advance. The results of the obtained optimal parameters with the objective errors and the corresponding number of evaluations (same as the number of simulations in this case) to convergence are summarized in Table 3.2. It can be seen that the GA has the fastest convergence speed, whereas the ABC has the slowest convergence speed. The DE has the strongest search ability but a slower

Table 3.2 Optimal parameters for different optimization methods with objective error and number of evaluations corresponding to convergence

Methods	E/kPa	$\phi/°$	$\psi/°$	c/kPa	Objective error/%	Number of evaluations to convergence
GA	29000	35.2	4.5	5.5	0.114	1381
PSO	29000	35.0	5.0	5.5	0.104	2716
SA	31000	39.0	2.0	3.0	0.145	1730
DE	29000	32.5	7.0	6.5	0.055	2041
ABC	30000	28.5	9.5	8.0	0.157	–

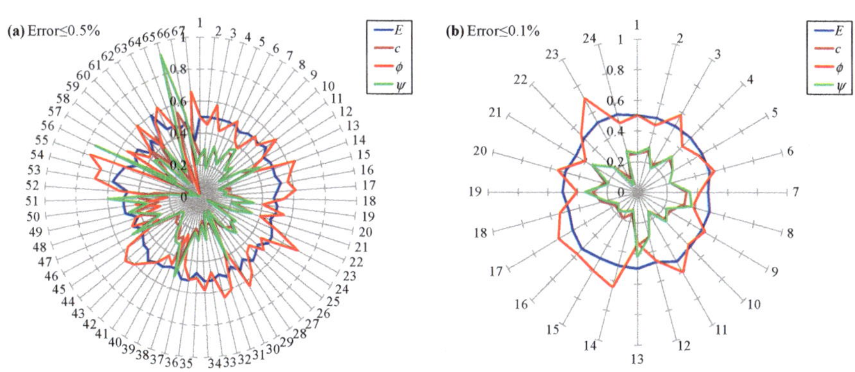

Fig. 3.5 Relationship between each parameter for similar simulation **a** error $\leq$ 0.5%; **b** error $\leq$ 0.1%

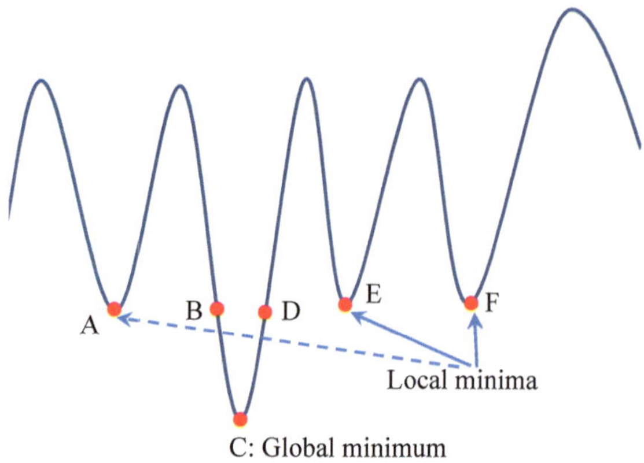

Fig. 3.6 Illustration of a multimodal optimization problem

convergence speed. However, none of the selected methods was able to find the pre-set parameters after the maximum of 3000 evaluations. In order to find the reason leading to the failure of the optimization methods, several sets of parameters with objective errors smaller than 0.5 and 0.1% were selected and plotted in Fig. 3.5. Note that all the parameters shown in Fig. 3.5 are normalized by using the ratio $(x - x_{min})/(x_{max} - x_{min})$ taking into account the upper and lower bounds. For those sets of parameters with an error smaller than 0.5%, it can be seen that a bigger ϕ with a smaller c or a smaller ϕ with a bigger c can result in similar objective errors, which indicates that a coupling effect exists among these two parameters apart from E. With a decrease in the value of the error, each parameter varies within a smaller range, which indicates a weak variability of each parameter. Thus, the identification of the MC parameters from a PMT can be considered to be a multimodal optimization problem, as illustrated in Fig. 3.6. Many existing local minima with similar objective errors would give a deceptive search direction, which could lead to a failure of the optimization process. This failure demonstrates also that the search ability of each selected algorithm is not satisfactory for solving conventional geotechnical optimization problems.

However, when solving a real engineering problem, the characteristics of the problem are usually unknown. Consequently, it is difficult to choose the most appropriate optimization method in advance. Therefore, an enhanced optimization algorithm accounting for finding a global minimum with faster convergence speed is the first choice for solving real engineering problems.

3.3 Case 2: Excavation

In order to further compare the performance of the above optimization algorithms, another identification procedure of the MC parameters was conducted on a synthetic

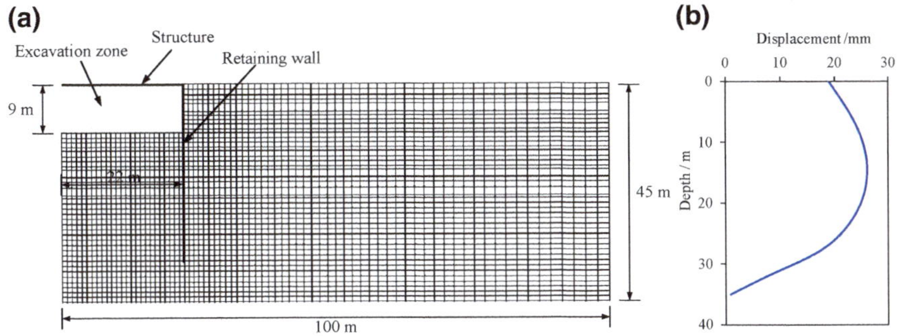

Fig. 3.7 **a** Geometry and finite element mesh of the synthetic excavation case in ABAQUS; **b** displacement of retaining wall in synthetic excavation

excavation. The same preset parameters as those used in the PMT case were adopted again to generate an objective test result. Figure 3.7a shows the geometry and the finite element mesh of the synthetic excavation. In the simulations, because of the geometric symmetry, only half of the excavation was modeled under a plain strain condition. The overall model size was 100 m long and 45 m high, which is considered large enough to avoid boundary constraints. The excavation was conducted in three steps of 3 m each. A strut was then installed at the level of the ground surface to support the retaining wall. The MC model was adopted for modeling the soil in the excavation. The retaining structure, including the retaining wall and the strut, was assumed to be linear elastic. The element type for the soil was a four-node bilinear rectangular element. The initial stress was set to K_0 condition according to Jaky's equation: $K_0 = 1 - \sin(\phi)$. Due to the relationship between ϕ and the K_0, it is not necessary to consider the K_0 as an independent parameter to be optimized in the identification procedure. A spring element and a two-node linear beam element were adopted for the strut and the retaining wall, respectively. The displacement of the retaining wall after the third excavation step is shown in Fig. 3.7b, and this result was taken as the objective test in the optimization process.

Prior to the optimization, a sensitivity analysis using a composite scaled sensitivity was also conducted. Figure 3.8 shows the normalized values of CSS_j of the MC parameters for the excavation case. A significant influence of Poisson's ratio on the wall deflection was found. This is reasonable since Poisson's ratio defined as the change of radial strain induced by the axial strain in tension test. Reflecting to the case of excavation, Poisson's ratio controls the ratio of deformations in vertical and horizontal directions. The influence of the other parameters are by decreasing order E, ϕ, c, and ψ. The effect of ψ is small and can be ignored in the optimization. The intervals of the MC parameters were determined according to the sensitivity analysis and are shown in Table 3.1. The settings for each selected optimization algorithm were the same as those used in the PMT case.

The optimal parameters and the number of evaluations to convergence for all the selected optimization algorithms are shown in Table 3.3. It can be seen that the

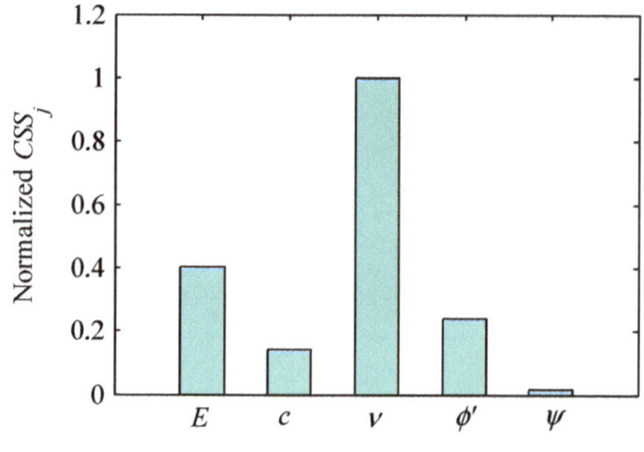

Fig. 3.8 Composite scaled sensitivity (CSS_j) of MC model parameters on excavation

Table 3.3 Optimal parameters for different optimization methods with objective error and number of evaluations corresponding to convergence

Methods	E/kPa	ϕ/°	ν	c/kPa	Objective error/%	Number of evaluations to convergence
GA	30000	28.0	5.0	8.0	0.033	1681
PSO	30000	35.0	0.3	5.0	0.0	1714
SA	30000	22.6	0.3	11.0	0.16	–
DE	30000	35.0	0.3	5.0	0.0	1259
ABC	30000	32.5	0.3	6.0	0.045	845

PSO and DE can eventually obtain the preset parameters, at a faster convergence speed for DE than for PSO. In terms of convergence speed, the ABC has the fastest convergence speed and the SA has the slowest convergence speed among all selected methods for this excavation test.

In order to evaluate the robustness of DE on identifying the parameters of different constitutive models in excavation, the proposed optimization procedure with five selected algorithms was applied to the same excavation using a more advanced constitutive model. For simulating the clay foundation, an anisotropic elastoplastic model S-CLAY1 by Wheeler et al. [6] was adopted in this case. Details of S-CLAY1 can be found in Wheeler et al. [6]. A set of parameters (initial void ratio $e_0 = 1.5$, Poisson's ratio $\nu = 0.3$, swelling index $\kappa = 0.02$, compression index $\lambda = 0.15$, slope of critical state line $M = 1.2$, and overburden pressure POP $= \sigma'_{p0} - \sigma'_{v0} = 1$ kPa) corresponding to a normally consolidated clay was preset to generate the objective curve. The number of initial population was set to 50, and the number of generations was set to 100. The search domain for each model parameter is shown in Table 3.4. Following the same procedure, all the parameters of S-CLAY1 were identified from the excavation using the five selected optimization algorithms, respectively. The optimal parameters with the objective error and the generation number of convergence obtained by different algorithms are summarized in Table 3.4. The results indicate that the DE can eventually obtain the preset parameters of S-CLAY1 model, the ABC

Table 3.4 Search domain for each S-CLAY1 parameter and optimal parameters obtained by different algorithms

Parameters	ν	κ	λ	M	POP/kPa	Objective error/%	Number of evaluations to convergence
Upper bound	0.4	0.1	0.5	2.0	5.0		
Lower bound	0.1	0.001	0.01	0.8	0.0		
Step size	0.01	0.001	0.01	0.01	0.1		
Preset	0.3	0.02	0.15	1.2	1.0	0	–
GA	0.3	0.02	0.13	1.17	0.8	0.0762	53
PSO	0.29	0.022	0.32	1.36	8.0	0.357	81
SA	0.29	0.022	0.29	1.45	3.0	0.533	83
DE	0.30	0.02	0.15	1.20	1.0	0	70
ABC	0.30	0.02	0.23	1.34	0.8	0.164	34

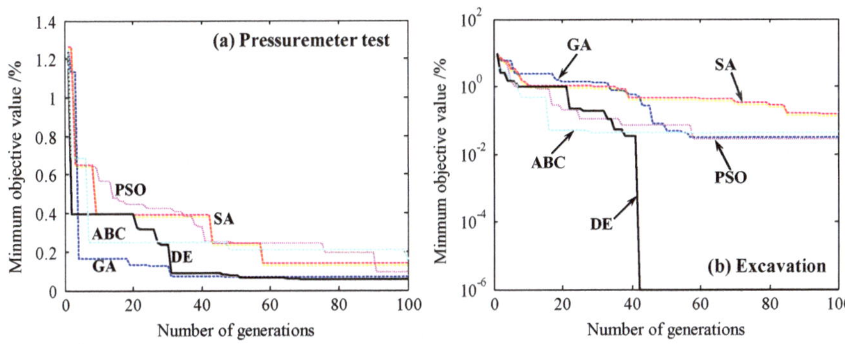

Fig. 3.9 Minimization process with increasing generation numbers

has the fastest convergence speed, and the SA has the slowest convergence speed among all selected methods.

Overall, all the results demonstrate that the classic DE performs better in identifying parameters even for the advanced soil model, as shown in Fig. 3.9, and is thus recommended. However, the classic DE still underperforms in convergence speed and search ability in the view of engineering application. Therefore, an enhancement of DE is worth trying.

3.4 Summary

A comparative study was performed for identifying the Mohr–Coulomb parameters from a synthetic PMT and an excavation tests, respectively. The GA, PSO, SA, DE, and ABC were selected to conduct the optimizations for both cases. All the comparisons demonstrated that the DE has the strongest search ability with the smallest objective error but has a relatively slow convergence speed. The performance of GA is slightly poorer than DE.

Other details can be found in Yin et al. [7].

References

1. Zentar R, Hicher P, Moulin G (2001) Identification of soil parameters by inverse analysis. Comput Geotech 28(2):129–144
2. Papon A, Riou Y, Dano C, Hicher PY (2012) Single-and multi-objective genetic algorithm optimization for identifying soil parameters. Int J Numer Anal Methods Geomech 36(5):597–618
3. Calvello M, Finno RJ (2004) Selecting parameters to optimize in model calibration by inverse analysis. Comput Geotech 31(5):410–424
4. Finno RJ, Calvello M (2005) Supported excavations: observational method and inverse modeling. J Geotech Geoenviron Eng 131(7):826–836
5. Hill MC (1998) Methods and guidelines for effective model calibration: US geological survey denver. CO, USA
6. Wheeler SJ, Näätänen A, Karstunen M, Lojander M (2003) An anisotropic elastoplastic model for soft clays. Can Geotech J 40(2):403–418
7. Yin Z-Y, Jin Y-F, Shen JS, Hicher P-Y (2018) Optimization techniques for identifying soil parameters in geotechnical engineering: comparative study and enhancement. Int J Numer Anal Methods Geomech 42(1):70–94

Chapter 4
Examples of Enhancing Optimization Algorithms

4.1 Introduction

The aim of this chapter is to show how to develop a new hybrid real-coded genetic algorithm (RCGA) and an enhanced differential evolution (DE) to identify soil parameters. As first example, under the framework of a classical GA, the hybrid RCGA is developed by combining two recently developed and efficient crossover operators with a hybrid strategy. A dynamic random mutation has also been incorporated into the new RCGA to maintain the diversity of the population. Additionally, in order to improve the convergence speed, a chaotic local search (CLS) has been adopted. Then, for the second example, an enhanced DE was developed by implementing the Nelder–Mead simplex method in a differential algorithm in order to accelerate the convergence speed with strong reliable search ability. The performance of both improved optimization algorithms is highlighted by identifying model parameters from pressuremeter tests and excavation observations.

4.2 Example 1: New Hybrid RCGA

4.2.1 Scope of the Proposed RCGA

In this section, a new hybrid RCGA is proposed. A flowchart showing the new hybrid RCGA is plotted in Fig. 4.1, where p_C, p_M, and p_S are the probabilities at which "offspring" are produced by "crossover1", "mutation", and "crossover2", respectively.

First, the evolution process starts with a set of initial individuals which are randomly generated. Then, the individuals are selected from the parent population using a tournament selection to perform the crossover and mutation, which is critical for maintaining the diversity of the population. The tournament selection is implemented for selecting the individuals to enter the mating pool, which has been demonstrated

Z. Yin and Y. Jin, *Practice of Optimisation Theory in Geotechnical Engineering*,
https://doi.org/10.1007/978-981-13-3408-5_4

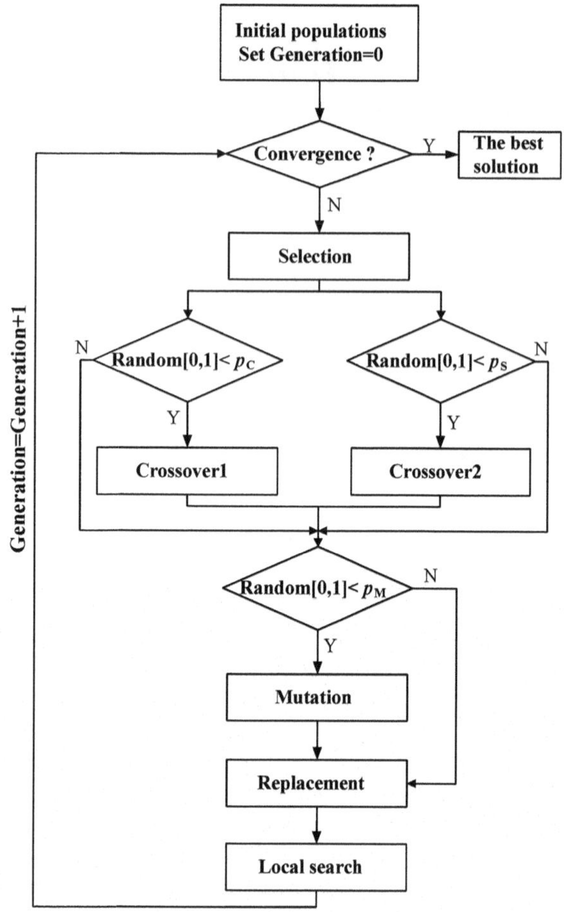

Fig. 4.1 Flowchart of the proposed RCGA

to perform well in RCGAs [1, 2]. In order to prevent the loss of diversity of the population, the chosen tournament size is two in the proposed algorithm.

In the new hybrid RCGA, a newly proposed hybrid strategy with two crossover operators is adopted to generate offspring. The selected crossover is determined by the probability of the crossover (p_C or p_S). In this hybrid RCGA, the simulated binary crossover (SBX) proposed by Deb and Agrawal [3] and the simplex crossover (SPX) developed by Da Ronco and Benini [4] is adopted. The SBX is a conventional outstanding operator, with an ability that has been highlighted by Deb and Agrawal [3] and other researchers in the optimization fields. According to Da Ronco and Benini [4], the experimental results with test functions used in their studies showed that SPX performed well on functions having multi-modality. Therefore, the search ability of the new RCGA can be improved by combining the advantages of each crossover operator.

In order to prevent the population to converge to a suboptimal solution, a newly developed mutation operator, the dynamic random mutation (DRM), proposed by Chuang et al. [5], was adopted to enlarge population diversity in the new RCGA. The DRM mutation is a self-adaptive operator, which can improve the search efficiency of the proposed genetic algorithm.

Since the population size is kept constant, the selection of survivor from both parent and offspring populations is critical to preserve the current best-found solution for subsequent evolution. Thus, the elitism strategy in NSGA-II proposed by Deb et al. [6] was implemented to perform the replacement process, which allows the parent and the offspring to compete after the crossover and mutation processes, ensuring better solutions.

Additionally, in order to increase the convergence speed, a chaotic local search (CLS) with a "shrinking strategy" proposed by Jia et al. [7] was adopted. At the beginning of the evolution process, the diversity of the population is rich, so that the convergence speed can be accelerated easily if the CLS is applied. As the generation number increases, the population converges to an optimal solution more gradually, so it is more difficult to make the population progress to the optimal solution. Therefore, in order to save computational time, only 1/3 of the total number of generations from the beginning was used in the CLS.

The pseudocode of new hybrid RCGA is given below:

1. *Begin*
2. *Set number of generation gen* $= 0$
3. *Initialization, generating the initial population P (gen)*
4. *Evaluate the fitness of P (gen)*
5. *Check the criterion of convergence. If yes, stop; if no, continue.*
6. *Select m individuals from the P (gen) using tournament selection to perform crossover*
7. *If (rand* $< p_C$*), apply SBX crossover; else if, (rand* $< p_S$*), apply SPX crossover*
8. *Apply DRM mutation on population generated after crossover according to probability* p_M *and get the new children* $P'(gen)$
9. *Put P(gen) and* $P'(gen)$ *to survive with elitism strategy, get new population* $P''(gen)$
10. *Apply the CLS conditionally on* $P''(gen)$*, and update the best individual*
11. *Gen* $=$ *gen* $+ 1$*, and P(gen* $+ 1) = P''(gen)$
12. *Back to step 5.*

4.2.2 Main Operators in the New Hybrid RCGA

4.2.2.1 Simulated Binary Crossover (SBX)

This operator simulates the behavior of the single-point crossover operator on binary strings in the sense that common interval schemata between the parents are repeated in the offspring. It works for generating the components of the offspring as follows:

$$\begin{aligned} \xi_i &= 0.5\left[(1+\beta_i)x_i^1 + (1-\beta_i)x_i^2\right] \\ \eta_i &= 0.5\left[(1-\beta_i)x_i^1 + (1+\beta_i)x_i^2\right] \end{aligned} \tag{4.1}$$

with,

$$\beta_i = \begin{cases} (2u)^{\frac{1}{\eta+1}} & \text{if } u \le 0.5 \\ \left(\frac{1}{2(1-u)}\right)^{\frac{1}{\eta+1}} & \text{if } u > 0.5 \end{cases} \tag{4.2}$$

where ξ_i and η_i are two offspring generated by SBX; β_i is spread factor; u is a uniformly distributed random variable within [0, 1]; and x_i^1 and x_i^2 are two parents selected by the tournament to create the offspring. In this case, the value of η is set to 20, as recommended by Deb et al. [6] and Zitzler and Thiele [8].

4.2.2.2 Simplex Crossover (SPX)

The offspring vector is formed as follows:

$$\xi_i = (1 + \text{Refl}) \cdot M - \text{Refl} \cdot x_i^2 \tag{4.3}$$

where M is the centroid of x_i^1, which can be calculated in the following manner:

$$M = \left(\frac{1}{n}\right) \cdot x_i^1 \tag{4.4}$$

where x_i^1 and x_i^2 are two parents selected by the tournament to create the offspring. It is assumed that x_i^1 is the best fitness individual among the two chosen parents to form the offspring. The Refl coefficient is set as equal to a random number [0, 1], n is the number of the remaining individuals, after the worst one is excluded, and $n = 2$ is employed, in this study, according to the test results conducted by Da Ronco and Benini [4].

4.2.2.3 Mutation Operator

The DRM applies the mutation rule of,

$$x_i^* = x_i + s_m \Phi_0\left(x_i^U - x_i^L\right) \tag{4.5}$$

where x_i^* is the offspring after the mutation; s_m is the mutation step size; and x_i^L and x_i^U are the lower and upper bounds of the variable in the chromosome. Φ_0 is a random perturbation vector in the n-dimensional cube $[-\phi_0, +\phi_0]^n$ of which ϕ_0 is a user-specified number chosen within the interval [0, 1].

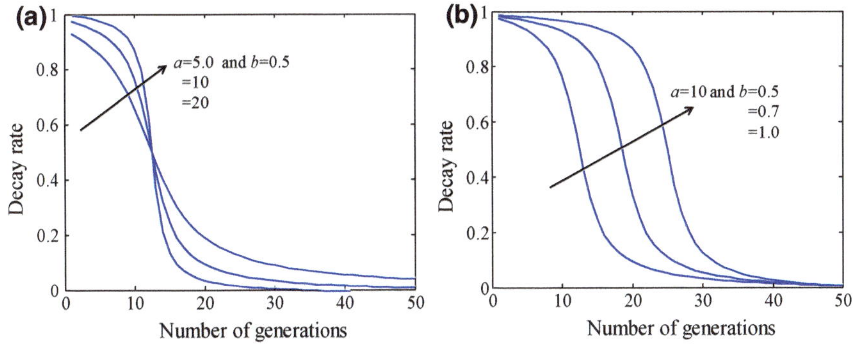

Fig. 4.2 Effects of **a** and **b** on the decay rate for the DRM

The step size was dynamically adjusted by the following update rule,

$$s_{\mathrm{m}} = (1 - k/k_{\max})^{b} \tag{4.6}$$

In this chapter, the dynamic adjustment step size can be adjusted as follows.

$$s_{\mathrm{m}} = \frac{1}{3} \cdot \arctan\left(a\left(1 - 2 \cdot \frac{k}{k_{\max}}\right)^{b}\right) + \frac{1}{2} \tag{4.7}$$

where the parameter $b > 0$ is used to control the decay rate of s_{m}; k and $k_{\max}$ denote the current generation number and the maximum number of generations, respectively. In this study, $\Phi_0 = 0.25$ and $b = 2$ are employed.

It can be seen that as the evolutionary algebra increases, the range of variation is gradually decreasing. Parameters a and b control the decay rate of s_{m}, the effect of which is shown in Fig. 4.2. It can be seen that a controls the decay rate and b locates the attenuation position. It should be noted that in the dynamic step control, a large step size can be retained in the initial generation of evolution, which will result in a larger variation range when the evolutionary algebra is increased. In the calculation, the step size needs to be gradually reduced to ensure that the result converges to the optimal solution. In addition, the DRM mutation algorithm is an adaptive operator, which also improves the search efficiency of the algorithm.

4.2.2.4 Chaotic Local Search (CLS)

The chaotic local search (CLS) with a "shrinking strategy" proposed by Jia et al. [7] was adopted as follows,

$$x_i^{\prime t} = (1 - \lambda)x_i^t + \lambda\beta_c \tag{4.8}$$

where $x_i^{\prime t}$ is a new vector of individual x_i^t in t generation produced by the chaotic local search; β_c is generated by the equation $\beta_c = x_i^L + \beta_j^t(x_i^U - x_i^L)$; and λ is the shrinking scale given by:

$$\lambda = 1 - \left|\frac{\text{FEs} - 1}{\text{FEs}}\right|^m \quad (4.9)$$

where FEs are the current function evaluations; m controls the shrinking speed. With higher m values, the shrinking speed is slower. In this study, the value of m was set at 1000, as suggested by Jia et al. [7].

β_j^t is a chaotic variable, which is obtained from the chaotic iteration. In this study, the logistic chaotic function was employed to construct a chaotic GA as follows:

$$\beta_j^{t+1} = \mu\beta_j^t\left(1 - \beta_j^t\right), \quad t = 1, 2, \ldots; \quad \beta_j \neq 0.25, 0.5 \text{ and } 0.75 \quad (4.10)$$

when $\mu = 4$, Eq. (4.10) reaches a complete chaotic state. Given β_j^1 an arbitrary initial value that is within the range of 0–1, but not equal to 0.25, 0.5, or 0.75, the chaos trajectory will finally search non-repeatedly any point within the range (0, 1).

4.2.3 Performance of the New Hybrid RCGA

To evaluate the performance of the proposed RCGA, six mathematical functions of different types were chosen as benchmark tests, which are usually used as benchmark tests to evaluate the performance of a new GA [9, 10]. Table 4.1 shows the selected benchmark tests with the optimum value corresponding to the best solution. In order to highlight the performance of the new RCGA, an extensive experimental study of various possible hybrid combinations of crossovers has been conducted. The other outstanding crossovers are arithmetical crossover (AC), Laplace crossover (LX), and bounded exponential crossover (BEX), which are described in Appendix I. Therefore, five different RCGAs have been defined and named as follows: AC + SPX + DRM, LX + SPX + DRM, BEX + SPX + DRM, SBX + SPX + DRM and SBX + SPX + DRM + Chaotic. For a fair comparison, the settings for each RCGA are the same except for the crossover, which has been changed according to a hybrid strategy. The settings for all the RCGAs are given in Table 4.2. In this study, six benchmark functions with 30 variables were adopted. The maximum number of generations was fixed at 100. For a uniform testing environment of all the RCGAs, the initial population size was taken to be ten times the number of decision variables. According to Poles et al. [11], using a well-distributed sampling can increase the robustness and avoid premature convergence. Thus, the initial populations for all the RCGAs were generated by Sobol.

The performance of a GA is usually measured on the basis of two criteria, efficiency and accuracy. The efficiency of a GA is a measure of the rate of conver-

Table 4.1 Selected benchmark tests for evaluating the new GA

Test function	x domain	Optimum
1. Ackley's problem: $f_1 = -20\exp\left(-0.02\sqrt{\frac{1}{n}\sum_{i=1}^{n} x_i^2}\right) - \exp\left(\frac{1}{n}\sum_{i=1}^{n}\cos(2\pi x_i)\right) + 20 + e$	$-30 \le x_i \le 30$	$\min f(x^*) = 0$ with $x^* = (0, 0, \ldots, 0)$
2. Exponential problem: $f_2 = -\left(\exp\left(-0.5\sum_{i=1}^{n} x_i^2\right)\right)$	$-1 \le x_i \le 1$	$\min f(x^*) = -1$ with $x^* = (0, 0, \ldots, 0)$
3. Griewank problem $f_3 = 1 + \frac{1}{4000}\sum_{i=1}^{n} x_i^2 - \prod_{i=1}^{n}\cos\left(\frac{x_i}{\sqrt{i}}\right)$	$-600 \le x_i \le 600$	$\min f(x^*) = 0$ with $x^* = (0, 0, \ldots, 0)$
4. Rosenbrock problem: $f_4 = \sum_{i=1}^{n-1}\left[100(x_{i+1} - x_i^2)^2 + (x_i - 1)^2\right]$	$-30 \le x_i \le 30$	$\min f(x^*) = 0$ with $x^* = (1, 1, \ldots, 1)$
5. Schwefel problem: $f_5 = -\sum_{i=1}^{n} x_i \sin\left(\sqrt{\lvert x_i\rvert}\right) + 418.98288n$	$-500 \le x_i \le 500$	$\min f(x^*) = 0$ with $x^* = (420.98, 420.98, \ldots, 420.98)$
6. De-Jong's function with noise $f_6 = \sum_{i=1}^{n}\left(x_i^4 + \text{rand}(0, 1)\right)$	$-10 \le x_i \le 10$	$\min f(x^*) = 0$ with $x^* = (0, 0, \ldots, 0)$

Table 4.2 Parameter settings for the five RCGAs

RCGA	Crossover (probability)	Mutation (probability)	CLS	Tournament size	Elitism
AC + SPX + DRM	AC (0.9) + SPX (0.5)	DRM (0.05)	No	2	Yes
LX + SPX + DRM	LX (0.9) + SPX (0.5)	DRM (0.05)	No	2	Yes
BEX + SPX + DRM	BEX(0.9) + SPX (0.5)	DRM (0.05)	No	2	Yes
SBX + SPX + DRM	SBX (0.9) + SPX (0.5)	DRM (0.05)	No	2	Yes
SBX + SPX + DRM + Chaotic	SBX (0.9) + SPX (0.5)	DRM (0.05)	Yes	2	Yes

gence, and the accuracy indicates the degree of precision in locating global minima. Figure 4.3 shows the evolution of the minimum objective value as the generation number increases. As shown in Fig. 4.3, for RCGAs without local search, SBX + SPX + DRM performs well on problems 1, 3, 4, and 5, which indicates that SBX + SPX + DRM has an outstanding ability in tackling complex problems; BEX + SPX + DRM performs well on problem 2, and AC + SPX + DRM performs well on problem 6. For problems 2 and 6, although the performance of SBX + SPX + DRM is not the best, the difference in performance between SBX + SPX + DRM and the best RCGA is relatively small and can be regarded as the same. Therefore, it can be seen that SBX + SPX + DRM has excellent search ability in detecting the optimal solution and a relatively faster convergence speed for various complex problems.

However, the convergence speed of SBX + SPX + DRM does not satisfy all the benchmark tests. In order to improve the convergence speed, but without consuming much more computational time, a chaotic local search (CLS) was added to enhance the performance of SBX + SPX + DRM, and the resulting RCGA is referred to as SBX + SPX + DRM + Chaotic. The CLS was only applied at the 30th generation from the beginning. As shown in Fig. 4.3, compared to SBX + SPX + DRM, the convergence speed was improved significantly by using the chaotic local search. For problems 4, 5, and 6, not only the convergence speed but also the accuracy of the optimal solution was further enhanced. All the comparisons demonstrate that the effectiveness of the CLS in accelerating convergence speed is excellent and should be adopted in the RCGAs.

Overall, the results demonstrate that the proposed hybrid RCGA (SBX + SPX + DRM + Chaotic) has an outstanding search ability, and the efficiency and accuracy of the proposed RCGA are ideal for tackling different problems.

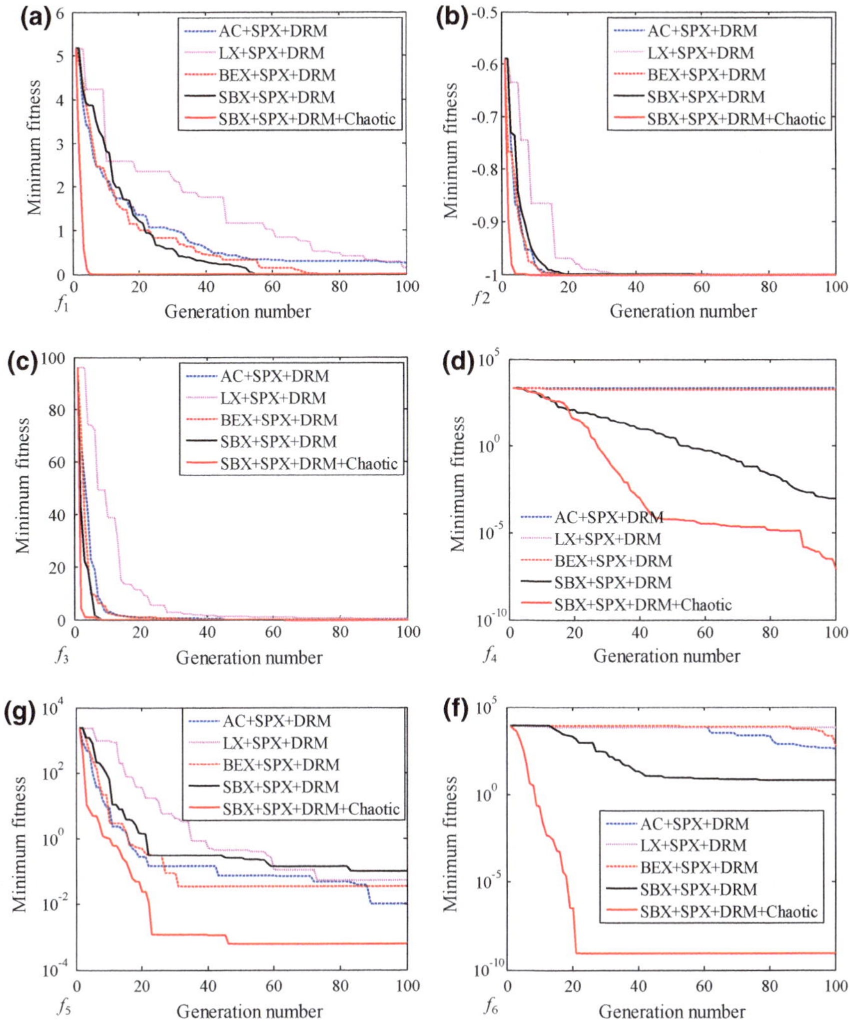

Fig. 4.3 Comparisons of performance between six RCGAs for different benchmark tests

4.3 Applications in the Identification of Soil Parameters

For further examining the ability of the proposed RCGA, the proposed hybrid RCGA has been applied to solve the problem of parameter identification.

4.3.1 Identification Methodology

The aim of an optimization is to find values for the model parameters that provide the best attainable fit between model predictions and corresponding observations. For this purpose, the error function is required, with "Error" based on the least square method as introduced by Levasseur et al. [12].

Generally, deformation and strength are two extremely important indicators for showing the mechanical behavior of soil. In a laboratory triaxial test, the isotropic or anisotropic compression test is conducted first, followed by the shear stage. During the whole process, the model parameters accounting for compression and shear behaviors are measured and obtained. For field tests, such as the pressuremeter test, cone penetration test, or vane shear test, the test results are usually displayed in the form of the displacement–pressure curve. The soil behavior (softening or hardening, contraction or dilation) are also implied in these curves, although behavior related to some parameters cannot be directly measured. In other words, the results of selected tests can provide information to optimize the model parameters. Therefore, in this study, a mono-objective framework was considered, which includes two objectives regarding the strength and deformation of soil, respectively:

$$\min[\text{Error}(x)] = \begin{bmatrix} \text{Error}(q) \\ \text{Error}(e)\,\text{or}\,\text{Error}(\Delta u) \end{bmatrix} \tag{4.11}$$

where Error(q) is the difference between the deviatoric stress from the simulations and that in the objectives; Error(e) is the difference between the void ratio from the simulations and that in the objectives for the drained tests; and Error(Δu) is the difference between the excess pore pressure from the simulations and that in the objectives for the undrained tests.

Figure 4.4 shows the identification procedure based on the successive use of two different codes: The code for the integration of the constitutive model is written in

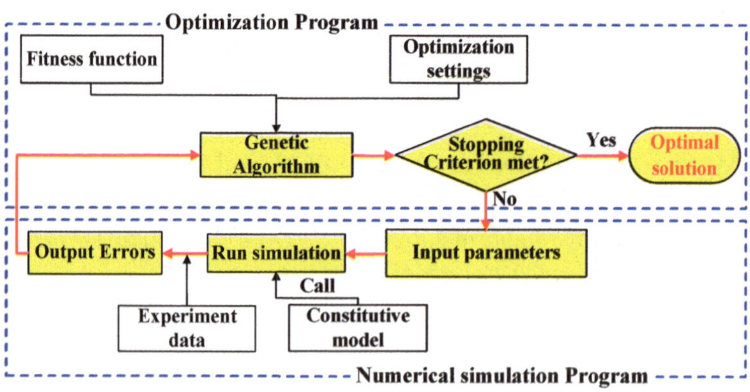

Fig. 4.4 Identification procedure

the FORTRAN language, and the code for the optimization process is written in the MATLAB language. To demonstrate cases in more general, some relatively simple constitutive models were adopted.

4.3.2 Identifying Parameters from Laboratory Testing

(1) For sand

The parameter identification for sand using the RCGA was performed first. A Mohr–Coulomb-like model with nonlinear elasticity and plastic hardening (NLMC), similar to the soil hardening model proposed by Schanz et al. [13] in PLAXIS, was developed to simulate the objective tests. The constitutive equations are shown in Table 4.3. For the NLMC model, Young's modulus is expressed as follows, according to Richart et al. [14]:

$$E = E_0 \cdot p_{\text{at}} \frac{(2.97 - e)^2}{(1 + e)} \left(\frac{p'}{p_{\text{at}}} \right)^{\zeta} \tag{4.12}$$

where E_0 is the reference value of Young's modulus; e is the void ratio; p' is the mean effective stress; p_{at} is the atmospheric pressure used as reference pressure (p_{at} = 101.325 kPa); and ζ is a constant.

The NLMC model has six parameters (1) elastic parameters: E_0 and ζ, which can be obtained from isotropic compression tests; and (2) plastic parameters: plastic modulus, k_p, friction angle, ϕ, dilatancy angle, ψ, and cohesion, c. Generally, a typical value of Poisson's ratio, $\upsilon = 0.2$ is assumed for the sand. All the parameters were then identified using the optimization method with the new hybrid RCGA from selected objective tests. The search domain and intervals of these parameters are

Table 4.3 Constitutive relations of selected soil models

Models	NLMC	MCC
Elasticity	$\dot{\varepsilon}_{ij}^{e} = \frac{1+\upsilon}{E}\sigma'_{ij} - \frac{\upsilon}{E}\sigma'_{kk}\delta_{ij}$	
Yield function	$f = \frac{q}{p'} - H = 0$	$f = \frac{q^2}{M^2} + p'(p' - p_c) = 0$
Potential function	$\frac{\partial g}{\partial p'} = M_{pt} - \frac{q}{p'}$ with $\frac{\partial g}{\partial q} = 1$ $M_{pt} = \frac{6\sin\phi_{pt}}{3 - \sin\phi_{pt}}$ with $\phi_{pt} = \phi_\mu - \psi$	$g = f$
Hardening law	$H = \frac{M\varepsilon_d^p}{k_p + \varepsilon_d^p}$ with $M = \frac{6\sin\phi_\mu}{3 - \sin\phi_\mu}$	$dp_c = p_c\left(\frac{1+e_0}{\lambda-\kappa}\right)d\varepsilon_v^p$
Number of parameters	6	5

Table 4.4 Search domain and intervals of parameters for NLMC and MCC

Model	NLMC					MCC			
Parameters	E_0	ζ	k_p	ϕ (°)	ψ (°)	κ	λ	M	p'_{c0} (kPa)
Lower bound	10	0.1	10^{-5}	10	0	10^{-4}	10^{-3}	0.5	10
Upper bound	500	1.0	0.1	50	20	0.1	0.5	2.0	200
Step	1.0	0.01	10^{-5}	0.1	0.1	10^{-4}	10^{-3}	0.01	0.5

given in Table 4.4 and are much larger than those corresponding to typical values. The initial population for the RCGA was generated in SOBOL within the search domain.

A series of standard drained triaxial compression tests [15] with an isotropic compression test [16] performed on Fontainebleau sand were selected as the objective tests during the optimization process. All the tests were isotropically consolidated to the corresponding consolidation pressure before shearing. The experimental results for these drained triaxial tests are shown in Fig. 4.5. Note that the cohesion, c, is taken equal to be zero, as dry Fontainebleau sand was used in the test. Thus, two elastic parameters together with other plastic-related parameters formed one set of model

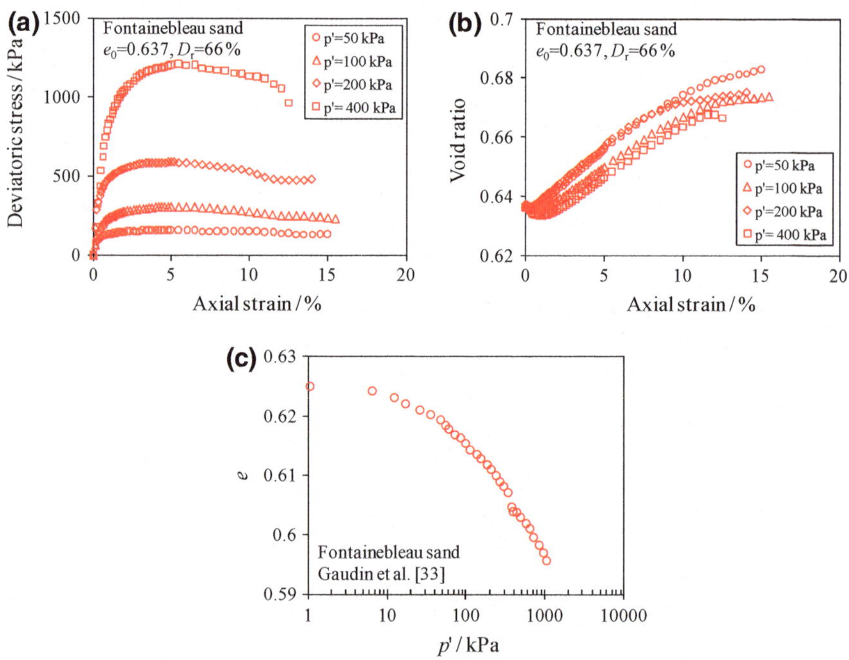

Fig. 4.5 Results of drained triaxial tests on Fontainebleau sand: **a** deviatoric stress versus axial strain; **b** void ratio versus axial strain; **c** isotropic compression test

parameters, which were used to simulate objective tests during the optimization process.

Following the proposed identification procedure, optimization involving the NLMC model was performed using the proposed hybrid RCGA. The optimization was performed many times following the new GA with different set values in order to select the best result. The optimal set of parameters obtained using the NLMC model corresponding to Fontainebleau sand is given in Table 4.5. Figure 4.6 shows the simulation results for Fontainebleau sand from the NLMC model when the optimal parameters are used. However, comparisons between simulated and experimental results indicate that the new hybrid RCGA has the ability to detect the optimal parameters. Note that due to the limitations of the NLMC model, e.g., disregarding the interlocking effect [17–20], the strain softening behavior cannot be reproduced.

Table 4.5 Optimal sets of parameters for NLMC for Fontainebleau sand

Parameters	E_0	ζ	k_p	ϕ_μ (°)	ψ (°)
Values	180	0.55	0.00031	36.4	5.9

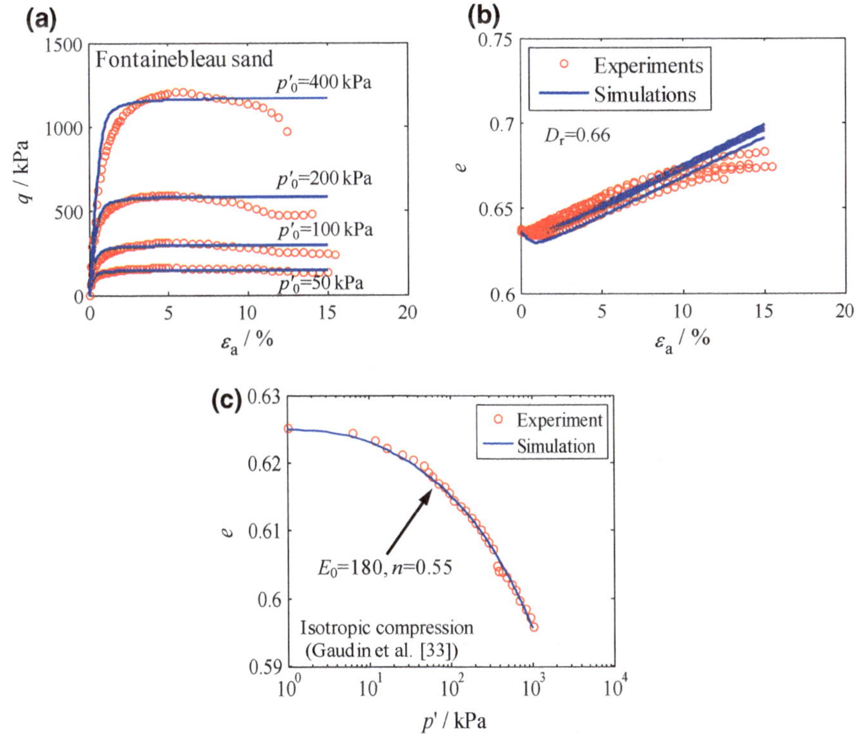

Fig. 4.6 Simulation results based on optimal parameters for Fontainebleau sand: **a** deviatoric stress versus axial strain; **b** void ratio versus axial strain; **c** isotropic compression test

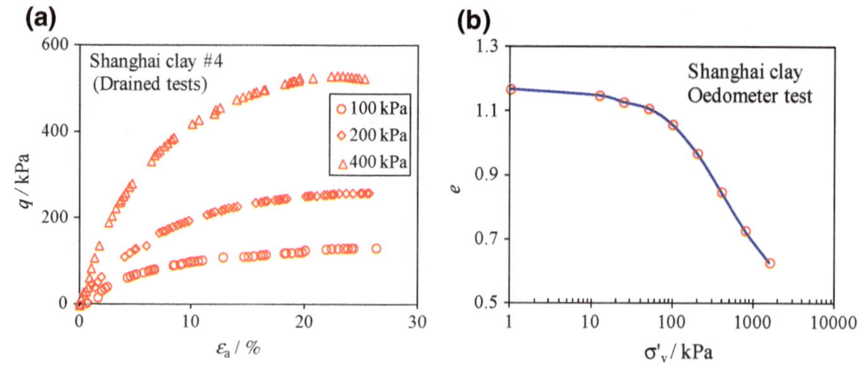

Fig. 4.7 Results of Shanghai clay: **a** stress–strain of drained triaxial test; **b** oedometer test

Table 4.6 Optimal sets of parameters of MCC for Shanghai soft clay

Parameters	κ	λ	M	p'_{c0} (kPa)
Value	0.032	0.171	1.34	100

(2) For clay

As previously described, the new RCGA can succeed in identifying the model parameters for sand. In order to show the ability of the new RCGA in identifying soil parameters for clay, the widely used modified Cam-Clay (MCC) model developed by Roscoe and Burland [21] and a series of tests on natural Shanghai clay conducted by Sheng [22] were selected.

Three drained compression triaxial tests and one oedometer test performed on Shanghai clay were selected as the objective to identify basic MCC parameters, as shown in Fig. 4.7. Thus, in this case, a mono-objective framework with three different criteria is considered. A typical value of Poisson's ratio $\upsilon = 0.25$ for Shanghai clay can be assumed. Thus, there are four parameters: swelling index, κ, compression index, λ, the slope of critical state line, M, and the size of yield surface p'_{c0}, which need to be determined by an optimization method with the new hybrid RCGA. The search domain and intervals of these parameters are given in Table 4.4, and they are much larger than those corresponding to typical values.

The same optimization procedure as that used for the NLMC model was conducted again. The optimal parameters of the MCC model are presented in Table 4.6. All the optimal parameters are in a reasonable range and were obtained from the experimental measurements made by Sheng [22]. Using the optimal parameters, objective tests were simulated in the MCC model, as shown in Fig. 4.8. The best fit between experiments and simulations by MCC model was detected. The big difference in volumetric strain is due to the limitation of the model in which the location of the

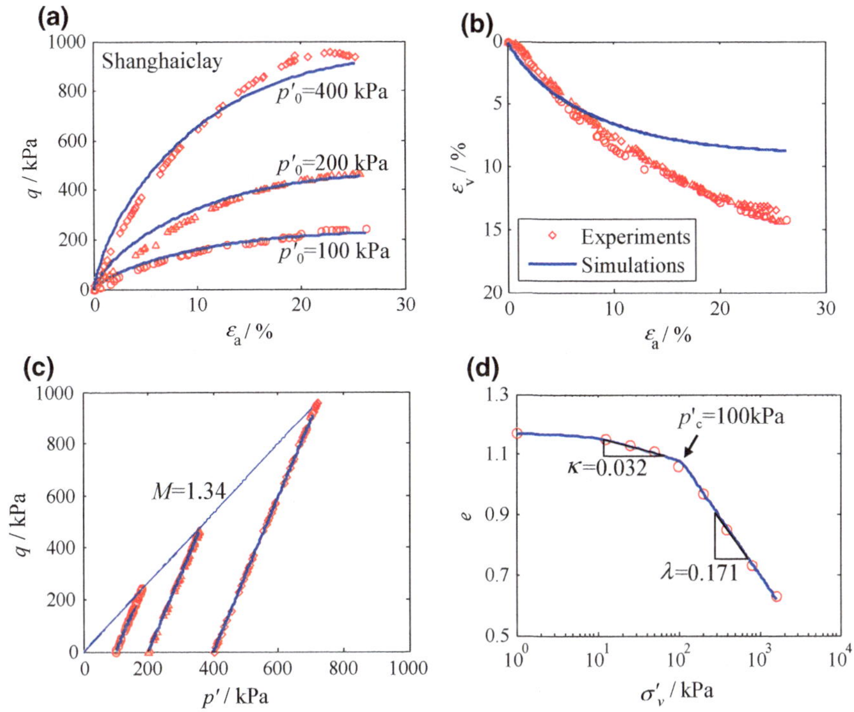

Fig. 4.8 Simulation results based on optimal parameters for Shanghai clay: **a** deviatoric stress versus axial strain; **b** void ratio versus axial strain; **c** stress path; **d** oedometer test

critical state line in e-log p' plane is fixed and not suitable for a given clay [23], or for a lack of considering anisotropy, destructuration, or time effect [24–29]. Therefore, all the comparisons between simulated and experimental results demonstrate that the optimal parameters from the MCC model corresponding to Shanghai clay detected by the hybrid RCGA are reasonable. Furthermore, this also demonstrates that the new hybrid RCGA performs well in identifying clay model parameters.

Overall, for the identification of parameters based on the laboratory tests, the proposed hybrid RCGA shows the strong and stable search ability for different constitutive models with various numbers of model parameters. Moreover, the parameters obtained from the optimization with the proposed hybrid RCGA are reliable and reasonable, which demonstrates that the optimization procedure is feasible and can be adopted as a tool in engineering practice.

4.4 Example 2: Enhancement of Differential Evolution Algorithm

As mentioned above, the simplex algorithm (NM simplex) has local convergence characteristics and fast convergence. It is suitable for use as a local search operator in a random algorithm with global search ability, thus forming a combined algorithm with high performance. Therefore, with the help of this idea, this section uses an improved simplex algorithm (Nelder–Mead method or downhill simplex method) to enhance the optimization ability of the traditional DE algorithm. The new improved optimization algorithm is named NMDE in this book.

In NMDE, in order to accelerate the convergence speed, the Nelder–Mead simplex presented previously is used. Before performing the DE mutation, all the individuals are sorted based on their fitness values, and the best $n + 1$ (n is the number of variables) individuals are selected to perform the Nelder–Mead simplex. Based on the results of the Nelder–Mead simplex, the best individual is updated. Then, this best individual was recombined with the $N - (n + 1)$ remaining individuals to perform the DE mutation. This process will be executed N times, and a new population with N individuals is obtained. Then, the obtained population is applied to the crossover operation. In order to avoid a rapid loss of diversity, an elitism strategy is adopted to perform the selection. In the selection, 10% of individuals with the highest fitness are selected from the parents and children to survive to the next generation. The remainders are chosen by tournament selection from the mating pool composed of parents and children other than the 10% individuals. The completion mechanism can help the NMDE find the better solutions (Fig. 4.9).

During the optimization process, the NMS could accelerate the convergence speed, but could easily get trapped in the local minimum. On the other hand, the DE process can overcome the local minimum problem, but the convergence speed is always slow. Therefore, the advantages of each process can be obtained through this hybrid strategy, resulting in better optimization performance.

For the NMS, a Nelder–Mead method with adaptive parameters for high-dimensional problems suggested by Gao and Han [30] was implemented, and a brief introduction to this method is presented in Appendix I. The tolerance for convergence in NMS was set to 10^{-4}, as recommended by Gao and Han [30].

For the adopted DE, a newly proposed mutation strategy by Zhang and Sanderson [31] was adopted as follows

$$v_i = \boldsymbol{x}_i + F^i\left(\boldsymbol{x}_{\text{best},i} - \boldsymbol{x}_i\right) + F^i(\boldsymbol{x}_{r1} - \boldsymbol{x}_{r2}) \tag{4.13}$$

where the indices r_1 and r_2 are distinct integers uniformly chosen from the set $\{1, 2, \ldots, N_p\}$; N_p is the number of individuals in one generation; $(\boldsymbol{x}_{r1} - \boldsymbol{x}_{r2})$ is a difference vector to mutate the corresponding parent $\boldsymbol{x}_i$; $\boldsymbol{x}_{\text{best},i}$ is the best vector in the current generation i, which is randomly chosen as one of the top $100p\%$ individuals in the current population with $p \in (0, 1]$, and in this case p was set to 0.1; F^i is the

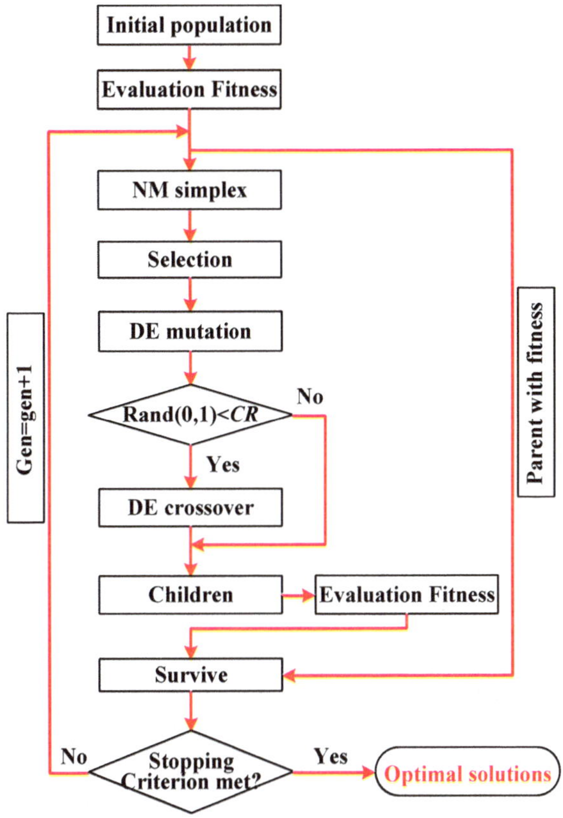

Fig. 4.9 Flowchart of NMDE

mutation factor that is associated with $\boldsymbol{x}_i$ and is regenerated at each generation by a randomly uniform distribution within [0.5, 1.0].

After mutation, a binomial crossover operation forms the final trial/offspring vector

$$\boldsymbol{u}_{i,j} = \begin{cases} \boldsymbol{v}_{i,j}, & \text{if } \mathrm{rand}(0,1) \le CR \text{ or } j = j_{\mathrm{rand}} \\ \boldsymbol{x}_{i,j}, & \text{otherwise} \end{cases} \tag{4.14}$$

where rand(a, b) is a uniform random number in the interval [a, b] and is independently generated for each j and each i; $j_{\mathrm{rand}} = \mathrm{randint}\ (1, D)$ is an integer randomly chosen from 1 to D and is newly generated for each i, D being the dimension of the problem; the crossover probability $CR \in [0, 1]$ corresponds roughly to the average fraction of the vector components that are inherited from the mutation vector; $CR =$ 1 was taken in this study.

In order to avoid a rapid loss of diversity, an elitism strategy was adopted to perform the selection. In the selection, 10% of individuals with the highest fitness were selected from the parents and children to survive to the next generation. The remainders were chosen by tournament selection from the mating pool composed of parents and children other than the 10% individuals. The enhanced algorithm is named as NMDE in the following sections for the simplicity.

The pseudocode of NMDE is given below.

1. Begin
2. Set number of generation gen = 1
3. Initialization, generating the initial population P (gen)
4. Evaluate the fitness of P(gen)
5. selecting n + 1 individuals with high fitness, then apply the NM simplex on the n + 1 individuals, other individuals N − (n + 1) keep unchanged. The obtained new population is P′(gen)
6. Select m individuals from the P′(gen) to perform mutation (Eq. 4.13), *the process will be executed N times, the obtained new population is P″(gen)*
7. Apply crossover (Eq. 4.14) *on population P″(gen) and get the new children P‴(gen), then calculating the fitness of P‴(gen)*
8. Put P(gen) and P‴(gen) to survive with elitism strategy, get new population P⁗(gen)
9. Check the criterion of convergence. If yes, stop; if no, continue
10. Gen = gen + 1, and P(gen + 1) = P⁗(gen)
11. Back to step 5

4.5 Validation by Synthetic Cases and Real PMTs

Then, the same identifications in the PMT and in the excavation tests were conducted again using the proposed algorithm. Table 4.7 shows the optimization results, and Fig. 4.10 shows the minimization process with increasing generation numbers, compared with other optimization methods. It can be seen that the preset solution was finally found by the proposed algorithm with a faster convergence speed, which demonstrates the high performance of the proposed algorithm.

Moreover, to further validate the performance of the enhanced algorithm, an inverse analysis of real PMTs investigated by Papon et al. [32] was conducted.

Table 4.7 Optimal parameters for different optimization methods with objective error and number of evaluations corresponding to convergence

Case	E (kPa)	ϕ (°)	ψ (°) (ν)	c (kPa)	Objective error (%)	Number of evaluations to convergence
PMT	30,000	35.0	5.0	5.0	0.0	1111
Excavation	30,000	35.0	0.3	5.0	0.0	661

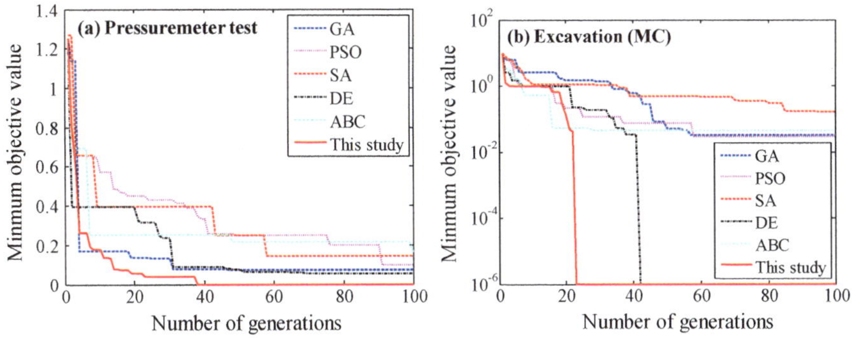

Fig. 4.10 Minimization process in identifying parameters for all selected optimization methods, **a** PMT; **b** excavation

Table 4.8 Optimal sets of parameters obtained by NMDE for real PMTs

PMT	Algorithm	E (MPa)	ϕ (°)	ψ (°)	c (kPa)	Error (%)
PMT1	NMDE	31.0	34.0	4.0	10.0	7.32
	GA	37.0	33.0	3.0	9.5	7.98
PMT2	NMGE	34.0	39.2	9.2	10	13.24
	GA	36.0	39.0	9.0	8.5	13.84

According to Papon et al. [32], two different PMTs (taken at a depth of 2 and 5.9 m) were performed in different soil layers under a spread footing. For reproducing the in situ conditions, the initial stress state of the soil was defined under K_0 condition with the initial vertical stress equal to 31 kPa and the horizontal stress equal to 22 kPa. The imposed PMT loading was displacement-controlled, and, at each step, the same displacement increment was applied all along the probe.

According to Papon et al. [32], the elastic modulus E, the friction angle ϕ, and the cohesion c are obtained by the identification procedure, whereas Poisson's ratio was taken equal to 0.33 and the correlation between the friction and the dilatancy angles $\psi = \phi - 30°$ was considered.

For the optimization, the initial population was set at 30 and was randomly generated using the algorithm SOBOL. The search domain for the linear elastic-perfectly plastic Mohr–Coulomb model is the same as the one proposed by Papon et al. [32]. All the parameters of NMDE have the recommended values discussed above. The optimal results are shown in Table 4.8, and the simulations of the PMTs using the optimal set of parameters are shown in Fig. 4.11. The poor simulation at the relatively small strain (between 0 and 5%) is found, which attributes to that the first part of the experimental curves (up to around $u(a)/a = 4\%$) was not taken into account in the calculation of the error function because of the unusual curvature at the beginning of the pressuremeter tests, due to the remolding of the soil along the cavity wall indicated by Papon et al. [32].

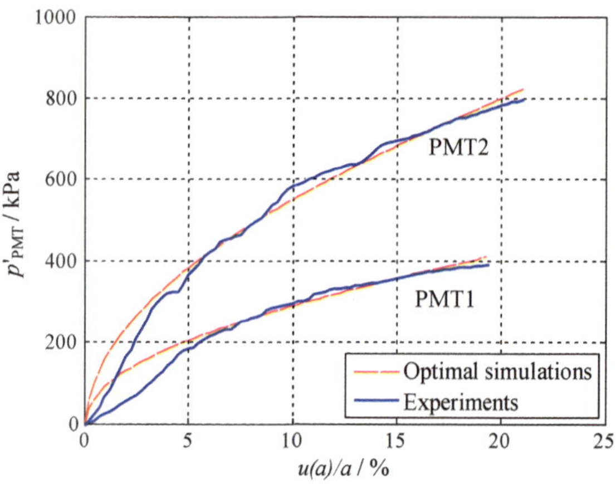

Fig. 4.11 Comparison between experimental results and NMDE-based optimal simulations for real PMTs

Since the genetic algorithm is superior to others except DE, as shown in the previous sections, only the genetic algorithm-based identification of the same case by Papon et al. [32] was compared here to highlight the performance of the proposed NMDE. For PMT1 at a depth of 2 m, a lower value of the error was obtained with the optimal set of parameters, compared to the error given by GA conducted by Papon et al. [32]. Note that the value of the friction angle given by NMDE is closer to the one obtained from the triaxial tests performed on specimens sampled at a depth of 0.6 m (34.2°) and is thus more reliable and accurate than that given by the genetic algorithms (Papon et al. [32]). For PMT2 at a depth of 5.9 m, a similar optimal set of parameters with a slightly lower error was obtained by the NMDE compared to the one obtained by Papon et al. [32]. For the convergence speed, the NMDE needs only 15 generations with 30 individuals, which is much faster than the genetic algorithm (39 generations with 200 individuals by Papon et al. [32]).

For further evaluating the efficiency and effectiveness of the enhanced algorithm on identifying parameters of complex constitutive models, two self-boring pressuremeter (SBP) tests performed in the Burswood Peninsula site at a depth of 5.25 m by Lee and Fahey [33] were selected. For two tests, the expansion phase was conducted at two different rates up to around 10% of cavity strain: 0.167 and 0.0185%/s, respectively. According to Lee and Fahey [33], the effective vertical stress σ'_{v0} was 31 kPa and the lateral earth coefficient K_0 was 0.9. The water pore pressure was 38.8 kPa. To simulate the rate-dependent characteristics of soft clay in the SBP tests, a newly developed elasto-viscoplastic model called "ANICREEP" by Yin et al. [25, 34] was adopted and implemented into ABAQUS to couple with the Biot's consolidation theory. To focus on identifying the rate-dependent parameters, the initial void ratio, e_0, the swelling index, κ, and the compression index, λ, with the permeability ($k = 3.3 \times 10^{-9}$ m/s) measured from oedometer tests by Lee and Fahey [33] were given in the optimization procedure. Poisson's ratio ν was assumed to be 0.3, which

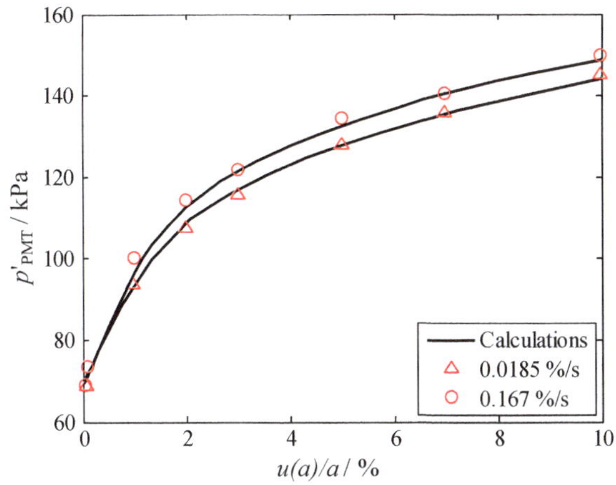

Fig. 4.12 Comparison between predicted and experimental results for SBP tests on Burswood clay

Table 4.9 Values of ANICREEP model for Burswood clay

Clay	ν	e_0	κ	λ	M	σ_{p0} (kPa)	$C_{\alpha e}$
Burswood clay	0.3	2.8	0.036	0.40	1.69	59	0.019

can be considered as a typical value assuming ideally elastic-perfectly plastic MC model. The other parameters (slope of critical state line M, the initial preconsolidation pressure σ_{p0}, and the secondary compression coefficient $C_{\alpha e}$) were determined by the proposed procedure of parameter identification. Figure 4.12 compares the predicted and measured results for SBP tests in terms of total pressure at the cavity wall versus cavity strain. The identified $M(=1.69)$ and $\sigma_{p0}(= 59\text{ kPa})$ are summarized in Table 4.9, which are close to values by Lee and Fahey [33]. The ratio of $C_{\alpha e}/C_c = 0.019/0.40 = 0.0475$ for Burswood clay is also in a reasonable range according to Mseri and Godlewski [35] for the silty clay. All the comparisons demonstrate that the proposed NMDE is found efficient for identifying soil parameters from in situ testing.

Overall, all these comparisons for the identification of soil parameters based on both synthetic cases and real PMTs demonstrate a better performance of the proposed NMDE than the above-mentioned algorithms.

4.6 Discussion

It should be pointed out that in this paper we only focus on investigating the performance of the proposed optimization-based inverse method on identifying parameters using several synthetic cases and two real PMTs. The presented results can give a clear evidence to show the ability of proposed algorithm on identifying model parameters. However, the two examples in this study are very common when dealing with evaluation of optimization techniques. Coupling effects that exist between parameters and their consequences on optimization process (due to the presence of many local minima) are well known. Furthermore, trying to identify one unique set of parameters for a geotechnical problem might be hopeless or at least not the most relevant manners to use optimization techniques in geotechnics considering uncertainties on models and measurements as well as heterogeneities of soils. Therefore, difficulties to identify parameters are accentuated when considering a real engineering problem. In practice, the performance of the proposed optimization procedure on parameter identification mainly depends on three aspects: (1) an appropriate soil model which can accurately describe the soil behavior; (2) the number of tests involving in the objective; (3) the accuracy of measurements for the field tests or the real engineering. In order to reduce the effects of the uncertainties of models, a soil model selection should be conducted to find a most "appropriate" one for solving this engineering problem. In order to reduce the effects of the uncertainties of measurements, the measuring devices with high accuracy should be used and more measurements should be involved in the optimization. In order to account for the effects of the heterogeneities of soils, the optimal results should contain many sets of parameters with the objective errors less than a "satisfactory" value instead of only the best optimal set of parameters (Papon et al. [32]). Then, considering all these measures, the proposed optimization technique for identifying soil parameters with further model selection can be applied to a real engineering problem.

4.7 Summary

A new combined hybrid real-coded genetic algorithm and an enhanced differential evolution have been developed in this study. In this RCGA, a hybrid strategy was adopted by using two outstanding crossovers: the simulated binary crossover (SBX) and the simplex crossover (SPX). A newly developed mutation operator, the dynamic random mutation (DRM), was adopted to maintain the diversity of the population. Additionally, in order to accelerate the convergence speed, a newly developed chaotic local search (CLS) was applied conditionally. In enhanced DE, the Nelder–Mead simplex for accelerating the convergence speed of the DE was selected for its good search ability.

The performance of the new RCGA was first estimated by comparing with other hybrid RCGAs on six mathematical benchmark functions and then further evaluated by applying the RCGA to inverse analysis in identifying soil parameters based on laboratory tests; finally, an inverse analysis of pressuremeter tests (PMT) was adopted to estimate the performance of the new hybrid RCGA. Based on all the results, the proposed RCGA is recommended for conducting an inverse analysis to identify soil parameters.

The performance of the enhanced DE algorithm was validated by identifying the MC parameters from synthetic PMT and excavation and was further validated by identifying MC parameters from two real PMTs in sand and ANICREEP parameters from two real PMTs in soft clay. All the comparisons demonstrated that the enhanced DE has the strongest search ability with the smallest objective error.

Other details can be found in Jin et al. [36] and Yin et al. [37].

References

1. Arumugam MS, Rao M, Palaniappan R (2005) New hybrid genetic operators for real coded genetic algorithm to compute optimal control of a class of hybrid systems. Appl Soft Comput 6(1):38–52
2. Mokhade AS, Kakde OG (2014) Overview of selection schemes in real-coded genetic algorithms and their applications. J Ind Intell Inf 2(1):71–77
3. Deb K, Agrawal RB (1995) Simulated binary crossover for continuous search space. Complex Syst 9(2):115–148
4. Da Ronco CC, Benini E (2013) A simplex crossover based evolutionary algorithm including the genetic diversity as objective. Appl Soft Comput 13(4):2104–2123
5. Chuang Y-C, Chen C-T, Hwang C (2015) A real-coded genetic algorithm with a direction-based crossover operator. Inf Sci 305(1):320–348
6. Deb K, Pratap A, Agarwal S, Meyarivan T (2002) A fast and elitist multiobjective genetic algorithm: NSGA-II. IEEE Trans Evol Comput 6(2):182–197
7. Jia D, Zheng G, Khan MK (2011) An effective memetic differential evolution algorithm based on chaotic local search. Inf Sci 181(15):3175–3187
8. Zitzler E, Thiele L (1998) Multiobjective optimization using evolutionary algorithms—a comparative case study. In: Parallel problem solving from nature—PPSN V. Springer, London, pp 292–301
9. Deep K, Thakur M (2007) A new crossover operator for real coded genetic algorithms. Appl Math Comput 188(1):895–911
10. Deep K, Thakur M (2007) A new mutation operator for real coded genetic algorithms. Appl Math Comput 193(1):211–230
11. Poles S, Fu Y, Rigoni E (2009) The effect of initial population sampling on the convergence of multi-objective genetic algorithms. In: Multiobjective programming and goal programming. Springer, Berlin, pp 123–133
12. Levasseur S, Malécot Y, Boulon M, Flavigny E (2008) Soil parameter identification using a genetic algorithm. Int J Numer Anal Methods Geomech 32(2):189–213
13. Schanz T, Vermeer P, Bonnier P (1999) The hardening soil model: formulation and verification. In: Brinkgreve RBJ (ed) Beyond 2000 in computational geotechnics. Balkema, Rotterdam, pp 281–296
14. Richart F, Hall J, Woods R (1970) Vibrations of soils and foundations. In: International series in Theoretical and applied mechanics. Prentice-Hall, Englewood Cliffs
15. Andria-Ntoanina I, Canou J, Dupla J (2010) Caractérisation mécanique du sable de Fontainebleau NE34 à l'appareil triaxial sous cisaillement monotone. Laboratoire Navier-Géotechnique (CERMES, ENPC/LCPC)
16. Gaudin C, Schnaid F, Garnier J (2005) Sand characterization by combined centrifuge and laboratory tests. Int J Phys Model Geotech 5(1):42–56
17. Yin ZY, Chang CS (2013) Stress–dilatancy behavior for sand under loading and unloading conditions. Int J Numer Anal Methods Geomech 37(8):855–870
18. Yin ZY, Chang CS, Hicher PY (2010) Micromechanical modelling for effect of inherent anisotropy on cyclic behaviour of sand. Int J Solids Struct 47(14–15):1933–1951

19. Yin Z-Y, Zhao J, Hicher P-Y (2014) A micromechanics-based model for sand-silt mixtures. Int J Solids Struct 51(6):1350–1363
20. Chang CS, Yin ZY (2011) Micromechanical modeling for behavior of silty sand with influence of fine content. Int J Solids Struct 48(19):2655–2667
21. Roscoe KH, Burland J (1968) On the generalized stress-strain behaviour of wet clay. In: Engineering plasticity. Cambridge University Press, Cambridge, pp 535–609
22. Sheng J (2012) Laboratory tests and constitutive modelling on the mechanical behavior of Shanghai clays. Shanghai Jiaotong University, Shanghai
23. Yin Z-Y, Xu Q, Hicher P-Y (2013) A simple critical-state-based double-yield-surface model for clay behavior under complex loading. Acta Geotech 8(5):509–523
24. Yin ZY, Chang CS, Hicher PY, Karstunen M (2009) Micromechanical analysis of kinematic hardening in natural clay. Int J Plast 25(8):1413–1435
25. Yin ZY, Chang CS, Karstunen M, Hicher PY (2010) An anisotropic elastic-viscoplastic model for soft clays. Int J Solids Struct 47(5):665–677
26. Yin ZY, Hattab M, Hicher PY (2011) Multiscale modeling of a sensitive marine clay. Int J Numer Anal Methods Geomech 35(15):1682–1702
27. Yin ZY, Xu Q, Chang CS (2013) Modeling cyclic behavior of clay by micromechanical approach. J Eng Mech 139(9):1305–1309
28. Yin Z-Y, Yin J-H, Huang H-W (2015) Rate-dependent and long-term yield stress and strength of soft Wenzhou marine clay: experiments and modeling. Mar Georesour Geotechnol 33(1):79–91
29. Yin Z-Y, Zhu Q-Y, Yin J-H, Ni Q (2014) Stress relaxation coefficient and formulation for soft soils. Géotech Lett 4(1):45–51
30. Gao F, Han L (2012) Implementing the Nelder-Mead simplex algorithm with adaptive parameters. Comput Optim Appl 51(1):259–277
31. Zhang J, Sanderson AC (2009) JADE: adaptive differential evolution with optional external archive. IEEE Trans Evol Comput 13(5):945–958
32. Papon A, Riou Y, Dano C, Hicher PY (2012) Single-and multi-objective genetic algorithm optimization for identifying soil parameters. Int J Numer Anal Methods Geomech 36(5):597–618
33. Lee Goh A, Fahey M (1991) Application of a 1-dimensional cavity expansion model to pressuremeter and piezocone tests in clay. In: Proceeding of the seventh international conference on computer methods and advances in geomechanics, Cairns, pp 255–260
34. Yin ZY, Karstunen M, Chang CS, Koskinen M, Lojander M (2011) Modeling time-dependent behavior of soft sensitive clay. J Geotech Geoenviron Eng 137(11):1103–1113
35. Mesri G, Godlewski PM (1977) Time-and stress-compressibility interrelationship. J Geotech Geoenviron Eng 103(ASCE 12910)
36. Jin Y-F, Yin Z-Y, Shen S-L, Zhang D-M (2017) A new hybrid real-coded genetic algorithm and its application to parameters identification of soils. Inverse Probl Sci Eng 25(9):1343–1366
37. Yin Z-Y, Jin Y-F, Shen JS, Hicher P-Y (2018) Optimization techniques for identifying soil parameters in geotechnical engineering: comparative study and enhancement. Int J Numer Anal Methods Geomech 42(1):70–94

Chapter 5
Optimization-Based Evolutionary Polynomial Regression

5.1 Introduction

This chapter aims to propose a robust and effective evolutionary polynomial regression (EPR) model for C_α of clay. First, a database covering various clays is formed, in which 120 data are randomly selected for training and the remaining data are used for testing. To avoid overfitting, a novel EPR procedure using a newly enhanced differential evolution (DE) algorithm is proposed with two enhancements: (1) a new fitness function is proposed using the structural risk minimization (SRM) with L_2 regularization that penalizes polynomial complexity, and (2) an adaptive process for selecting the combination of involved variables and size of polynomial terms is incorporated. By comparing the predictive ability, model complexity, robustness and monotonicity, the EPR formulation for C_α involving clay content, plasticity index and void ratio with three terms is selected as the optimal model. A parametric study is then conducted to assess the importance of each input in the proposed model. All results demonstrate that the proposed model of C_α is simple, robust, and reliable for applications in engineering practice.

5.2 Differential Evolution-Based EPR Modeling

5.2.1 General EPR Procedure

The evolutionary polynomial regression (EPR) is a data-driven method based on evolutionary computing, aiming to search for polynomial structures representing a system, which was first introduced by Giustolisi and Savic [1] with applications in the hydroinformatics and environment-related problems. A general EPR expression can be mathematically formulated as:

Z. Yin and Y. Jin, *Practice of Optimisation Theory in Geotechnical Engineering*,
https://doi.org/10.1007/978-981-13-3408-5_5

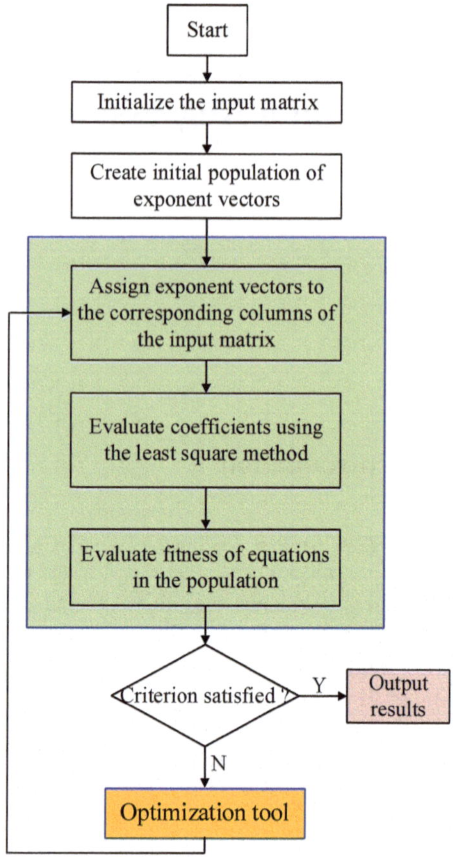

Fig. 5.1 Typical flowchart of EPR procedure

$$y = \sum_{j=1}^{m} F\left(\mathbf{X}, f(\mathbf{X}), a_j\right) + a_0 \tag{5.1}$$

where y is the estimated vector of output of the process; a_0 is an optional bias; a_j is an adjustable parameter for the jth term; F is a function constructed by the process; $\mathbf{X}$ is the matrix of input variables; f is a function defined by the user; and m is the number of terms of the target expression.

According to Giustolisi and Savic [1], the first step in identifying the model structure is to transfer Eq. (5.1) to the following vector form:

$$\mathbf{Y}_{N\times1}(\boldsymbol{\theta}, \mathbf{Z}) = \left[\mathbf{I}_{N\times1}\ \mathbf{Z}^j_{N\times m}\right] \times \left[a_0\ a_1\ \ldots\ a_m\right]^{\mathrm{T}} = \mathbf{Z}_{N\times d} \times \boldsymbol{\theta}^{\mathrm{T}}_{d\times1} \tag{5.2}$$

where $\mathbf{Y}_{N\times1}(\boldsymbol{\theta}, \mathbf{Z})$ is the least squares (LS) estimator vector of N target values; $\boldsymbol{\theta}_{d\times1}$ is the vector of d $(= m + 1)$ parameters a_j and a_0 ($\boldsymbol{\theta}^{\mathrm{T}}$ is the transposed vector); and

$\mathbf{Z}_{N \times d}$ is a matrix formed by $\mathbf{I}$ (unitary vector) for bias a_0, with m vectors of variables $\mathbf{Z}^j$. More details about the EPR can be found in Giustolisi and Savic [1].

Figure 5.1 shows a typical flowchart for the EPR procedure [1]. The general functional structure represented by $f(\mathbf{X}, a_j)$ in Eq. (5.1) is constructed from elementary functions by EPR using an optimization algorithm strategy (such as genetic algorithm). Note that any optimization algorithm guaranteeing the global optimal solution can be employed in the EPR procedure. The building blocks (elements) of the structure are defined by the user based on understanding of the physical process. The selection of feasible structures to be combined is conducted through an evolutionary process, while the parameters a_j in Eq. (5.2) are estimated by the least squares method.

5.2.2 Implementation of NMDE in EPR Modeling

In order to improve the efficiency of EPR modeling, the newly developed NMDE by Yin et al. [2] was employed to select the useful input vectors from $\mathbf{X}$ to formulate the EPR. In NMDE, the Nelder–Mead simplex (NMS) is used to accelerate the convergence speed. Before performing the DE mutation, all individuals are sorted based on their fitness value, and the best $n + 1$ (n is the number of variables) is selected to perform the NMS. Based on the results of the NMS, the best individual is updated and then recombined with the N-(n + 1) remaining individuals to perform the DE mutation. This process will be executed N times, resulting in a new population of N individuals. Then, the obtained population is applied to the crossover operation. To avoid a rapid loss of diversity, an elitism strategy is adopted when performing the selection, in which the 10% of individuals with the highest fitness are selected from the parents and children to survive to the next generation. The remainders are chosen by tournament selection from the mating pool composed of parents and children. The completion mechanism can help the NMDE to identify better solutions.

5.2.3 New Fitness Function Considering $\mathbf{L}_2$ Regularization

The performance of an EPR procedure mainly depends on fitness function. A widely used fitness function is structural risk minimization (SRM) [3], which involves the addition of model complexity term (size of model) in the empirical error and punishes the model fitness based on its size. Another problem is that a relatively small amount of data will increase the risk to cause the model overfitting, making the training error small and the testing error particularly large, which would weaken the generalization ability of an EPR model. Then, the use of regularization/penalty functions (e.g., L_0, L_1 and L_2 regularizations) to avoid overfitting is suggested [4]. Among various regularizations, the L_2 regularization is usually adopted [5]. Therefore, a modified mathematical formulation of SRM considering the L_2 regularization was adopted in this study, given as:

$$\mathrm{SRM} = \frac{\mathrm{SSE}}{N}\left(1 - \sqrt{\left(\frac{n}{N} - \left(\frac{n}{N}\log\left(\frac{n}{N}\right)\right) + \left(\frac{\log\left(\frac{n}{N}\right)}{2N}\right)\right)}\right)^{-1} + \lambda\|\boldsymbol{\omega}\|_2^2 \quad (5.3)$$

with

$$\mathrm{SSE} = \sum_{i=1}^{N}\left(\mathbf{Y_m} - \mathbf{Y_p}\right)^2 \text{ and } \|\boldsymbol{\omega}\|_2^2 = \sum_{j=1}^{n}\boldsymbol{\omega}_j = \boldsymbol{\omega}^{\mathrm{T}}\boldsymbol{\omega} \quad (5.4)$$

where N is the number of data points on which the SRM is computed; $\mathbf{Y}_{\mathrm{m}}$ is the vector of measured values; $\mathbf{Y}_{\mathrm{p}}$ is the vector of predicted values; $\boldsymbol{\omega}$ is the vector of model coefficients; λ is regularization parameter that requires manual adjustment to find an appropriate value.

5.2.4 Adaptive Selection of Correlating Variables and Term Size

A reliable EPR model should have a reasonable trade-off between predictive ability and generalization ability. As stated by Wood [6], simple yet adequate models are favored on the basis of practicality. Therefore, an EPR procedure combining with model selection process should be proposed to ensure the model "simple" enough based on minimizing the training error. Then the model could also have a good generalization performance (e.g., the testing error is also small). In this case, the model selection involves two aspects: selecting the suitable combination of correlating variables and the appropriate size of terms.

Figure 5.2 presents the proposed procedure, where $\boldsymbol{\theta}$ is the decision variables corresponding to the exponents of EPR model; Comb represents the number of combination of correlating variables; m is the size of terms. Compared to the common EPR process, two additional variables Comb (an integer number) and m (an integer number) are added to the vector of optimization variable in the proposed procedure. Firstly, all variables in initial generation are generated randomly within their domains. Next, the possible combination of correlating variables is selected according to the value of Comb and then a possible term size is chosen according to the value of m. Subsequently, a generated EPR model with unknown coefficients according to Eq. (5.6) is attained. Then, the vector of coefficients $\boldsymbol{a}$ is determined by regression between the measurements and predictions. Finally, the fitness SRM with L_2 regularization is computed to evaluate the performance of EPR model, which determines whether the formula can survive to next generation in the DE-evolution.

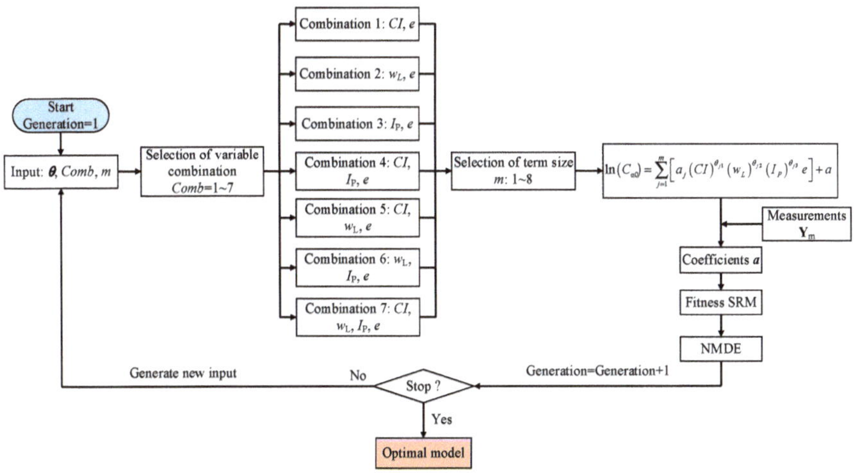

Fig. 5.2 Procedure of model selection combined with EPR process

Once the stop criterion (e.g., the maximum number of generation) is reached, the whole process is exited; otherwise, the process will continue to the next generation.

With increasing the number of generation, the appropriate combination of correlating variables and term size will be automatically selected among numerous calculations. Moreover, through adjusting the regularization parameter, the most appropriate EPR model in terms of model complexity and generalization ability can be finally found.

5.2.5 Suggestion of Regularization Parameter

To find an appropriate value of regularization parameter λ to trade-off the performance between prediction and generalization, several attempts to assign different values of λ are needed in the proposed EPR procedure. Firstly, the $\lambda = 0$ is tested to check the general predictive performance of the EPR model without regularization. Then, a value of λ (e.g., 10^{-4}) that closes to zero is assigned to evaluate the effect of regularization on the selection of model. Note that the value of first tried λ is different for various concerned cases because the value of λ is related to the coefficients of obtained equation. Next, based on investigated results, a series of calculation attempts using different values of λ are conducted based on which several formulas can be achieved. Among them, the formula that the predictive performance is similar to one of the models with $\lambda = 0$ is saved, otherwise discarded. The finally retained formulas have different number of term sizes. The most appropriate model will be eventually selected through the comparison of predictive ability and the number of term sizes, even other criteria depending on the problem on which for instance the robustness is more important or the accuracy.

5.3 EPR Modeling of Creep Index

Natural soft clays exhibit significant creep under both laboratory and in situ conditions after primary consolidation, which significantly influences the long-term safety of infrastructures in various fields, such as tunneling [7–10], excavation [11–13], embankment [14–20], urban land subsidence [21–25], etc. Usually, the creep property of soft clays is represented by the creep index $C_\alpha = \Delta e/\Delta \log(t)$, where e is void ratio and t is time during secondary compression. The creep index is a key parameter for most viscoplastic constitutive models applicable to the engineering practice [26–37], which is usually obtained by a conventional oedometer test. According to studies of [29, 30, 34], the C_α corresponding to intact clays is not constant because of the effect of destructuration to the creep. In contrast, the C_α of reconstituted clay is an intrinsic property, which is the base for understanding the creep characteristic and thus more suitable to be adopted in practice [38]. Because of this, the attention is paid to the C_α of reconstituted clay in this study.

The creep property should relate to the microstructure of clay [28, 30, 39, 40]. Unfortunately, the microstructure of clay is expensive to measure which may lead to a practical obstacle. Physical properties can somehow reflect the microstructure of clay. Thus, practically, it is very convenient to get ideas of the intrinsic value of C_α only based on physical properties of clay. Some attempts have been made to correlate the C_α to some physical properties of soils (such as water content, void ratio, Atterberg limits) [41–47]. However, these correlations are only applicable for few clays and thus not enough reliable for soft clay engineering practice. Therefore, a robust and effective correlation between C_α and physical properties of clay is worth investigating.

Numerical regression is the most powerful and commonly applied form of regression used to solve the problem of finding the best model to fit the observed data [48–52]. Evolutionary polynomial regression (EPR) is a recently developed hybrid regression method [1] that has advantages in modeling the nonlinear complex problems. Applications in geotechnics include stability prediction of slopes [49, 53, 54], modeling of clay compressibility [55, 56], modeling of permeability and compaction characteristics of soils [57], evaluation of liquefaction potential of sand [58, 59], prediction of soil saturated water content [60], settlement prediction of foundations [61–63], evaluation of pile bearing capacity [64–66], pipeline failure prediction [67], and modeling of soil behaviors [68–71]. These successful applications have demonstrated that the EPR technique is superior to other soft computing techniques, such as artificial neural networks (ANNs) [72], genetic programming (GP) [48, 58]. More recently, the development of optimization algorithms [2, 31, 55, 73–78] can improve the EPR technique in a more adaptive way. Thus, the optimization combined EPR technique is worth trying for the C_α of clay. However, in current EPR modeling most fitness functions are only based on training data error, such as the sum of squared errors (SSE) or coefficient of determination (COD). As a result, the proposed models are usually overfitting and weak in terms of the generalization ability. Therefore, the EPR technique needs to be improved to avoid overfitting and with good generalization ability.

5.3.1 Database

5.3.1.1 Statistics and Basic Correlation Analysis

Massive experimental data from various studies [43, 44, 47, 79, 80] were collected and used to develop the EPR model of intrinsic C_α. The clay content (CI), liquid limit (w_L), plastic limit (w_P), plasticity index (I_P) and void ratio (e) were treated as correlating variables of interest. Table 5.1 summarizes those physical properties and C_α for all selected reconstituted clays. Based on the data, the statistical analysis (maximum value, minimum value, mean value, and standard deviation) for each property was conducted, summarized in Table 5.2.

The linear correlations between C_α and each main basic physical property (CI, w_L, w_P, I_P and e) are presented in Fig. 5.3. It is found that the C_α relatively highly correlates to e, followed by I_p and w_L, and very poorly correlates to w_P and CI. It is obvious that none of the basic correlations involving single physical property for predicting C_α is satisfactory for engineering purposes ($R^2 < 0.8$), and thus, the performance of the correlation still needs to be improved. Note that the increasing of the ratio of clay to silt or sand can increase significantly the creep index (Yin [43]), the CI should be considered in correlation.

5.3.2 Discrepancy of Current Correlation Formula

Table 5.3 summarizes five existing empirical correlations of C_α, which were used to fit the database presented in Table 5.1. Figure 5.4 shows the comparison between predictions and measurements for these empirical correlations. For the correlations only involving I_P [41, 43], their performances are poor with a low value of correlation coefficient R^2. For the correlations proposed by Zeng et al. [44] and Zhu et al. [47], their performances are almost the same, but the R^2 is still smaller than 0.8. Therefore, it is still recommended and should be practically useful to improve the correlation of C_α to physical properties.

Therefore, a novel correlation approach to find a reliable and reasonable correlation between C_α and physical properties was proposed, as presented below.

Table 5.1 Summary of physical properties and creep index for all selected clays

Clay	CI (%)	w_L (%)	w_P (%)	I_P (%)	C_α	e	Reference
Haarajoki clay	65	88	26	62	0.0461	2.283	[47, 79]
	65	88	26	62	0.0299	2.006	
	65	88	26	62	0.0226	1.733	
	65	88	26	62	0.0154	1.486	
	65	88	26	62	0.0140	1.273	
	65	88	26	62	0.0297	2.044	
	65	88	26	62	0.0329	1.779	
	65	88	26	62	0.0244	1.545	
	65	88	26	62	0.0173	1.338	
	65	88	26	62	0.0113	1.141	
	65	88	26	62	0.0332	2.058	
	65	88	26	62	0.0258	1.831	
	65	88	26	62	0.0187	1.602	
	65	88	26	62	0.0154	1.395	
	65	88	26	62	0.0108	1.196	
Suurpelto clay	80	80	23	57	0.0467	1.881	[47, 79]
	80	80	23	57	0.0267	1.558	
	80	80	23	57	0.0166	1.092	
	80	80	23	57	0.0484	1.862	
	80	80	23	57	0.0203	1.635	
	80	80	23	57	0.0196	1.430	
	80	80	23	57	0.0143	1.272	
	80	80	23	57	0.0161	1.127	
	80	80	23	57	0.0548	2.080	
	80	80	23	57	0.0279	1.831	
	80	80	23	57	0.0196	1.579	
	80	80	23	57	0.0159	1.356	
	80	80	23	57	0.0131	1.147	
	80	80	23	57	0.0516	2.035	
	80	80	23	57	0.0288	1.782	
	80	80	23	57	0.0138	1.311	
	80	80	23	57	0.0117	1.104	
Mixed clay	78	45	26	19	0.0069	1.040	[47, 79]
	78	45	26	19	0.0053	0.963	
	78	45	26	19	0.0062	0.883	
	78	45	26	19	0.0041	0.803	
	78	45	26	19	0.0037	0.729	
	78	45	26	19	0.0071	1.063	

(continued)

Table 5.1 (continued)

Clay	CI (%)	w_L (%)	w_P (%)	I_P (%)	C_α	e	Reference
	78	45	26	19	0.0053	0.809	
	78	45	26	19	0.0076	1.057	
	78	45	26	19	0.0062	0.982	
	78	45	26	19	0.0069	0.918	
	78	45	26	19	0.0051	0.843	
	78	45	26	19	0.0051	0.761	
	78	45	26	19	0.0058	0.997	
	78	45	26	19	0.0055	0.929	
	78	45	26	19	0.0051	0.858	
	78	45	26	19	0.0051	0.790	
Vanttila clay	53	98	30	68	0.0447	1.926	[47, 79]
	53	98	30	68	0.0336	1.498	
	53	98	30	68	0.0315	1.287	
	53	98	30	68	0.0265	1.165	
	53	98	30	68	0.0272	1.075	
	53	98	30	68	0.0451	1.908	
	53	98	30	68	0.0230	1.677	
	53	98	30	68	0.0345	1.460	
	53	98	30	68	0.0249	1.252	
	53	98	30	68	0.0235	1.062	
	53	98	30	68	0.0357	1.942	
	53	98	30	68	0.0338	1.726	
	53	98	30	68	0.0239	1.518	
	53	98	30	68	0.0246	1.313	
	53	98	30	68	0.0196	1.121	
	53	98	30	68	0.0212	1.019	
	53	98	30	68	0.0302	1.999	
	53	98	30	68	0.0253	1.825	
Murro clay	26	88	34	54	0.0359	1.684	[47, 79]
	26	88	34	54	0.0302	1.536	
	26	88	34	54	0.0269	1.385	
	26	88	34	54	0.0286	1.231	
	26	88	34	54	0.0233	1.099	
	26	88	34	54	0.0207	0.954	
	26	88	34	54	0.0311	1.691	
	26	88	34	54	0.0283	1.536	
	26	88	34	54	0.0281	1.4101	
	26	88	34	54	0.0237	1.260	

(continued)

Table 5.1 (continued)

Clay	CI (%)	w_L (%)	w_P (%)	I_P (%)	C_α	e	Reference
	26	88	34	54	0.0214	1.123	
	26	88	34	54	0.0233	1.394	
	26	88	34	54	0.0228	1.246	
	26	88	34	54	0.0184	1.107	
	26	88	34	54	0.0184	0.980	
	26	88	34	54	0.0269	1.446	
	26	88	34	54	0.0235	1.294	
	26	88	34	54	0.0228	1.145	
	26	88	34	54	0.0249	0.997	
	26	88	34	54	0.0258	1.411	
	26	88	34	54	0.0262	1.331	
	26	88	34	54	0.0214	1.183	
	26	88	34	54	0.0196	1.042	
Kaolin clay	83	40	20	20	0.0060	0.913	[47]
	83	40	20	20	0.0065	0.841	
	83	40	20	20	0.0062	0.766	
	83	40	20	20	0.0060	0.689	
	83	40	20	20	0.0058	0.598	
Shanghai clay-1	33	51	26.4	24.6	0.0086	0.949	[80]
	33	51	26.4	24.6	0.0083	0.857	
	33	51	26.4	24.6	0.0076	0.746	
	33	51	26.4	24.6	0.0072	0.680	
Shanghai clay-2	26	42.5	22.5	20	0.0076	0.861	[47]
	26	42.5	22.5	20	0.0074	0.763	
	26	42.5	22.5	20	0.0069	0.671	
	26	42.5	22.5	20	0.0062	0.585	
Nanjing clay-9 m	34	52	26	26	0.0154	1.015	[44]
	34	52	26	26	0.0142	0.915	
	34	52	26	26	0.0124	0.805	
	34	52	26	26	0.0116	0.699	
	34	52	26	26	0.0107	0.605	
	34	52	26	26	0.0093	0.518	
Wenzhou clay-10 m	47.6	65	28	37	0.0228	1.334	[44]
	47.6	65	28	37	0.0212	1.207	
	47.6	65	28	37	0.0200	1.077	
	47.6	65	28	37	0.0184	0.938	
	47.6	65	28	37	0.0173	0.799	

(continued)

Table 5.1 (continued)

Clay	CI (%)	w_L (%)	w_P (%)	I_P (%)	C_α	e	Reference
	47.6	65	28	37	0.0152	0.654	
Lianyungang clay-12 m	38	63	27	36	0.0179	1.101	[44]
	38	63	27	36	0.0147	0.834	
	38	63	27	36	0.0142	0.700	
Zhoushan clay	37	40.7	26.7	20	0.0071	0.882	[81]
	37	40.7	26.7	20	0.0074	0.778	
	37	40.7	26.7	20	0.0069	0.669	
	37	40.7	26.7	20	0.0069	0.574	
	37	40.7	26.7	20	0.0058	0.482	
HKMC	27.5	60	28	32	0.0071	1.158	[43]
	27.5	60	28	32	0.0046	0.894	
	27.5	60	28	32	0.0036	0.730	
	27.5	60	28	32	0.0034	0.610	
	27.5	60	28	32	0.0024	0.466	
Nanjing clay-7 m	11.5	44	23	21	0.0111	0.901	[44]
	11.5	44	23	21	0.0100	0.846	
	11.5	44	23	21	0.0091	0.786	
	11.5	44	23	21	0.0089	0.704	
	11.5	44	23	21	0.0086	0.644	
	11.5	44	23	21	0.0074	0.562	
Wenzhou clay-4 m	40.7	60	28	32	0.0208	1.404	[44]
	40.7	60	28	32	0.0192	1.289	
	40.7	60	28	32	0.0176	1.169	
	40.7	60	28	32	0.0164	1.032	
	40.7	60	28	32	0.0153	0.901	
	40.7	60	28	32	0.0144	0.786	
	40.7	60	28	32	0.0134	0.639	
Lianyungang clay-4 m	40	86	31	55	0.0347	1.704	[44]
	40	86	31	55	0.0314	1.502	
	40	86	31	55	0.0284	1.289	
Kyuhoji clay	70	1.02	78.3	28.2	0.0320	1.02	[82]
Illite clay	61	1.354	58	26	0.0092	1.354	[83]
Merville clay	26	1.223	99	40	0.0158	1.223	[84]
Kawasaki clay	22.3	1.07	55.3	25.9	0.0134	1.07	[85]
Boston blue clay	57.6	1.181	45.4	21.7		1.1801	[86]

Table 5.2 Statistics of properties in the database

Properties	CI (%)	w_L (%)	w_P (%)	I_p (%)	e	C_α
Maximum value	83	98	34	68	2.284	0.0548
Minimum value	11.5	40	20	19	0.466	0.0037
Mean value	50.3	70.6	27.4	43.4	1.175	0.0182
Standard deviation	22.0	20.5	3.8	18.3	0.415	0.0110

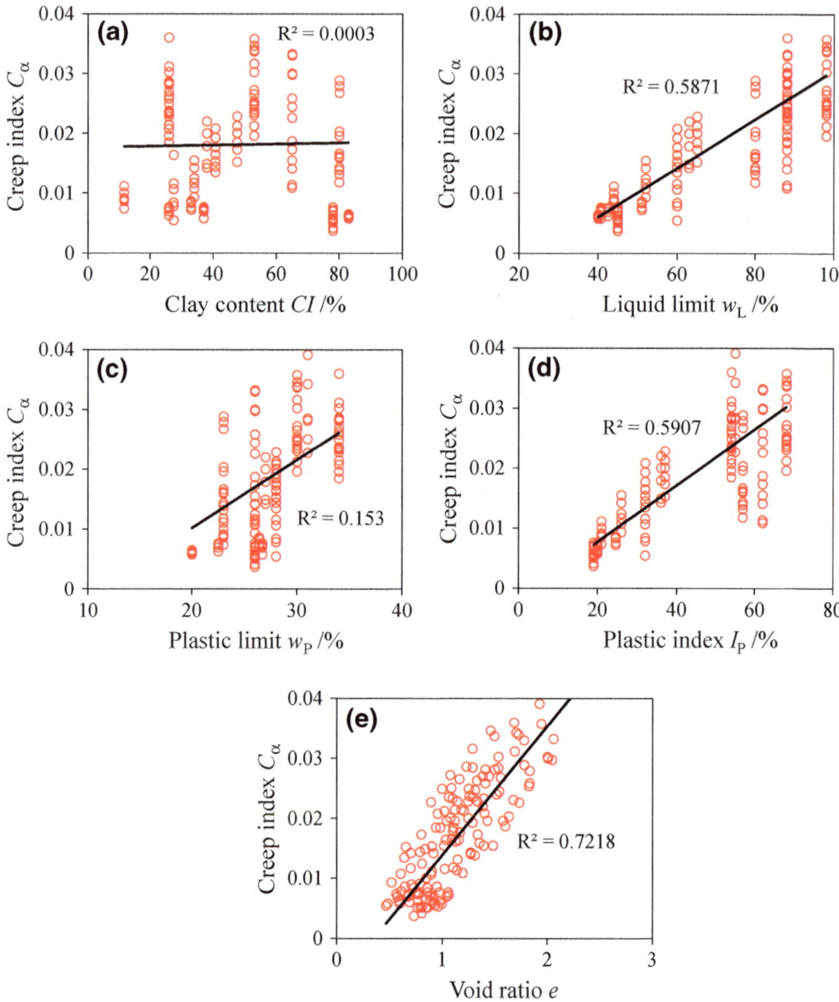

Fig. 5.3 Basic correlations between C_α and each physical property of soils

Table 5.3 Summary of existing empirical correlations for C_α

Empirical correlations	Applicability	Reference
$C_\alpha = 0.00168 + 0.00033I_p$	Remold clays	[41]
$C_\alpha = 0.000369I_p - 0.00055$	Remold clays	[43]
$C_\alpha = \left(-0.0067 + 0.0115e_L - 0.0016(e_L)^2\right)(1+e)$	Remold clays	[44]
$C_\alpha =$ $\left(-0.0274 + 0.0011w_L - 0.00048I_p\right)\left(\frac{w}{w_L}\right)^{0.7872-0.0369w_L+0.0619I_p}$	Remold clays	[47]
$C_\alpha = (0.0007w_L - 0.0223)\left(\frac{w}{w_L}\right)^{0.014978w_L-0.23031}$	Remold clays	[47]

Where e_L is the void ratio corresponding to liquid limit; w is water content, equivalent to e for a given clay

5.3.3 EPR Modeling Process for C_α

Since selecting (w_L, I_p) or (w_p, I_p) is physically the same for evaluating the C_α, based on the statistics results of database, four physical properties (CI, w_L, I_P, and e) were selected as the correlating variables of interest to training the EPR model. To attain the nonlinear creep behavior with a consecutively decreasing creep index C_α that fully relates to the soil density [47, 79], a general structure of EPR expression for C_α was proposed as:

$$\ln(C_\alpha) = \sum_{j=1}^{m}\left[f(\mathrm{CI}, w_L, I_P)e\right] + a_0 \tag{5.5}$$

which was further expressed as:

$$\ln(C_\alpha) = \left(\sum_{j=1}^{m}\left[a_j(\mathrm{CI})^{\theta_{j1}}(w_L)^{\theta_{j2}}(I_P)^{\theta_{j3}}\right]\right)e + a_0 \tag{5.6}$$

where a_0 is a constant in the EPR equation; a_j is the coefficient corresponding to j term and θ_j is the vector of exponent. Note that the use of logarithm in C_α can guarantee the positiveness of C_α.

To obtain an accurate and reasonable correlation, 120 data randomly selected in the prepared database were used for training and the remaining data were used for testing. For simplicity, the value of exponent was constrained to [−2, 2] with a step size to 1. Also, the maximum number of terms was set to 8 for restricting the model complexity. For NMDE, the number of initial population was set to ten times of decision variables and the maximum generation was set to 200. The probability of crossover CR is 0.3. For NMS, the tolerance for convergence was set to 10^{-4}. Independent multiple runs were performed to avoid randomness.

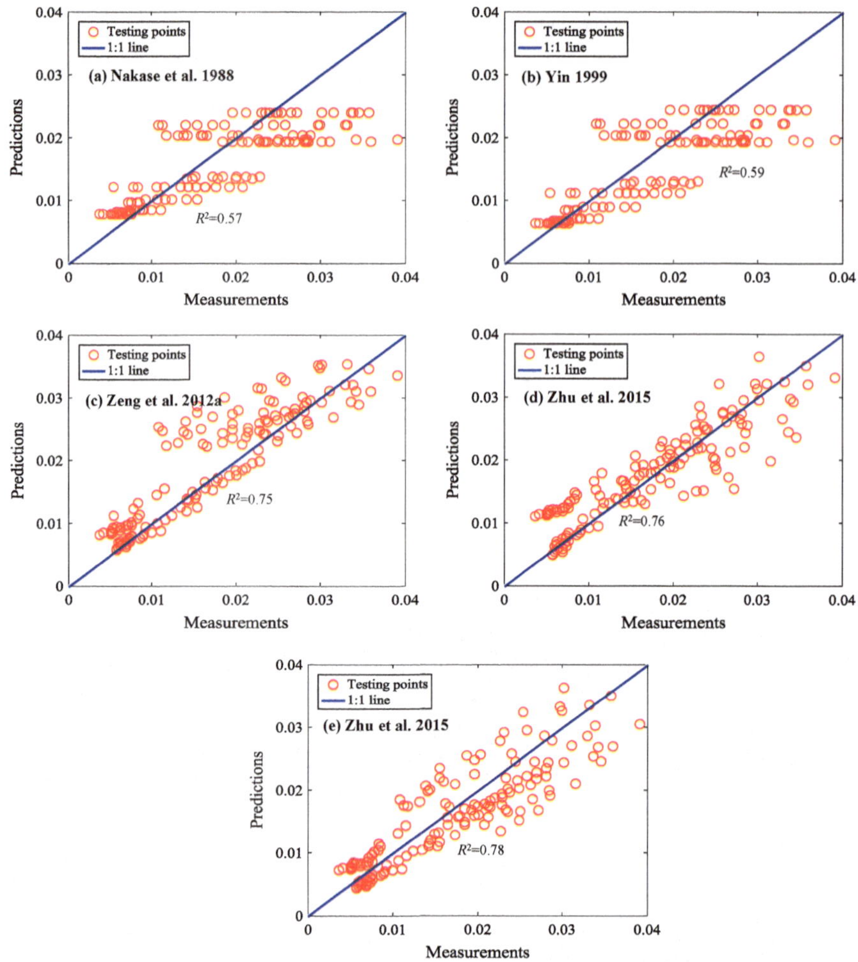

Fig. 5.4 Comparison between predictions and measurements for five empirical correlations

As shown in Fig. 5.2, for modeling C_α, the variable e is fixed to keep the unique relationship between C_α and e. A total of seven combinations ($= C_3^1 + C_3^2 + C_3^3$) that each one contains different physical properties are obtained. Thus, the total number of combinations Comb is 7 and the total number of term size m is 8. Following the proposed EPR procedure, the most appropriate EPR model for C_α in terms of model complexity and generalization ability can be finally found.

5.3.4 Analysis of Results

To evaluate the performance of the obtained EPR model, five indicators are used. Besides the mean value u and standard deviation value σ, coefficient of determination (R^2), root mean square error (RMSE) index and mean absolute error (MAE) are expressed as:

$$R^2 = \frac{\sum_{i=1}^{N} (\mathbf{Y}_{\mathrm{m}})^2 - \sum_{i=1}^{N} \left(\mathbf{Y}_{\mathrm{m}} - \mathbf{Y}_{\mathrm{p}}\right)^2}{\sum_{i=1}^{N} (\mathbf{Y}_{\mathrm{m}})^2} \tag{5.7}$$

$$\mathrm{RMSE} = \sqrt{\frac{1}{N}\sum_{i=1}^{N} \left(\mathbf{Y}_{\mathrm{m}} - \mathbf{Y}_{\mathrm{p}}\right)^2} \tag{5.8}$$

$$\mathrm{MAE} = \frac{1}{N}\sum_{i=1}^{N} \left|Y_{\mathrm{m}} - Y_{\mathrm{p}}\right| \tag{5.9}$$

Higher R^2 or lower RMSE and MAE values represent better model performance. Meanwhile, both the mean value "u" and the standard deviation value "σ" of $\mathbf{Y}_{\mathrm{p}}/\mathbf{Y}_{\mathrm{m}}$ were calculated. A "u" value greater than 1.0 indicates over-estimation and under-estimation otherwise.

Followed the suggestion of selecting λ, a series of calculation attempts using different values of λ (i.e., $\lambda = 0, 0.0001, 0.001, 0.01, 0.05$, and 0.1) were carried out. Figure 5.5 shows the evolution of model selection in terms of variable combination and size of terms for different values of λ. The results show that all models compete with each other to keep the diversity of population during the initial stage of EPR; then, with increasing the number of generations, the models having higher fitness (small training error) survive to next generation and the others with lower fitness are discarded; finally, the percentage of the models with good performance continues to rise to 100%. Therefore, the most appropriate model represented by variable combination and size of terms are automatically selected using the proposed EPR procedure. The results demonstrate that the proposed EPR procedure combined with model selection is efficient.

Table 5.4 summarizes the expressions of all the proposed EPR models. It is found that all obtained EPR expressions contain the I_{P}, which demonstrates that the I_{P} has an important effect on the C_{α}. Figure 5.6 shows the comparison of C_{α} between measurements and different EPR predictions for training and testing data. All obtained results are summarized in Table 5.5. With increasing the value of λ, a trend that the term size of obtained model decreases is found. Apart from the model with $\lambda = 0$, the performance of other models is similar in terms of R^2, RMSE, MAE, u and σ. The performance of all EPR models is acceptable. Based on preliminary results, it seems that the performance of all EPR models is acceptable. However, considering the less complexity of an appropriate model, only EPR models with three terms (Eqs. 5.11 and 5.12) can be considered as the optimal models.

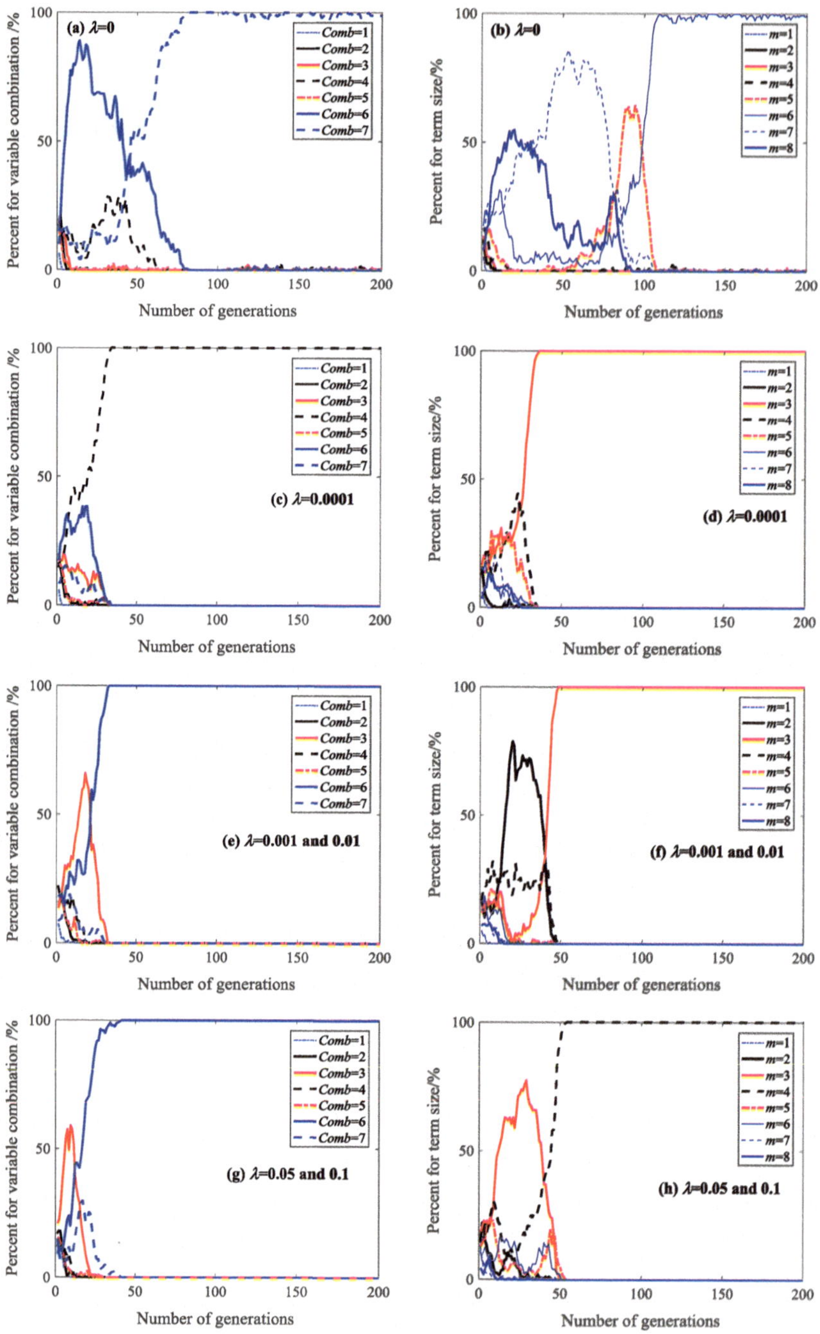

Fig. 5.5 Evolution of model selection in terms of variable combination and size of terms

Table 5.4 Optimal correlations of C_α for different values of λ

λ	Proposed optimal correlation
0	$\ln(C_\alpha) = \left(0.9092\frac{\mathrm{CI}^2}{w_L I_P} + 0.6283\frac{\mathrm{CI}^2}{w_L^2 I_P} - 10.3212(I_P\mathrm{CI})^2 - 0.1963\left(\frac{\mathrm{CI}}{w_L I_P}\right)^2 + 3.9741(I_P)^2\right)e - 5.1618$ (5.10)
0.0001	$\ln(C_\alpha) = \left(0.3114\frac{I_P^2}{\mathrm{CI}} - 0.1229\frac{1}{I_P^2} + 0.6455\frac{1}{I_P}\right)e - 5.1308$ (5.11)
0.001 and 0.01	$\ln(C_\alpha) = \left(0.1265\frac{1}{w_L^2 I_P} - 0.2463\frac{1}{I_P^2} + 0.6264\left(\frac{w_L}{I_P}\right)^2\right)e - 5.1098$ (5.12)
0.05 and 0.1	$\ln(C_\alpha) = \left(-0.0877\frac{1}{w_L I_P^2} + 0.1998\left(\frac{w_L}{I_P}\right)^2 + 0.1497\frac{1}{w_L^2 I_P} + 0.3846\frac{(w_L)^2}{I_P}\right)e - 5.1660$ (5.13)

Remark CI, w_L and I_P are in real number, not in percentage

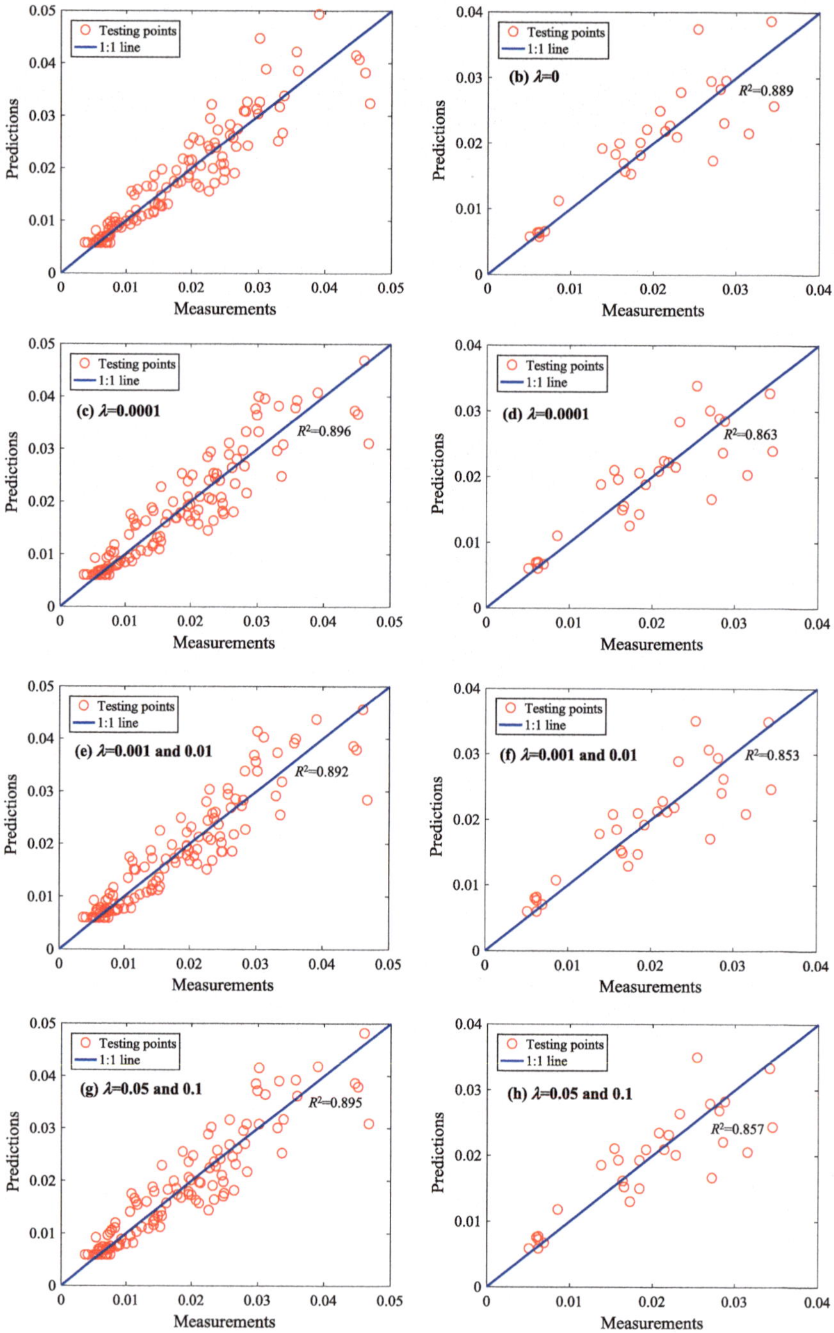

Fig. 5.6 Comparison of C_α between measurements and EPR predictions for different values of λ

Table 5.5 Summary of indicators for all calculations of C_α with different values of λ

λ	Comb	Training					Testing				
		R^2	RMSE	MAE	u	σ	R^2	RMSE	MAE	u	σ
0	7	0.924	0.0042	0.0028	1.0464	0.0042	0.889	0.0053	0.0035	1.0284	0.1983
0.0001	4	0.896	0.0045	0.0032	1.0226	0.2204	0.863	0.0055	0.0036	1.0035	0.2218
0.001	6	0.892	0.0047	0.0032	1.0233	0.2219	0.853	0.0057	0.0038	1.0215	0.2219
0.01	6	0.892	0.0047	0.0032	1.0233	0.2219	0.853	0.0057	0.0038	1.0215	0.2219
0.05	6	0.895	0.0045	0.0031	1.0229	0.2221	0.857	0.0055	0.0037	1.013	0.2209
0.1	6	0.895	0.0045	0.0031	1.0229	0.2221	0.857	0.0055	0.001	1.013	0.2209

5.3.5 Robustness Testing for Proposed EPR Models

An appropriate model has not only a good predictive ability and less complexity but also good robustness. The latter indicates the predicted values are always guaranteed reasonable for reasonable input values. To assess the robustness of each EPR model, the robustness tests were performed and a criterion representing the robustness was defined as:

$$\text{Robustness ratio} = \frac{\text{Samples located in reasonable range}}{\text{Total samples}} \tag{5.14}$$

The reasonable range for C_α in this case is [0.001–0.1] according to the statistical results, which is also applicable for most reconstituted soft clays [28, 30, 47, 79]. To generate the testing samples, it supposes that variables (CI, w_L, I_p and e) are independent of each other and meet the multivariable lognormal distribution according to various studies [87–89]. According to the statistic results of used database shown in Table 5.2, the values of mean and standard deviation for each variable were employed to randomly generate 10,000 samples from its lognormal distribution. Note that for the robustness testing, the samples should be adjusted according to the specific problem and the related variables, not 10,000 samples for all cases.

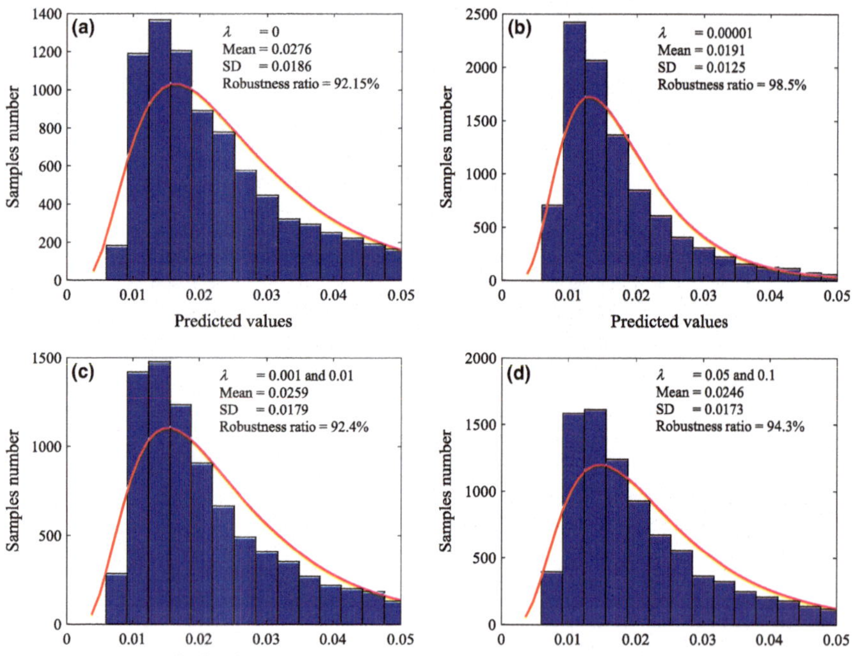

Fig. 5.7 Distribution of C_α located in reasonable range in robustness testing

Then, the C_α was predicted by each proposed EPR model and the robustness ratio was then computed for each model. For each robustness test, the mean and standard deviation for samples locating in the reasonable range were also calculated.

Figure 5.7 presents the results of robustness tests for four potential EPR models. It can be seen that the proposed EPR model (Eq. 5.11) involving CI and I_P with 3 terms has the highest robustness ratio, and the mean and standard deviation values (0.0191 ± 0.0125) predicted by this model are very close to the values (0.0182 ± 0.0110) of the used database. Thus, this EPR model can most probably give more reliable prediction of C_α on unseen data. Therefore, Eq. (5.11) is the optimal model in terms of robustness, followed by Eq. (5.12). These two formulas will be further examined.

5.3.6 Monotonicity and Sensitivity Analysis

The mathematical characteristics (e.g., monotonicity) of a formula can somehow imply whether it is physically correct or not. Thus, to deeply understand the mathematical characteristics of two proposed EPR models and select one of them as the optimum, a parametric study was conducted on the involving physical properties. Note that when one variable is being studied, the other two variables were fixed to their common values (CI = 50%, I_P = 40% and e = 1.0) for Eq. (5.11) and (w_L = 50%, I_P = 40% and e = 1.0) for Eq. (5.12). Figures 5.8 and 5.9 show the results of parametric study for Eq. (5.11) and Eq. (5.12), respectively. For Eq. (5.11), it is found that with increasing the value of CI, the predicted values of C_α has a slight decrease; with increasing the value of I_P, the value of C_α increases quickly up to a point then increases slowly. However, for Eq. (5.12), the C_α decreases firstly and then increases with increasing w_L; with increasing the I_p, the C_α decreases, which is different from the investigation shown in Fig. 5.3 so that it is unreasonable. With increasing the void ratio, the C_α increases for both EPR models, which is in accordance with the findings by Yin et al. [79]. Therefore, the EPR model involving CI and I_P with 3 terms for C_α is better in the monotonicity. Overall, this model was finally selected in terms of the predictive ability, model complexity, robustness, and monotonicity.

To assess the importance of each input of the proposed EPR model on C_α, a sensitivity analysis was performed. The composite scaled sensitivity (CSS_j) analysis proposed by Hill [90] was adopted, which indicates the amount of information provided by the ith observations for the estimation of jth parameter and is defined as:

$$\mathrm{CSS}_j = \sqrt{\left(\frac{1}{N}\sum_{i=1}^{N}\left(\left(\frac{\partial y_i}{\partial x_j}\right)\cdot x_j\sqrt{\omega_i}\right)^2\right)} \tag{5.15}$$

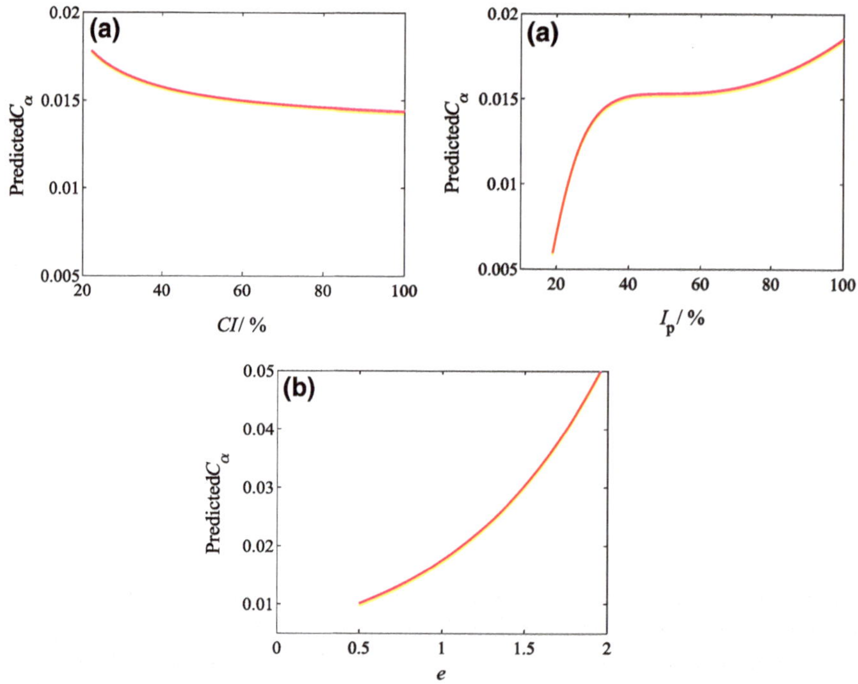

Fig. 5.8 Results of the C_α computed by Eq. (5.11) against **a** clay content, **b** plasticity index, and **c** void ratio

where y_i is the ith simulated value; x_j is the jth estimated parameter; $\partial y_i / \partial x_j$ is the sensitivity of the ith simulated value with respect to the jth parameter; N is the number of observations; ω_i is the weighting factor, which is related to the ith observation and can be evaluated based on the statistics (i.e., variance, or standard deviation, or coefficient of variation of the error of the observations). The composite scaled sensitivities indicate the total amount of information provided by the observations for the estimation of parameter j and measure the relative importance of the input parameters being simultaneously estimated.

To obtain a reliable sensitivity, the CSS_j was calculated on three different points for each involved variable and thus an average value was finally given. Figure 5.10 shows the results of sensitivity analysis for C_α based on the proposed EPR model. The variable having the most significant influence on C_α is I_p, which has been highlighted by Yin [43] and Nakase et al. [41]. The *e* and CI have a relatively minor important influence on predicting C_α. A slight higher sensitivity of CI well reveals the experimental results of Yin [43].

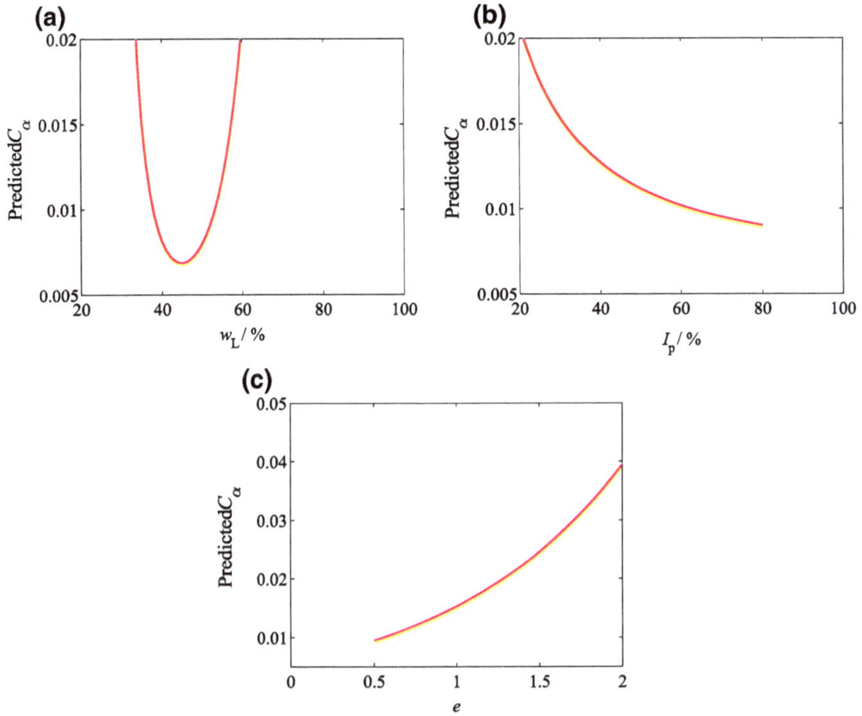

Fig. 5.9 Results of the C_α computed by Eq. (5.12) against **a** liquid limit, **b** plasticity index, and **c** void ratio

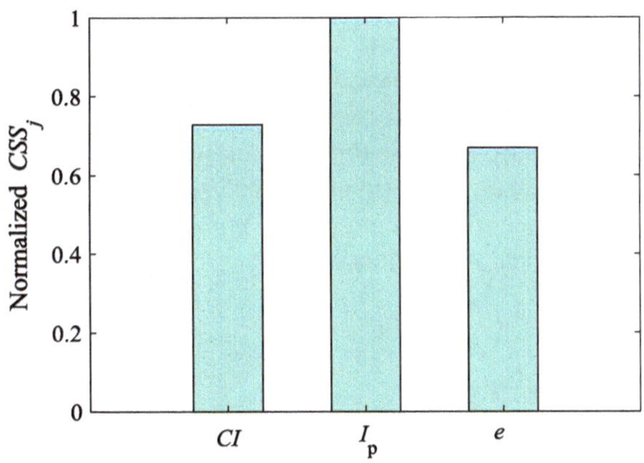

Fig. 5.10 Results of sensitivity analysis for EPR model of C_α

5.4 Discussion

The CI is involved in the proposed EPR model, which implies the need of measurement of CI in laboratory. Comparing to Atterberg limits and void ratio, the measurement of CI is less conventional. Thus, the need of CI will reduce the utility of the proposed model. Therefore, when the data of CI is not available, the EPR correlation (Eq. 5.12) only involving Atterberg limits (w_L and I_p) can also be an alternative choice for predicting C_α although its monotonicity is worse.

In contrast to other techniques to obtain nonlinear creep parameters of clays, such as trust-region-reflective least squares algorithm [91–93], Simplex [94], genetic algorithm [31, 78], the proposed EPR model only requires the basic physical information of soil samples and no additional laboratory tests (e.g., oedometer test, triaxial test) are needed. Moreover, the computational cost is less compared to numerous calculations on obtaining the fitness, sorting and selection for optimizations.

Since EPR is a data mining technique which is heavily dependent on the amount of data used, especially the range covered by the data, the formula obtained will be more applicable if more experimental data with wide range can be found and used. Currently, both EPR models are trained on limited data. Thus, their further performance needs more unseen data to verify. Moreover, since the EPR models are polynomial, it is inevitable to predict very unreasonable values on few special cases. As both models have high robustness ratios, this unreasonable probability should be slight. However, the robustness ratio is always smaller than 1, which means that the proposed ERP model fails in generating a reasonable value on some samples. Therefore, it is still necessary to pay attention when the predicted values are out of the proposed range of C_α.

For the applicability of proposed EPR model in real engineering practice, the basic physical properties (e.g., e, I_p and CI) of soil samples from the dominated soil layer can be easily measured in laboratory. Then, the intrinsic C_α (corresponding to reconstituted state) can be obtained using the proposed EPR model. This C_α is a key input parameter for many elasto-viscoplastic models (e.g. Kimoto and Oka [82]; Yin et al. [36]; Yin et al. [26, 29]), and then long-term performance of various real engineering structures (e.g., embankment, slope and tunnel) can be estimated by numerical simulations.

5.5 Summary

A simple, robust, and accurate EPR model for modeling C_α of reconstituted clays using physical properties has been proposed. Prior to EPR procedure, the database for training the EPR model was built, which contains clay content (CI), liquid limit (w_L), plastic limit (w_P), plasticity index (I_P), void ratio (e), and C_α. Based on the database, the statistical analysis and basic correlations between C_α and each physical property have then been conducted. The C_α is relatively highly correlated to e, followed by I_p and w_L, and very poorly correlated to w_P and CI.

To avoid overfitting and reduce the model complexity, a novel EPR procedure using a newly enhanced DE algorithm was proposed with two enhancements: (1) a new fitness function was proposed using the structural risk minimization (SRM) with L_2 regularization factor that penalizes polynomial complexity; (2) an adaptive process for selecting the combination of involved variables and size of polynomial terms was incorporated. The selection of regularization parameter was suggested.

To attain the nonlinear creep behavior that the C_α consecutively decreases with increasing the soil density, a general structure of EPR expression for C_α was proposed, in which the CI, w_L, and I_P were chosen as the dynamic correlating variables and the e was considered as a fixed variable. 120 data randomly selected in database were used for training and the remaining data were used for testing. The maximum size of terms was set to 8 and the total number of possible combinations of involved variables was 7.

Six EPR models with different variable combinations and size of model terms corresponding to different values of regularization parameter λ were firstly achieved. The performance of each model was compared using five indicators. Based on preliminary results, two EPR models with three terms were temporarily suggested as the optimal models. Then, a robustness testing was conducted on all obtained EPR models, in which the EPR model involving CI with three terms was selected as the optimal model. To deeply understand the mathematical characteristics of two proposed EPR models and select one of them as the optimum, a monotonicity analysis was conducted. Overall, the EPR model involving CI and I_P with 3 terms is finally recommended in terms of the predictive ability, model complexity, robustness, and monotonicity. Hereafter, the sensitivity analysis of CI, I_p and e for the optimal model was carried out. The analysis results indicated that the I_p has the most significant influence on predicting C_α.

The proposed EPR procedure can be applied to other properties of soils. More details can be found in Yin et al. [55] and Jin et al. [95].

References

1. Giustolisi O, Savic D (2006) A symbolic data-driven technique based on evolutionary polynomial regression. J Hydroinform 8(3):207–222
2. Yin Z-Y, Jin Y-F, Shen JS, Hicher P-Y (2018) Optimization techniques for identifying soil parameters in geotechnical engineering: comparative study and enhancement. Int J Numer Anal Methods Geomech 42(1):70–94
3. Garg A, Lam JSL, Panda B (2017) A hybrid computational intelligence framework in modelling of coal-oil agglomeration phenomenon. Appl Soft Comput 55:402–412
4. Coelho F, Neto JP (2017) A method for regularization of evolutionary polynomial regression. Appl Soft Comput 59:223–228
5. Ng AY (2004) Feature selection, L 1 vs. L 2 regularization, and rotational invariance. In: Proceedings of the twenty-first international conference on machine learning. ACM, p 78
6. Wood DM (2003) Geotechnical modelling. CRC Press

7. Ren D-J, Shen S-L, Arulrajah A, Wu H-N (2018) Evaluation of ground loss ratio with moving trajectories induced in DOT tunnelling. Can Geotech J 55(6):894–902
8. Shen S-L, Wu H-N, Cui Y-J, Yin Z-Y (2014) Long-term settlement behaviour of metro tunnels in the soft deposits of Shanghai. Tunn Undergr Space Technol 40:309–323
9. Wu H-N, Shen S-L, Liao S-M, Yin Z-Y (2015) Longitudinal structural modelling of shield tunnels considering shearing dislocation between segmental rings. Tunn Undergr Space Technol 50:317–323
10. Wu H-N, Shen S-L, Yang J (2017) Identification of tunnel settlement caused by land subsidence in soft deposit of Shanghai. J Perform Constructed Facil 31(6):04017092
11. Feng J, Chuhan Z, Gang W, Guanglun W (2003) Creep modeling in excavation analysis of a high rock slope. J Geotech Geoenviron Eng 129(9):849–857
12. Zhang J-F, Chen J-J, Wang J-H, Zhu Y-F (2013) Prediction of tunnel displacement induced by adjacent excavation in soft soil. Tunn Undergr Space Technol 36:24–33
13. Wang J, Xu Z, Wang W (2009) Wall and ground movements due to deep excavations in Shanghai soft soils. J Geotech Geoenviron Eng 136(7):985–994
14. Chai J-C, Shen JS-L, Liu MD, Yuan D-J (2018) Predicting the performance of embankments on PVD-improved subsoils. Comput Geotech 93:222–231
15. Karstunen M, Yin ZY (2010) Modelling time-dependent behaviour of Murro test embankment. Geotechnique 60(10):735–749
16. Rezania M, Bagheri M, Nezhad MM, Sivasithamparam N (2017) Creep analysis of an earth embankment on soft soil deposit with and without PVD improvement. Geotext Geomembr 45(5):537–547
17. Yin Z, Karstunen M, Wang J, Yu C (2011) Influence of features of natural soft clay on behaviour of embankment. J Central South Univ Technol 18(5):1667–1676
18. Shen SL, Chai JC, Hong ZS, Cai FX (2005) Analysis of field performance of embankments on soft clay deposit with and without PVD-improvement. Geotext Geomembr 23(6):463–485
19. Zhu Q-Y, Yin Z-Y, Xu C-J, Yin J-H, Xia X-H (2015) Uniqueness of rate-dependency, creep and stress relaxation behaviors for soft clays. J Central South Univ 22:296–302
20. Zhu Q-Y, Wu Z-X, Li Y-L, Xu C-J, Wang J-H, Xia X-H (2014) A modified creep index and its application to viscoplastic modelling of soft clays. J Zhejiang Univ Sci A 15(4):272–281
21. Shen SL, Xu YS (2011) Numerical evaluation of land subsidence induced by groundwater pumping in Shanghai. Can Geotech J 48(9):1378–1392
22. Shen S-L, Ma L, Xu Y-S, Yin Z-Y (2013) Interpretation of increased deformation rate in aquifer IV due to groundwater pumping in Shanghai. Can Geotech J 50(11):1129–1142
23. Xu YS, Ma L, Du YJ, Shen SL (2012) Analysis of urbanisation-induced land subsidence in Shanghai. Nat Hazards 1–13
24. Xu YS, Ma L, Shen SL, Sun WJ (2012) Evaluation of land subsidence by considering underground structures that penetrate the aquifers of Shanghai, China. Hydrol J 1–12
25. Xu Y-S, Shen S-L, Ren D-J, Wu H-N (2016) Analysis of factors in land subsidence in Shanghai: a view based on a strategic environmental assessment. Sustainability 8(6):573
26. Yin ZY, Chang CS, Karstunen M, Hicher PY (2010) An anisotropic elastic-viscoplastic model for soft clays. Int J Solids Struct 47(5):665–677
27. Yin Z-Y, Yin J-H, Huang H-W (2015) Rate-dependent and long-term yield stress and strength of soft Wenzhou marine clay: experiments and modeling. Mar Georesour Geotechnol 33(1):79–91
28. Yin Z-Y, Zhu Q-Y, Yin J-H, Ni Q (2014) Stress relaxation coefficient and formulation for soft soils. Géotech Lett 4(January–March):45–51
29. Yin ZY, Karstunen M, Chang CS, Koskinen M, Lojander M (2011) Modeling time-dependent behavior of soft sensitive clay. J Geotech Geoenviron Eng 137(11):1103–1113
30. Yin Z-Y, Zhu Q-Y, Zhang D-M (2017) Comparison of two creep degradation modeling approaches for soft structured soils. Acta Geotech 1–19
31. Yin Z-Y, Jin Y-F, Shen S-L, Huang H-W (2017) An efficient optimization method for identifying parameters of soft structured clay by an enhanced genetic algorithm and elastic–viscoplastic model. Acta Geotech 12(4):849–867

32. Yin ZY, Wang JH (2012) A one-dimensional strain-rate based model for soft structured clays. Sci China-Technol Sci 55(1):90–100
33. Yin ZY, Karstunen M, Hicher PY (2010) Evaluation of the influence of elasto-viscoplastic scaling functions on modelling time-dependent behaviour of natural clays. Soils Found 50(2):203–214
34. Yin ZY, Karstunen M (2011) Modelling strain-rate-dependency of natural soft clays combined with anisotropy and destructuration. Acta Mech Solida Sin 24(3):216–230
35. Yin JH, Cheng CM (2006) Comparison of strain-rate dependent stress-strain behavior from K_0-consolidated compression and extension tests on natural Hong Kong marine deposits. Mar Georesour Geotechnol 24(2):119–147
36. Yin J-H, Zhu J-G, Graham J (2002) A new elastic viscoplastic model for time-dependent behaviour of normally and overconsolidated clays: theory and verification. Can Geotech J 39(1):157–173
37. Zhu G, Yin JH (2000) Elastic visco-plastic consolidation modelling of clay foundation at Berthierville test embankment. Int J Numer Anal Methods Geomech 24(5):491–508
38. Mesri G, Godlewski PM (1977) Time and stress-compressibility interrelationship. J Geotech Eng Div 103(5):417–430
39. Yin ZY, Chang CS (2009) Microstructural modelling of stress-dependent behaviour of clay. Int J Solids Struct 46(6):1373–1388
40. Yin Z, Chang C, Hicher P, Karstunen M (2008) Microstructural modeling of rate-dependent behavior of soft soil. In: Proceeding of 12th IACMAG, Goa, 2008, pp 862–868
41. Nakase A, Kamei T, Kusakabe O (1988) Constitutive parameters estimated by plasticity index. J Geotech Eng 114(7):844–858
42. Suneel M, Park LK, Im JC (2008) Compressibility characteristics of Korean marine clay. Mar Georesour Geotechnol 26(2):111–127
43. Yin J-H (1999) Properties and behaviour of Hong Kong marine deposits with different clay contents. Can Geotech J 36(6):1085–1095
44. Zeng L, Hong Z, Liu S, Chen F (2012) Variation law and quantitative evaluation of secondary consolidation behavior for remolded clays. Chin J Geotech Eng 34(8):1496–1500
45. Zeng L, Liu S (2010) A calculation method of secondary compression index for natural sedimentary clays using void index. In: Geo-Shanghai international conference Shanghai, 2010, pp 14–21
46. Anagnostopoulos C, Grammatikopoulos I (2011) A new model for the prediction of secondary compression index of soft compressible soils. Bull Eng Geol Environ 70(3):423–427
47. Zhu Q-Y, Yin Z-Y, Hicher P-Y, Shen S-L (2016) Nonlinearity of one-dimensional creep characteristics of soft clays. Acta Geotech 11(4):887–900
48. Alemdag S, Gurocak Z, Cevik A, Cabalar A, Gokceoglu C (2015) Modeling deformation modulus of a stratified sedimentary rock mass using neural network, fuzzy inference and genetic programming. Eng Geol 203:70–82
49. Gurocak Z, Alemdag S, Zaman MM (2008) Rock slope stability and excavatability assessment of rocks at the Kapikaya dam site, Turkey. Eng Geol 96(1):17–27
50. Gurocak Z, Solanki P, Alemdag S, Zaman MM (2012) New considerations for empirical estimation of tensile strength of rocks. Eng Geol 145:1–8
51. Zhang W, Goh ATC, Zhang Y, Chen Y, Xiao Y (2015) Assessment of soil liquefaction based on capacity energy concept and multivariate adaptive regression splines. Eng Geol 188:29–37
52. Zhou C, Yin K, Cao Y, Ahmed B (2016) Application of time series analysis and PSO–SVM model in predicting the Bazimen landslide in the Three Gorges Reservoir, China. Eng Geol 204:108–120
53. Ahangar-Asr A, Faramarzi A, Javadi AA (2010) A new approach for prediction of the stability of soil and rock slopes. Eng Comput 27(7):878–893
54. Doglioni A, Crosta GB, Frattini P, Melidoro NL, Simeone V (2015) Predicting landslide displacements by multi-objective evolutionary polynomial regression. In: Engineering geology for society and territory, vol 5. Springer, Berlin, pp 651–654

55. Yin Z-Y, Jin Y-F, Huang H-W, Shen S-L (2016) Evolutionary polynomial regression based modelling of clay compressibility using an enhanced hybrid real-coded genetic algorithm. Eng Geol 210:158–167
56. Wu Z, Ji H, Yu C, Zhou C (2018) EPR-RCGA-based modelling of compression index and RMSE-AIC-BIC-based model selection for Chinese marine clays and their engineering application. J Zhejiang Univ Sci A 19(3):211–224
57. Ahangar-Asr A, Faramarzi A, Mottaghifard N, Javadi AA (2011) Modeling of permeability and compaction characteristics of soils using evolutionary polynomial regression. Comput Geosci 37(11):1860–1869
58. Rezania M, Javadi AA, Giustolisi O (2010) Evaluation of liquefaction potential based on CPT results using evolutionary polynomial regression. Comput Geotech 37(1):82–92
59. Rezania M, Faramarzi A, Javadi AA (2011) An evolutionary based approach for assessment of earthquake-induced soil liquefaction and lateral displacement. Eng Appl Artif Intell 24(1):142–153
60. Khoshkroudi SS, Sefidkouhi MAG, Ahmadi MZ, Ramezani M (2014) Prediction of soil saturated water content using evolutionary polynomial regression (EPR). Arch Agron Soil Sci 60(8):1155–1172
61. Shahnazari H, Shahin MA, Tutunchian MA (2014) Evolutionary-based approaches for settlement prediction of shallow foundations on cohesionless soils. Geotech Eng 12(1):55–64
62. Ghorbani A, Firouzi Niavol M (2017) Evaluation of induced settlements of piled rafts in the coupled static-dynamic loads using neural networks and evolutionary polynomial regression. Appl Comput Intell Soft Comput
63. Shahin MA (2014) State-of-the-art review of some artificial intelligence applications in pile foundations. Geosci Front 7(1):33–44
64. Ebrahimian B, Movahed V (2013) Evaluation of axial bearing capacity of piles in sandy soils by CPT results. Evaluation 29:31
65. Ebrahimian B, Movahed V (2017) Application of an evolutionary-based approach in evaluating pile bearing capacity using CPT results. Ships Offshore Struct 12(7):937–953
66. Ahangar-Asr A, Javadi AA, Johari A, Chen Y (2014) Lateral load bearing capacity modelling of piles in cohesive soils in undrained conditions: an intelligent evolutionary approach. Appl Soft Comput 24:822–828
67. Kakoudakis K, Behzadian K, Farmani R, Butler D (2017) Pipeline failure prediction in water distribution networks using evolutionary polynomial regression combined with K-means clustering. Urban Water J 14(7):737–742
68. Javadi AA, Faramarzi A, Ahangar-Asr A (2012) Analysis of behaviour of soils under cyclic loading using EPR-based finite element method. Finite Elem Anal Des 58:53–65
69. Shahnazari H, Tutunchian MA, Rezvani R, Valizadeh F (2013) Evolutionary-based approaches for determining the deviatoric stress of calcareous sands. Comput Geosci 50:84–94
70. Faramarzi A, Alani AM, Javadi AA (2014) An EPR-based self-learning approach to material modelling. Comput Struct 137:63–71
71. Nassr A, Javadi A, Faramarzi A (2018) Developing constitutive models from EPR-based self-learning finite element analysis. Int J Numer Anal Methods Geomech 42(3):401–417
72. Kalinli A, Acar MC, Gündüz Z (2011) New approaches to determine the ultimate bearing capacity of shallow foundations based on artificial neural networks and ant colony optimization. Eng Geol 117(1):29–38
73. Jin Y-F, Yin Z-Y, Shen S-L, Hicher P-Y (2016) Selection of sand models and identification of parameters using an enhanced genetic algorithm. Int J Numer Anal Methods Geomech 40(8):1219–1240
74. Jin Y-F, Yin Z-Y, Shen S-L, Hicher P-Y (2016) Investigation into MOGA for identifying parameters of a critical-state-based sand model and parameters correlation by factor analysis. Acta Geotech 11(5):1131–1145
75. Jin Y-F, Wu Z-X, Yin Z-Y, Shen JS (2017) Estimation of critical state-related formula in advanced constitutive modeling of granular material. Acta Geotech 12(6):1329–1351

76. Jin Y-F, Yin Z-Y, Shen S-L, Zhang D-M (2017) A new hybrid real-coded genetic algorithm and its application to parameters identification of soils. Inverse Prob Sci Eng 25(9):1343–1366
77. Jin Y-F, Yin Z-Y, Wu Z-X, Zhou W-H (2018) Identifying parameters of easily crushable sand and application to offshore pile driving. Ocean Eng 154:416–429
78. Jin Y-F, Yin Z-Y, Riou Y, Hicher P-Y (2017) Identifying creep and destructuration related soil parameters by optimization methods. KSCE J Civil Eng 21(4):1123–1134
79. Yin Z, Xu Q, Yu C (2015) Elastic-viscoplastic modeling for natural soft clays considering nonlinear creep. Int J Geomech 15(5):A6014001
80. Li Q, Ng CWW, Liu G (2012) Low secondary compressibility and shear strength of Shanghai Clay. J Central South Univ 19(8):2323–2332
81. Zhu Q-Y, Jin Y-F, Yin Z-Y, Hicher P-Y (2013) Influence of natural deposition plane orientation on oedometric consolidation behavior of three typical clays from southeast coast of China. J Zhejiang Univ Sci A 14(11):767–777
82. Kimoto S, Oka F (2005) An elasto-viscoplastic model for clay considering destructuralization and consolidation analysis of unstable behavior. Soils Found 45(2):29–42
83. Yin J-H, Graham J (1989) Viscous–elastic–plastic modelling of one-dimensional time-dependent behaviour of clays. Can Geotech J 26(2):199–209
84. Han J, Yao Y, Yin Z (2018) Influences of overconsolidation ratio on undrained creep behavior of overconsolidated saturated clay. Chin J Geotech Eng 40(3):426–430
85. Nakase A, Kamei T (1986) Influence of strain rate on undrained shear characteristics of K_0-consolidated cohesive soils. Soils Found 26(1):85–95
86. Qu G, Hinchberger S, Lo K (2010) Evaluation of the viscous behaviour of clay using generalised overstress viscoplastic theory. Geotechnique 60(10):777–789
87. Cao Z, Wang Y (2014) Bayesian model comparison and selection of spatial correlation functions for soil parameters. Struct Saf 49:10–17
88. Zhang J, Zhang L, Tang WH (2009) Bayesian framework for characterizing geotechnical model uncertainty. J Geotech Geoenviron Eng 135(7):932–940
89. Zhang L, Li D-Q, Tang X-S, Cao Z-J, Phoon K-K (2017) Bayesian model comparison and characterization of bivariate distribution for shear strength parameters of soil. Comput Geotech 95:110–118
90. Hill MC (1998) Methods and guidelines for effective model calibration. US Geological Survey Denver, CO, USA
91. Le TM, Fatahi B (2016) Trust-region reflective optimisation to obtain soil visco-plastic properties. Eng Comput 33(2):410–442
92. Le TM, Fatahi B, Khabbaz H (2015) Numerical optimisation to obtain elastic viscoplastic model parameters for soft clay. Int J Plast 65:1–21
93. Le TM, Fatahi B, Khabbaz H, Sun W (2017) Numerical optimization applying trust-region reflective least squares algorithm with constraints to optimize the non-linear creep parameters of soft soil. Appl Math Model 41:236–256
94. Ye L, Jin Y-F, Shen S-L, Sun P-P, Zhou C (2016) An efficient parameter identification procedure for soft sensitive clays. J Zhejiang Univ Sci A 17(1):76–88
95. Jin Y-F, Yin Z-Y, Zhou W-H, Yin J-H, Shao J-F (2019) A single-objective EPR based model for creep index of soft clays considering L2 regularization. Eng Geol

Chapter 6
Parameter Identification for Soft Structured Clays

6.1 Introduction

Soft structured clays usually exhibit complex behaviors, which can lead to difficulties in the determination of parameters and high testing costs. This chapter aims to propose an efficient optimization method for identifying the parameters of advanced constitutive model for soft structured clays from only limited conventional triaxial tests. First, the developed real-coded genetic algorithm (RCGA) is adopted. A newly developed elastic–viscoplastic model accounting for anisotropy, destructuration, and creep features is enhanced with the cross-anisotropy of elasticity and is adopted for test simulations during optimization. Laboratory tests on soft Wenzhou marine clay are selected, with three of them being used as objectives for optimization and others for validation. The optimization process, using the RCGA with a uniform sampling initialization method, is carried out to obtain the soil parameters. A classic genetic algorithm (NSGA-II)-based optimization is also conducted and compared to the RCGA for estimating the performance of the RCGA. Finally, the optimal parameters are validated by comparing with other measurements and test simulations on the same clay. All comparisons demonstrate that a reliable solution can be obtained by the RCGA optimization combined with the appropriate soil model, which is practically useful with a reduction in testing costs.

Z. Yin and Y. Jin, *Practice of Optimisation Theory in Geotechnical Engineering*,
https://doi.org/10.1007/978-981-13-3408-5_6

6.2 Difficulties in Determining Parameters of Soft Structured Clay

6.2.1 *Experimental Observations on Coupling of Creep and Destructuration*

Experimental observations show a strong coupling of creep and destructuration (Leroueil et al. [1]; Rocchi et al. [2]; Yin and Karstunen [3]; Yin and Wang [4]; Yin et al. [5]). For instance, Fig. 5.4 shows a schematic plot of a typical 1D compression curve and the evolution of a secondary compression coefficient with the vertical stress for both intact and reconstituted soft clays. By extending the compression curve of the reconstituted sample, an intersection point with the compression curve of the intact sample can be obtained. The difference between the intact and the reconstituted samples is due to the state of the soil structure which can be influenced by cementation or chemical bonding during the natural deposition of the clay. The initial value of the post-yield secondary compression coefficient of the intact sample is large, and it then decreases with increasing vertical stress due to the debonding process induced by the plastic strain, and it finally approaches the value of the secondary compression coefficient of the reconstituted sample. Thus, creep and destructuration are strongly coupled. This coupling can also be found in the triaxial condition and can significantly influence the stability of geotechnical structures.

6.2.2 *Discrepancy in Standard Parameter Determination*

Since the secondary compression coefficient changes with the state of bonding for soft structured clays, the coupling effect links two soil properties: the secondary compression coefficient of soil without bonding (corresponding to the reconstituted clay) $C_{\alpha ei}$, and the soil sensitivity S_t. As shown by Zhu et al. [6], the global secondary compression coefficient $C_{\alpha e}$ of intact samples depends on two soil properties: the intrinsic secondary compression coefficient $C_{\alpha ei}$ of its reconstituted samples and the state of bonding represented by the soil sensitivity S_t or the bonding ratio χ_0 (note that $C_{\alpha ei}$ and S_t or χ_0 are independent). In numerical or analytical methods for geotechnical applications, the $C_{\alpha ei}$ and the bonding ratio χ_0 reflecting S_t (see Fig. 5.4) are simultaneously estimated with the soil constants controlling the debonding rate (e.g., ξ and ξ_d in constitutive models of structured clays by Yin et al. [7] and Gens and Nova [8]).

For determining $C_{\alpha ei}$, a conventional consolidation test on a reconstituted sample is usually required for most soft structured clays in which the interparticle bonding is generally not fully destroyed during mechanical loading. Thus, reconstituted samples need to be tested, which requires additional time (about one month) and considerable cost. For determining the initial bonding ratio, χ_0, 1D compression tests on both intact

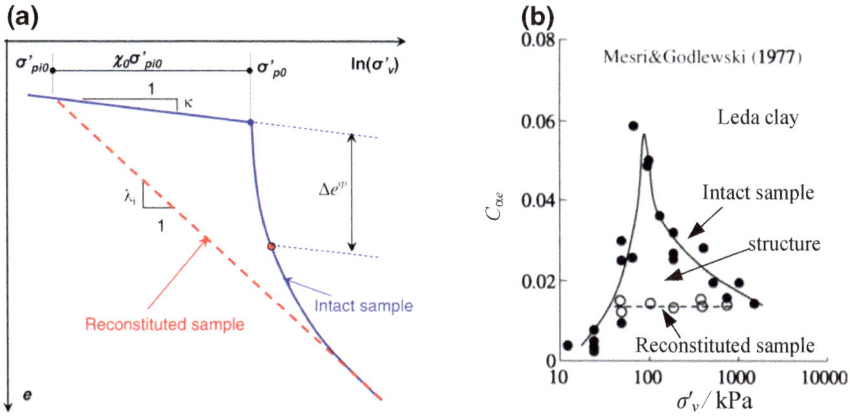

Fig. 6.1 Typical results of oedometer test for intact and reconstituted soft clays

and reconstituted samples are also needed (see Fig. 6.1). The determination of the destructuration constants, ξ and ξ_d requires both 1D and isotropic compression tests on intact and reconstituted samples, as they control different debonding mechanisms (Yin et al. [7]).

Thus, 1D and 3D tests on both intact and reconstituted samples of soft structured clay are currently required. This leads to high demands on time and testing costs, and can lead to difficulties in engineering practice. Finding a more efficient way to determine these parameters with conventional tests on intact clay alone is essential.

6.2.3 *Necessity of Optimization-Based Parameter Identification*

If there is a way to obtain such parameter values by simulating conventional laboratory or field tests and minimizing the difference between experimental and theoretical results, (so-called optimization), this will be practically useful. Different optimization techniques for identifying soil parameters have been successfully used in the geotechnical field in the last few decades. The existing optimization techniques can be divided into two categories: (1) deterministic optimization techniques, such as gradient-based algorithms and simplex (Lecampion et al. [9]; Calvello and Finno [10]; Yin and Hicher [11]; Papon et al. [12]) work with a single solution and are focused on reaching local minima, because they begin the search procedure with a first guess solution (often chosen randomly in the search space). If this guess solution is not close enough to the global minimum solution, it is likely to be trapped in a local minimum solution. (2) Stochastic optimization techniques, such as evolutionary algorithms (Pal et al. [13]; Javadi et al. [14]; Levasseur et al. [15]; Rokonuzzaman and

Sakai [16]; Papon et al. [12]; Vardakos et al. [17]; Moreira et al. [18]) and simulated annealing (Yepes et al. [19]), rely significantly on computational power.

All these optimization techniques are usually applied to laboratory tests, in situ testing or field measurements. Among these, evolutionary algorithms are found to be very promising global optimizers. Genetic algorithms (GA) are perhaps the most commonly used evolutionary algorithms (Deb [20]). However, the application of GAs for soil parameters accounting for combined anisotropy, creep and destructuration has not been reported to date. Therefore, enhancing a more efficient GA based on recent developments and applying it to tackle the problem of parameter identification for soft structured clays may be a useful approach.

6.3 Brief Introduction of Laboratory Tests and Adopted Constitutive Model

The Wenzhou clay deposit is a marine clay characterized as slightly organic and highly plastic. A relatively homogenous layer of Wenzhou clay from 10.5 to 11.5 m was selected for this study. Some common physical properties are presented in Table 6.1.

Intensive laboratory tests were carried out along various stress paths, focusing on the rate-dependent mechanical properties of Wenzhou clay (Yin et al. [5]). The tests selected for this study were three conventional undrained triaxial tests on K_0-consolidated Wenzhou clay under different confining stresses. The K_0-consolidation was performed over 2 days up to a vertical stress of 75 kPa, for another 2 days up to 150 kPa, and finally for another 2 days up to 300 kPa. The average stress rate was 2.08 kPa/h. For the undrained triaxial shearing stage, a strain rate of 2%/h was applied during tests in accordance with the ASTM standard, and this was also adopted in the simulations. The results of these undrained triaxial tests with their K_0-consolidation curves are shown in Fig. 6.2.

Since the stress–strain–time relationship is uniquely controlled by the secondary compression coefficient and the applied strain rate of test (see Yin et al. [21]), and the destructuration behavior is also uniquely controlled by a test with a stress path of varying stress ratio, an undrained triaxial shearing test with its consolidation stage is theoretical enough for identifying related parameters. Then for higher accuracy, three triaxial tests under different confining stresses with their consolidation stages as recommended in engineering design, usually for simple elastoplastic models, were adopted in this study. If successful, the application of the design-based test requirement can be directly extended to more advanced constitutive models.

Table 6.1 Typical physical properties of Wenzhou clay

Depth (m)	γ (kN/m^3)	e_0	w (%)	w_L (%)	w_P (%)	σ'_{p0} (kPa)	σ'_{v0} (kPa)
10.5–11.5	15.5	1.895	67.5	63.4	27.6	81.3	75.4

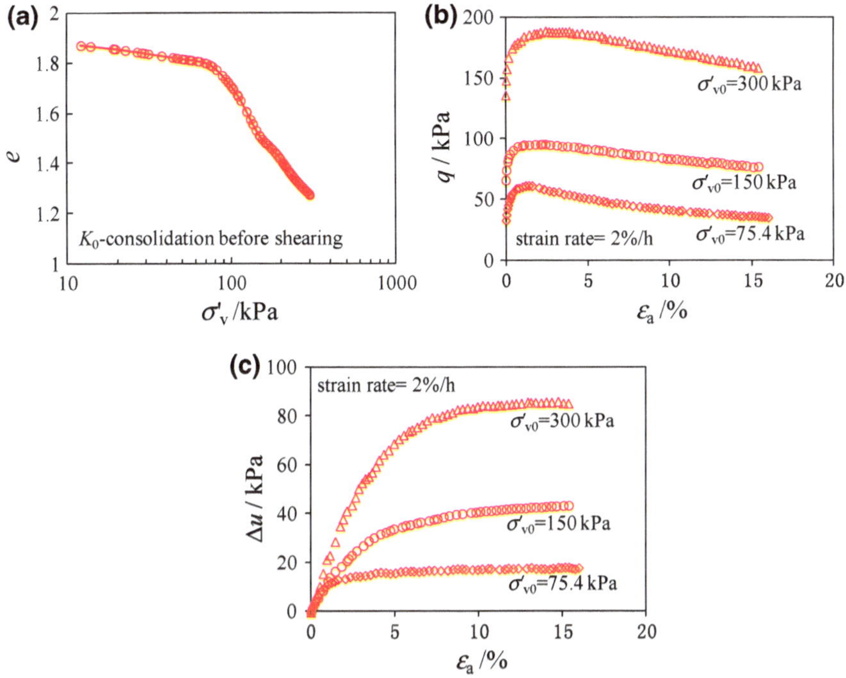

Fig. 6.2 Results of triaxial tests on Wenzhou clay: **a** K_0-consolidation stage; **b** deviatoric stress versus axial strain; and **c** excess pore pressure versus axial strain

For soft structured clays, creep- and destructuration-related parameters are of importance not only for constitutive modeling but also for engineering practice (e.g., directly relating to secondary compression coefficient and sensitivity). A newly developed elastic–viscoplastic model (Yin et al. [7]) accounting for the main features of a natural soft clay (e.g., soil viscosity, anisotropy, and destructuration) can reproduce the decrease of the secondary compression coefficient with the interparticle debonding, and is thus adopted to simulate all selected tests in this study. A brief introduction to this model with its associated parameters can be found in the Appendix. Due to its natural deposition, the soil exhibits naturally inherent cross-anisotropy of elasticity (see Yin and Chang [22]; Chang and Yin [23]; Yin and Karstunen [3]; Yin et al. [24]; Yin and Hicher[25]). This anisotropic elastic behavior was considered to enhance the model of Yin et al. [7] by adopting the following matrix of elastic stiffness:

$$\begin{bmatrix} \dot{\varepsilon}_x \\ \dot{\varepsilon}_y \\ \dot{\varepsilon}_z \\ \dot{\varepsilon}_{xy} \\ \dot{\varepsilon}_{yz} \\ \dot{\varepsilon}_{zx} \end{bmatrix} = \begin{bmatrix} 1/E_\text{v} & -\upsilon'_\text{vv}/E_\text{v} & -\upsilon'_\text{vv}/E_\text{v} & 0 & 0 & 0 \\ -\upsilon'_\text{vv}/E_\text{v} & 1/E_\text{h} & -\upsilon'_\text{vh}/E_\text{h} & 0 & 0 & 0 \\ -\upsilon'_\text{vv}/E_\text{v} & -\upsilon'_\text{vh}/E_\text{h} & 1/E_\text{h} & 0 & 0 & 0 \\ 0 & 0 & 0 & 1/2G_\text{vh} & 0 & 0 \\ 0 & 0 & 0 & 0 & \left(1+\upsilon'_\text{vh}\right)/E_\text{h} & 0 \\ 0 & 0 & 0 & 0 & 0 & 1/2G_\text{vh} \end{bmatrix} \begin{bmatrix} \dot{\sigma}'_x \\ \dot{\sigma}'_y \\ \dot{\sigma}'_z \\ \dot{\sigma}'_{xy} \\ \dot{\sigma}'_{yz} \\ \dot{\sigma}'_{zx} \end{bmatrix} \tag{6.1}$$

with $E_\text{h} = nE_\text{v}$, $\upsilon'_\text{vh} = \sqrt{n}\upsilon'_\text{vv}$, where E_v and E_h are the vertical and horizontal Young's modulus, respectively, υ'_vv and υ'_vh are the vertical and horizontal Poisson's ratio, respectively and G_vh is the shear modulus (see Graham and Houlsby [26]). For stress-controlled isotropic compression with incremental stress of $\dot{\sigma}'_x = \dot{\sigma}'_y = \dot{\sigma}'_z = \dot{p}'$, then

$$\dot{\varepsilon}_\text{v} = \dot{\varepsilon}_x + \dot{\varepsilon}_y + \dot{\varepsilon}_z = \left(1 - 4\upsilon'_\text{vv} + 2/n - 2\,\upsilon'_\text{vv}/\sqrt{n}\right)\frac{\dot{p}'}{E_\text{v}} \tag{6.2}$$

Based on the definition of bulk modulus, $K = \delta p'/\delta\varepsilon_\text{v}$, the vertical Young's modulus can be obtained as follows, with the shear modulus, G_vh:

$$E_\text{v} = \left(1 - 4\upsilon'_\text{vv} + 2/n - 2\upsilon'_\text{vv}/\sqrt{n}\right)\left(\frac{1+e_0}{\kappa}\right)p' \tag{6.3}$$

$$G_\text{vh} = \frac{\sqrt{n}E_\text{v}}{2\left(1+\sqrt{n}\upsilon'_\text{vv}\right)} \tag{6.4}$$

Then for input parameters $\left(\kappa, \upsilon'_\text{vv}, n\right)$, one additional parameter, n, varying between 0 and 1, needs to be identified for anisotropic elasticity unlike the isotropic elasticity.

Overall, the recently developed "ANICREEP" model considering soil viscosity, anisotropy, and destructuration was adopted and enhanced with cross-anisotropy of elasticity. Apart from the Poisson's ratio (varying from 0.1 to 0.35 for clays, and the stress–strain response in 1D and triaxial conditions is not sensitive to these values (see Biarez and Hicher [27]; Yin and Hicher [11]), taken as 0.25, a typical value for clays) and the initial void ratio e_0, all other input parameters of the model are set for optimization in this study. The intervals of these parameters given in Table 6.2 are much larger than those corresponding to their typical values (see Biarez and Hicher [27]; Yin et al. [7, 21]; Zhu et al. [28]; Yao et al. [29, 31]; Yao and Sun [30]). Note that this is necessary to make sure the real solutions are within the range, and then no measurement or prejudgment is needed based on stress–strain or stress path curves. If successful, the high performance of the proposed identification procedure using the proposed RCGA can be highlighted.

Table 6.2 Search domain for creep and destructuration parameters of ANICREEP model

Parameters	κ	λ_i	M	$C_{\alpha ei}$	χ_0	ξ	ξ_d	n
Lower bound	0.001	0.001	0.5	0.0001	0	0	0	0.1
Upper bound	0.2	0.5	2.0	0.1	50	20	0.5	2.0
Step size	0.001	0.001	0.01	0.0001	0.5	0.1	0.05	0.05

6.4 Identification Procedure Based on RCGA

6.4.1 Error Function

In order to carry out an inverse analysis, the user must define a function that can evaluate the error between the experimental and the numerical results, and then minimize this function. In order to make the error independent of the type of test and the number of measurement points, an advanced error function proposed by Levasseur et al. [15] has been adopted. The difference between the measured and the simulated results is expressed in the form of the least square method,

$$\text{Error}(x) = \sqrt{\frac{\sum_{i=1}^{N}\left(\frac{U_{\text{exp}}^{i}-U_{\text{num}}^{i}}{U_{\text{exp}}^{i}} \times 100\right)^{2}}{N}}\,\% \tag{6.5}$$

where x is a vector of the parameters; N is the number of values; U_{exp}^{i} is the value of measurement point i; U_{num}^{i} is the value of calculation at point i.

The scale effects on the fitness between the experimental and the simulated results can be eliminated by this normalized formula. Additionally, the objective error calculated by this function is a dimensionless variable; thus, any difference in error can be avoided for different objectives with different variables.

6.4.2 Identification Methodology

Generally, deformation and strength are two extremely important indicators to illustrate the mechanical behavior of soil. For a laboratory triaxial test, the isotropic or anisotropic compression test has been conducted first, followed by the shearing stage. During the whole process, the model parameters accounting for the compression behavior and the shear behavior are measured and obtained. At the same time, some other parameters concerning features such as sensitivity or destructuration and creep are also implied in the stress–strain–time relationship (softening or hardening and contraction or dilation), although these variables cannot be directly measured. Therefore, the results of selected laboratory tests can provide information to optimize the model parameters. In this study, a mono-objective framework with three different criteria is considered:

$$\min[\text{Error}(x)] = \begin{bmatrix} \text{Error}(K_0) \\ \text{Error}(q) \\ \text{Error}(\Delta u \text{ or } \Delta e) \end{bmatrix} \tag{6.6}$$

Note that the errors can be calculated between measurements and simulations based on strains along the stress level or stresses along the strain level. It will be more convenient in strain-controlled tests to compute the errors based on p' for K_0-compression test (marked as Error (K_0)), and based on q and Δu or Δe (marked as Error (q) and Error (Δu or Δe)) for undrained or drained triaxial tests.

The total error function is expressed as:

$$\text{Error}(x) = \sum_{\text{i}=1}^{m} (l_i \cdot \text{Error}(x)_i) \tag{6.7}$$

where m is the number of objectives involved in the optimization; $\text{Error}(x)_i$ is the value of error corresponding to the objective, i. l_i is the weight factor with $\sum(l_i) = 1$; The weight factor, l_i is taken as 1/3 in this case, as each test plays the same important role in evaluating soil behavior. Then the average error is taken as being equal to (Error (q) + Error (Δu or Δe) + Error (K_0))/3, with Error (q), Error (Δu), Error (Δe) and Error (K_0) representing the difference between the experimental results and the numerical simulation values for deviatoric stress, excess pore pressure for undrained tests, the change in void ratio for drained tests and vertical stress for the K_0-consolidation stage, respectively.

6.4.3 Numerical Validation by Identifying Soil Parameters

To evaluate the performance of the adopted RCGA, a set of synthetic objective tests (one oedometer test and one undrained triaxial test on isotropically consolidated clay) was first generated by the modified Cam-Clay model [32] using a set of typical parameters ($\upsilon = 0.25$, $e_0 = 1.0$, $\kappa = 0.02$, $\lambda = 0.20$, $M_\text{c} = 1.2$, and $p_{\text{c}0}' = 100$), as shown in Fig. 6.3. Then, except for υ and e_0, the remaining of the parameters were optimized by the new RCGA and NSGA-II, respectively. The initial population size and the maximum number of generations were taken to 50 and 100. The interval of each parameter is shown in Table 6.3. Figure 6.4 shows the evolution of the minimum objective value as the generation number increases in identifying MCC parameters for both RCGA and NSGA-II. The Procedure with RCGA converges faster than that with NSGA-II, which demonstrates an improved search ability by RCGA.

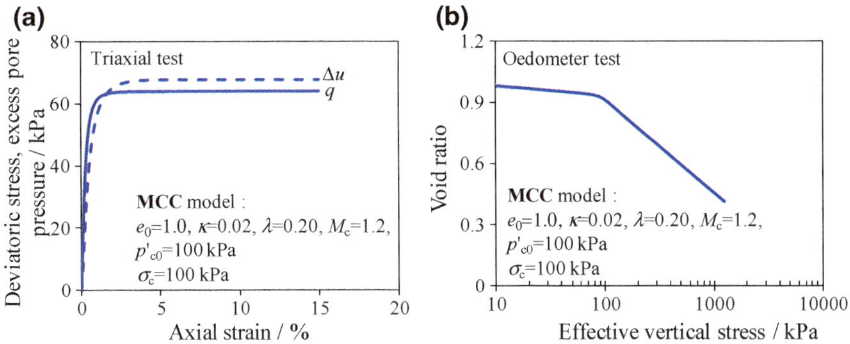

Fig. 6.3 Results of synthetic objective tests generated by MCC model

Table 6.3 Search domain and intervals of parameters for MCC model

Parameters	κ	λ	M_c	p_{c0}'
Lower bound	0.001	0.1	0.5	50
Upper bound	0.1	0.5	1.5	150
Step	0.001	0.01	0.01	1.0

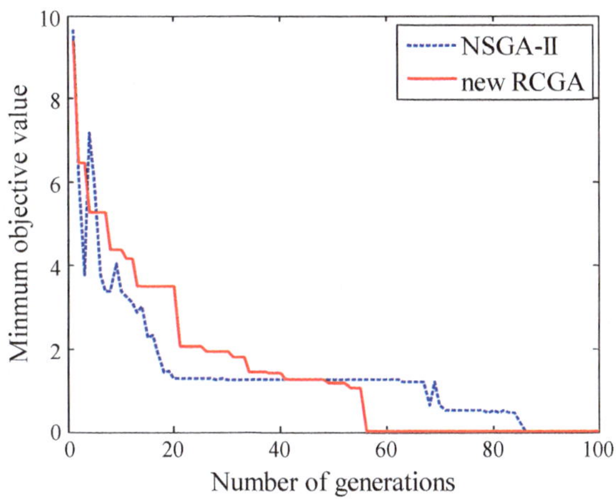

Fig. 6.4 Comparison of optimized results in identifying MCC parameters for RCGA and NSGA-II

Therefore, the excellent performance of the proposed RCGA was validated by conducting four benchmark mathematic tests and one constitutive model test, based on which the proposed RCGA is recommended for tackling more complex problems.

6.5 Optimization Results and Validation

6.5.1 *Optimization Results and Discussion*

The optimization procedure was initially conducted using the new RCGA. In order to estimate the performance of the new RCGA, the classic and widely adopted non-dominated sorting genetic algorithm (NSGA-II) proposed by Deb et al. [33] was selected to carry out the same optimization analysis. The controlling parameters of NSGA-II have the same values as the new RCGA, as shown in Table 6.4. Note that these parameter settings are typical and common, as recommended by many researchers (Deb et al. [33]; Deep and Thakur [34]; Jin et al. [35, 36]), the sensitivity of these parameters will not be presented here.

According to Deep and Thakur [34], computational effectiveness and efficiency are two important aspects for assessing an optimization algorithm. The computational effectiveness signifies the degree of precision in locating global minima, and the efficiency of a GA is the measure of the rate of convergence. Following the same optimization procedure, the optimal solution with minimum average error was respectively obtained by RCGA and NSGA-II. Generally, GA provides a population of individuals, which has to be selected according to a satisfaction criterion. All individuals, whose error value is lower than a reference value, are called "satisfactory" [12]. In this paper, the optimal solution is selected based on the minimum average error (average error = (Error (q) + Error (Δu) + Error (K_0))/3) from a thousand set of parameters generated during optimization, as shown in Table 6.5. This optimal solution is unique for different optimization runs when the number of generations and initial individuals are big enough in GA [12]. The significant difference between the optimal parameters obtained using both algorithms was found, which indicates that the two GAs used in this study have different search abilities in finding the best solution. Based only on the value of the objective error, the set of parameters obtained using the new RCGA is much more satisfying than that of NSGA-II. In other words, the new RCGA has a perfect search effectiveness and is more suitable than NSGA-II for identifying soil parameters.

For evaluating the search efficiency, the number of generations to convergence is a key criterion. Figure 6.5 shows the evolution of the minimum objective error in each generation with the number of generations. It can be seen that the new RCGA

Table 6.4 Parameters of selected algorithms

Algorithm	PopSize	NumGens	Selection	Probability of crossover	Probability of mutation	Elitism strategy
New RCGA	100	50	Tournament	0.7	0.05	Yes
NSGA-II	100	50	Tournament	0.7	0.05	Yes

Table 6.5 Simulation errors with two optimal sets of parameters optimized by GAs for Wenzhou clay

Algorithm	Convergence generation	Optimal parameters								Objective error (%)
		κ	λ_i	M	$C_{\alpha ei}$	χ_0	ξ	ξ_d	n	
RCGA	50	0.052	0.225	1.18	0.0081	5	11.5	0.425	0.80	7.94
NSGA-II	49	0.059	0.250	1.13	0.0041	3.5	8.5	0.50	0.70	9.69

Remark Objective error representing minimum average error of (Error (q) + Error (Δu) + Error (K_0))/3 during optimization

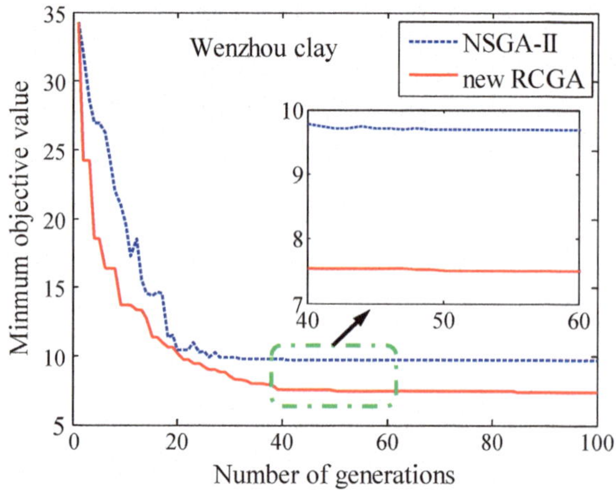

Fig. 6.5 Evolution of minimum objective value in each generation with the increase of the number of generations

in general has a smaller error than the NSGA-II, which demonstrates that a higher search efficiency is obtained with the new RCGA.

It is apparent that the advanced search mechanism and the maintenance of population diversity can lead to a good performance for the GAs. Unlike in the NSGA-II, the mutation operators adopted in the RCGA are self-adaptive. The DRM provides a greater chance of population variation by producing a relatively large allowable step size for the mutation at every initial evolution period, which can result in a higher probability of escaping from the local traps. When the population is gradually converging to the optimum solution, a small mutation region produced by the DRM can enhance the precision of the obtained solution.

All the comparisons demonstrate that the new RCGA is robust and suitable for identifying parameters of soft structured clay, in terms of computational effectiveness and efficiency. With regard to the reasonableness and reliability of optimal parameters, this will be further validated in the following sections.

6.5.2 Validation Based on Experimental Measurements

Additional test data on the same Wenzhou marine clay (Zeng [37]; Wang and Yin [38]; Yin et al. [5]) are used to determine which set of parameters obtained by different GAs is the most appropriate.

For the parameter concerning the compression behavior, the intrinsic compression index λ_i (corresponding to the reconstituted clay) cannot be directly measured for intact samples except under very high stress levels. According to Biarez and Hicher [27], the compression index $\lambda_i = 0.197$ can be estimated using the empirical form $C_c = 0.009\ (w_L - 13)$. Based on a study by Zeng [37] on reconstituted clay from the same location, a compression index of $\lambda_i = 0.202$ was obtained. In the comparison,

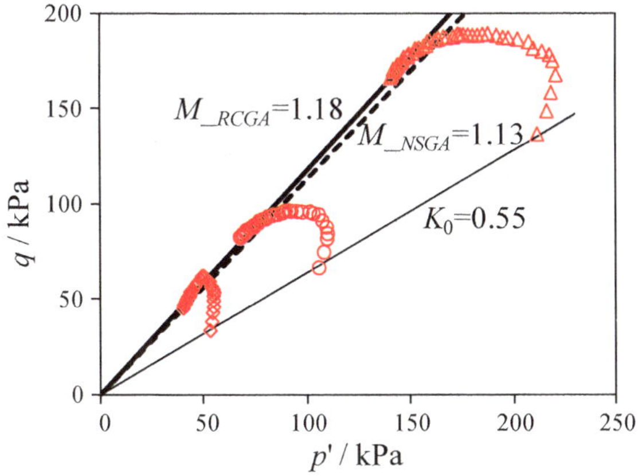

Fig. 6.6 Comparisons of M_c obtained by RCGA and NSGA-II between simulated and experimental results

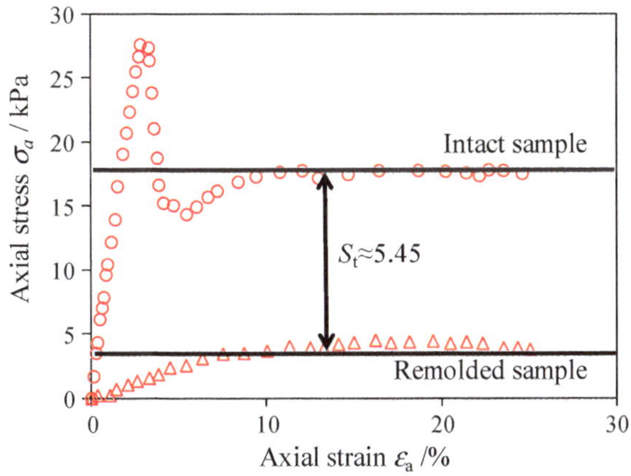

Fig. 6.7 Unconfined compression tests on intact and remolded Wenzhou marine clay

the value given by RCGA is closer to the actual measurement than that given by NSGA-II. Likewise, for the slope of the critical state line M, as shown in Fig. 6.6, the value of M given by RCGA ($M_{\_RCGA} = 1.18$) appears more reasonable than that given by NSGA-II($M_{\_NSGA} = 1.13$).

The measured value of sensitivity, S_t, for Wenzhou marine clay is approximately 5.45, as shown in Fig. 6.7. From the optimization analysis, a S_t value of 6 was obtained using the new RCGA, and a value of 4.5 using NSGA-II. Compared to the measured value, the one obtained using the RCGA appears more reasonable than the one obtained using NSGA-II. For this reason, the performance of the new RCGA is suitable for identifying soil parameters.

For $C_{\alpha ei}$, a higher value of $C_{\alpha ei} = 0.0081$ is obtained using the new RCGA and a smaller value of $C_{\alpha ei} = 0.0041$ is obtained using NSGA-II. Compared with the measurements given by Zeng [37] based on reconstituted clay from the same location, the value determined using RCGA is closer to the average measured value ($C_{\alpha ei} =$

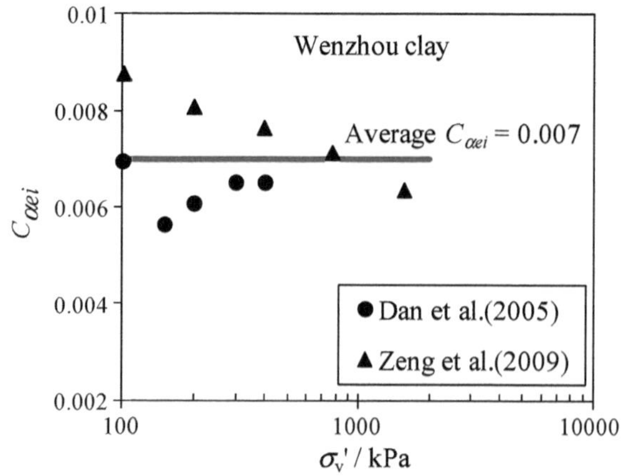

Fig. 6.8 Evolution of $C_{\alpha ei}$ with vertical stress for Wenzhou marine clay

0.007) than that given using NSGA-II, as shown in Fig. 6.8. It can be seen that the value obtained using the new RCGA appears more reasonable.

Considering the uniqueness of the solution and the comparisons between measurements and optimization results in terms of the intrinsic compression index λ_i, the critical state line M, the initial bonding χ_0, and the intrinsic secondary compression coefficient $C_{\alpha ei}$, the new RCGA appears more robust and suitable than the NSGA-II for identifying parameters. Finally, the optimal set of parameters using the new RCGA was selected, shown in Table 6.5, for further validation by simulating other tests on the same clay.

6.5.3 Validation Based on Test Simulations

One-dimensional multi-staged constant rate of strain (CRS) tests, undrained triaxial tests in compression and extension, and undrained creep tests on the same Wenzhou clay (Yin et al. [5]) were simulated using an enhanced ANICREEP model with the optimized parameters shown in Table 6.5.

6.5.4 Oedometer Tests at Constant Rate of Strain

Two one-dimensional multi-staged CRS tests with strain rates varying between 0.2 and 20%/h were simulated and compared with experimental results, as shown in Fig. 6.9. A good agreement between experiments and simulations was achieved for the two tests. This demonstrates that the enhanced ANICREEP model can predict the 1D rate-dependent behavior of Wenzhou marine clay, and that the soil parameters optimized by the new RCGA are suitable.

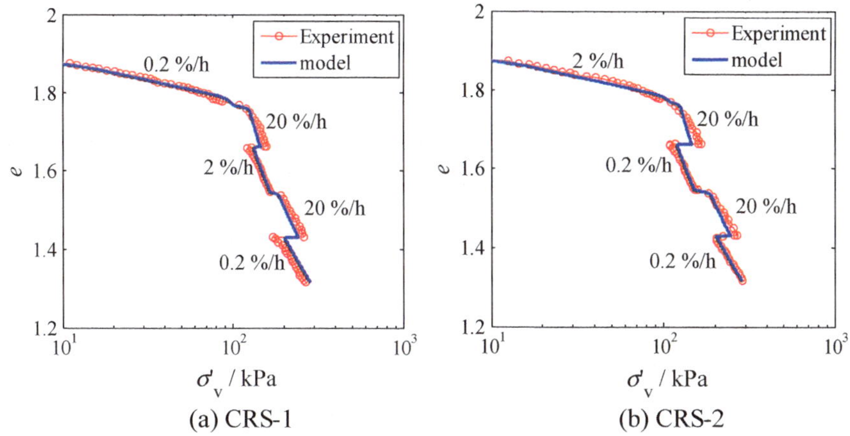

Fig. 6.9 Comparisons between simulated and experimental results of multi-staged one-dimensional tests with axial strain rate varying between 0.2 and 20%/h

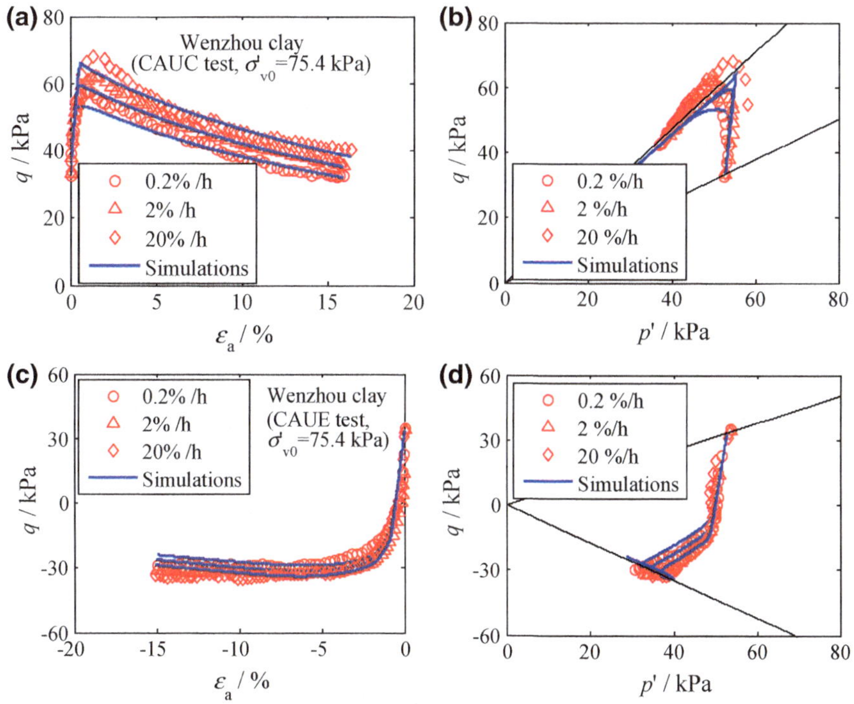

Fig. 6.10 Comparisons between simulated and experimental results of undrained triaxial CRS tests on samples K_0-consolidated at a vertical stress of 75.4 kPa: **a**, **b** in compression and **c**, **d** in extension

6.5.4.1 Undrained Triaxial Tests at Constant Rate of Strain

Three sets of undrained triaxial tests in compression and extension on K_0-consolidated samples under three vertical effective stresses ($\sigma'_{v0} = 75.4$, 150, and 300 kPa) at strain rates of 0.2, 2, and 20%/h were simulated. Figures 6.10, 6.11, and 6.12 show the comparisons between the predicted and the measured results for deviatoric stress versus axial strain and effective stress path. Good agreement between experimental results and simulations was generally achieved by the enhanced ANI-CREEP model with the set parameters optimized using the new RCGA method. This demonstrates that (1) the enhanced model has a good ability to reproduce the 3D rate-dependent and destructuration behavior of soft structured clay; and (2) the parameters obtained from the new RCGA are representative of the 3D behavior of Wenzhou clay.

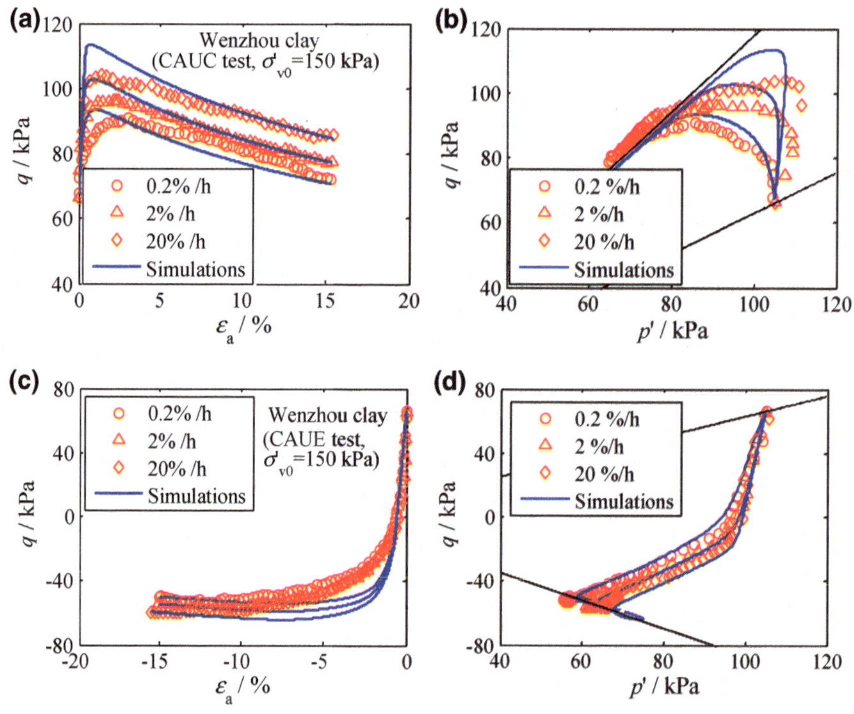

Fig. 6.11 Comparisons between simulated and experimental results of undrained triaxial CRS tests on samples K_0-consolidated at a vertical stress of 150 kPa: **a**, **b** in compression and **c**, **d** in extension

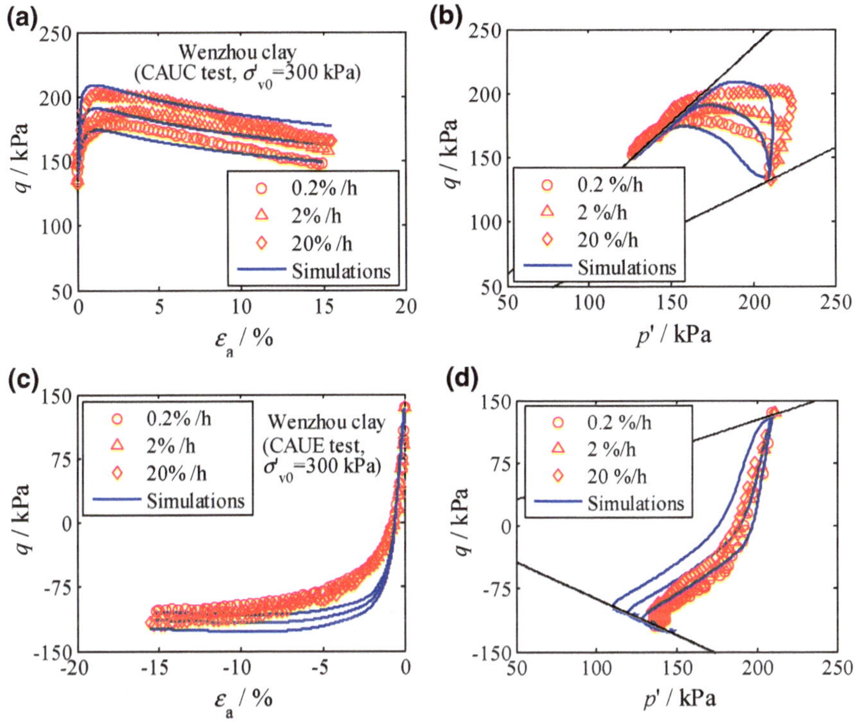

Fig. 6.12 Comparisons between simulated and experimental results of undrained triaxial CRS tests on samples K_0-consolidated at a vertical stress of 300 kPa: **a**, **b** in compression and **c**, **d** in extension

6.5.4.2 Undrained Triaxial Creep Tests

Four undrained triaxial creep tests on K_0-consolidated samples ($\sigma'_{v0} = 150$ kPa) under different applied stress levels ($\Delta\sigma'_{v0} = 12.6$, 16.8, 20.5, and 25.6 kPa) were simulated using the enhanced ANICREEP model with parameters obtained using the new RCGA. The comparison between experimental and simulation results is shown in Fig. 6.13, demonstrating again the good predictive ability of the model and the good quality of the optimization procedure of the new RCGA. Note that the simulation performance of mean effective stress–time relations is less good than that of axial strain, which can be improved by revising the stress–dilatancy relation in the model according to Wang and Yin [38].

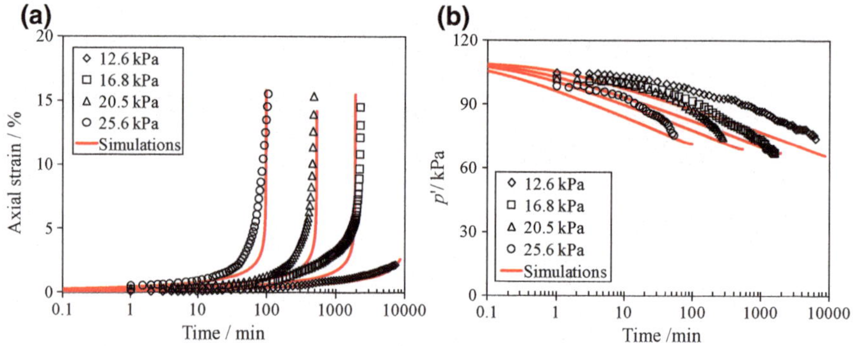

Fig. 6.13 Comparisons between simulated and experimental results of undrained triaxial creep tests: **a** axial strain versus time; **b** mean effective stress versus time

6.6 Discussion

As shown in the previous validation, the error between experiments and simulations for tests with different stress paths and loading rates is different. Then, if different tests are combined as objectives, the optimized parameters will be different. In this part, a comparative study is presented for the choice of the loading rates and load paths during the optimization-based parameters identification.

The influence of the loading rate on parameter identification was first evaluated by adopting the undrained triaxial compression tests with strain rates of 0.2, 2, and 20%/h under a same vertical effective stresses ($\sigma_{v0}' = 150$ kPa) in the optimization (marked as "Comb-1"). Then, for investigating the influence of loading path, three undrained triaxial extension tests ($\sigma_{v0}' = 75.4$, 150, and 300 kPa) at a same strain rate of 2%/h were selected as objectives (marked as "Comb-2") and two undrained triaxial compression tests ($\sigma_{v0}' = 75.4$ and 300 kPa) with one undrained triaxial extension test ($\sigma_{v0}' = 150$ kPa) at a same strain rate of 2%/h were selected as objectives (marked as "Comb-3"). The same K_0-compression test was used together for all three combinations. The same identification procedure was used for the above three combinations unlike the one based on undrained triaxial compression tests.

All the optimized parameters were summarized in Table 6.6. Together with the previous case in Table 6.6 (marked as "Std"), the combination with different loading rates gives a smallest objective error, followed by the "Std", "Comb-3", and "Comb-2". To evaluate the performance of each choice or combination, optimized values of $\lambda_i, M, C_{\alpha ei,}$ and χ_0 were compared to measurements, respectively, shown in Fig. 6.14. It can be observed that the combination with different loading rates gives a smallest objective error, followed by the "Std", "Comb-3", and "Comb-2".

For each combination, all optimized parameters were used to simulate other oedometer and triaxial tests. The total average errors between experimental data and simulations were calculated and plotted in Fig. 6.15 for comparison. The combination with different loading rates gives the smallest objective error, followed by the

Table 6.6 Three sets of optimal parameters with objective errors for Wenzhou clay based on different objective combinations

Combinations	Optimal parameters								Objective error (%)
	κ	λ_i	M	$C_{\alpha ei}$	χ_0	ξ	ξ_d	n	
Comb-2	0.062	0.294	1.16	0.0078	10	5.5	0.35	0.75	6.96
Comb-3	0.070	0.221	0.99	0.0071	5.0	11.0	0.50	0.90	10.41
Comb-4	0.050	0.201	1.17	0.0066	9.0	13.5	0.275	0.9	9.46

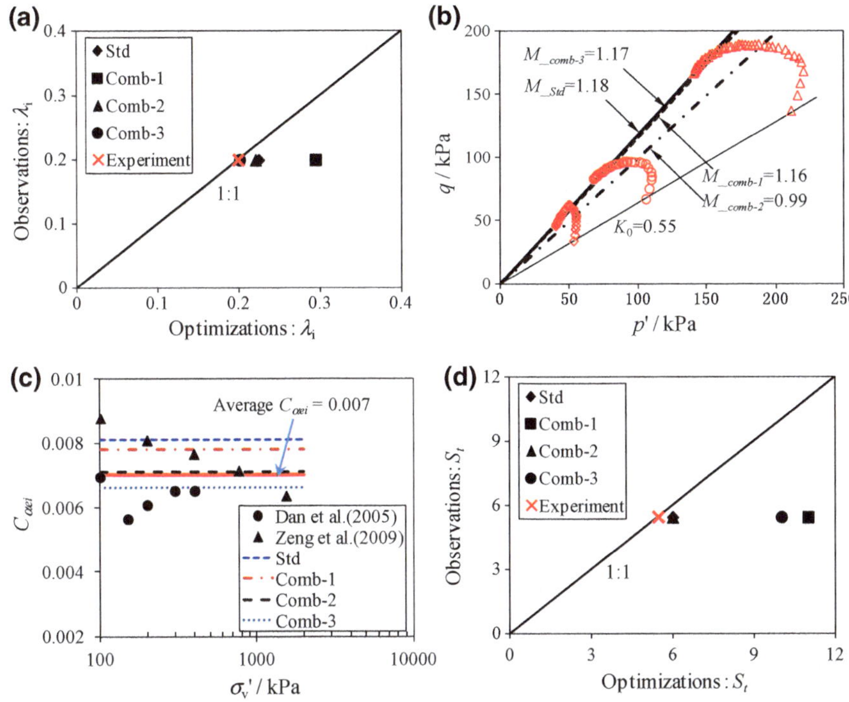

Fig. 6.14 Comparisons of different optimal parameters obtained from different combinations of objective tests

"Std", "Comb-3", and "Comb-2". The extension test combining with compression tests involved in the objective can result in more accurate parameters.

Overall, the "Std" and "Comb-3" have relatively better performance compared to others. Note that the undrained triaxial extension test is not a conventional test in laboratory and more difficult to be conducted, because by adding the extension test into objective can increase the difficulty in identifying model parameters, which is not our original intention. Thus, the proposed optimization method using three

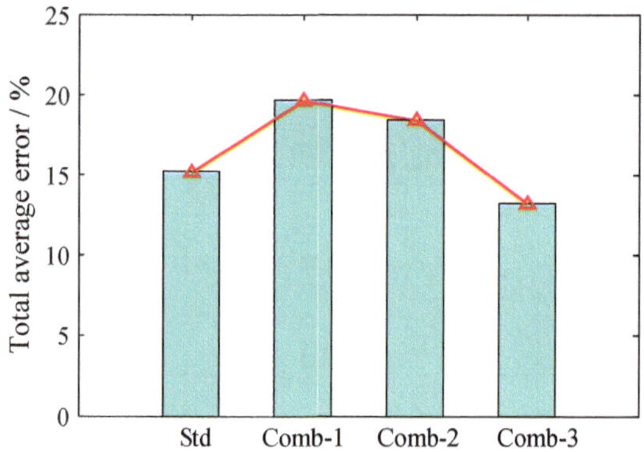

	Vertical stress (kPa)	Strain rate (% /h)	Compression (C) or Extension (E)
Std:	75.4, 150 and 300	2.0	3C
Comb-1:	150	0.2, 2.0 and 20	3C
Comb-2:	75.4, 150 and 300	2.0	3E
Comb-3:	75.4, 150(E) and 300	2.0	2C+1E

Fig. 6.15 Comparisons of total average errors simulated by optimal parameters for different combinations of objective tests

undrained triaxial compression tests as the objective is more suitable in terms of the accuracy of parameters and practical convenience.

6.7 Summary

An efficient optimization method for identifying parameters of soft structured clay using standard experimental tests has been proposed, in which an appreciation of genetic algorithms and constitutive models are required. A newly developed elastic–viscoplastic model accounting for soil viscosity, anisotropy, and destructuration was adopted and enhanced with cross-anisotropy of elasticity for simulating laboratory tests on soft structured clays.

The new RCGA with uniform samplings was first examined and discussed in terms of the optimization performance. The results demonstrate that an optimal solution can be guaranteed by the new RCGA. The computational effectiveness and efficiency of the new RCGA were considered to be better compared to the commonly used NSGA-II. The optimization performance of the new RCGA was further examined by comparing the optimized values for the intrinsic compression coefficient, the slope of the critical state line, the initial bonding ratio, and the intrinsic secondary compression coefficient with specific experimental measurements. The results demonstrate that the new RCGA solution is more suitable than the NSGA-II. The new RCGA solution was then further validated by simulating other tests on the same clay with different stress paths: 1D CRS tests with various strain rates, 3D CRS tests in compression and extension with various strain rates, and 3D undrained creep tests, which demonstrate that the new RCGA solution is reliable. Therefore, it can be concluded that the new RCGA optimization is a suitable and efficient way to identify parameters of soft structured clays. All the results demonstrate that the determination of the whole set of parameters of an advanced elastic viscoplastic model for natural structured clays can be determined by simply using a limited number of conventional soil tests, if an appropriate identification procedure is undertaken.

More details can be found in Yin et al. [39].

References

1. Leroueil S, Kabbaj M, Tavenas F, Bouchard R (1985) Stress-strain-strain relation for the compressibility of sensitive natural clays. Geotechnique 35(2):159–180
2. Rocchi G, Fontana M, Da Prat M (2003) Modelling of natural soft clay destruction processes using viscoplasticity theory. Geotechnique 53(8):729–745
3. Yin ZY, Karstunen M (2011) Modelling strain-rate-dependency of natural soft clays combined with anisotropy and destructuration. Acta Mech Solida Sin 24(3):216–230
4. Yin ZY, Wang JH (2012) A one-dimensional strain-rate based model for soft structured clays. Sci China-Technol Sci 55(1):90–100
5. Yin Z-Y, Yin J-H, Huang H-W (2015) Rate-dependent and long-term yield stress and strength of soft Wenzhou marine clay: experiments and modeling. Mar Georesour Geotechnol 33(1):79–91
6. Zhu Q-Y, Yin Z-Y, Xu C-J, Yin J-H, Xia X-H (2015) Uniqueness of rate-dependency, creep and stress relaxation behaviors for soft clays. J Central South Univ 22:296–302
7. Yin ZY, Karstunen M, Chang CS, Koskinen M, Lojander M (2011) Modeling time-dependent behavior of soft sensitive clay. J Geotech Geoenviron Eng 137(11):1103–1113
8. Gens A, Nova R (1993) Conceptual bases for a constitutive model for bonded soils and weak rocks. Geotech Eng Hard Soils-Soft Rocks 1(1):485–494
9. Lecampion B, Constantinescu A, Nguyen Minh D (2002) Parameter identification for lined tunnels in a viscoplastic medium. Int J Numer Anal Methods Geomech 26(12):1191–1211
10. Calvello M, Finno RJ (2004) Selecting parameters to optimize in model calibration by inverse analysis. Comput Geotech 31(5):410–424
11. Yin ZY, Hicher PY (2008) Identifying parameters controlling soil delayed behaviour from laboratory and in situ pressuremeter testing. Int J Numer Anal Methods Geomech 32(12):1515–1535
12. Papon A, Riou Y, Dano C, Hicher PY (2012) Single-and multi-objective genetic algorithm optimization for identifying soil parameters. Int J Numer Anal Methods Geomech 36(5):597–618
13. Pal S, Wije Wathugala G, Kundu S (1993) Calibration of a constitutive model using genetic algorithms. Comput Geotech 19(4):325–348
14. Javadi A, Farmani R, Toropov V, Snee C (1999) Identification of parameters for air permeability of shotcrete tunnel lining using a genetic algorithm. Comput Geotech 25(1):1–24
15. Levasseur S, Malécot Y, Boulon M, Flavigny E (2008) Soil parameter identification using a genetic algorithm. Int J Numer Anal Methods Geomech 32(2):189–213
16. Rokonuzzaman M, Sakai T (2010) Calibration of the parameters for a hardening–softening constitutive model using genetic algorithms. Comput Geotech 37(4):573–579
17. Vardakos S, Gutierrez M, Xia C (2012) Parameter identification in numerical modeling of tunneling using the Differential Evolution Genetic Algorithm (DEGA). Tunn Undergr Space Technol 28:109–123

18. Moreira N, Miranda T, Pinheiro M, Fernandes P, Dias D, Costa L et al (2013) Back analysis of geomechanical parameters in underground works using an Evolution Strategy algorithm. Tunn Undergr Space Technol 33:143–158
19. Yepes V, Alcala J, Perea C, González-Vidosa F (2008) A parametric study of optimum earth-retaining walls by simulated annealing. Eng Struct 30(3):821–830
20. Deb K (2001) Multi-objective optimization using evolutionary algorithms. Wiley, Hoboken
21. Yin ZY, Chang CS, Karstunen M, Hicher PY (2010) An anisotropic elastic-viscoplastic model for soft clays. Int J Solids Struct 47(5):665–677
22. Yin ZY, Chang CS (2009) Microstructural modelling of stress-dependent behaviour of clay. Int J Solids Struct 46(6):1373–1388
23. Chang CS, Yin ZY (2010) Micromechanical modeling for inherent anisotropy in granular materials. J Eng Mech-Asce 136(7):830–839
24. Yin ZY, Chang CS, Hicher PY (2010) Micromechanical modelling for effect of inherent anisotropy on cyclic behaviour of sand. Int J Solids Struct 47(14–15):1933–1951
25. Yin Z, Hicher PY (2013) Micromechanics-based model for cement-treated clays. Theor Appl Mech Lett 3(2):021006
26. Graham J, Houlsby G (1983) Anisotropic elasticity of a natural clay. Geotechnique 33(2):165–180
27. Biarez J, Hicher P-Y (1994) Elementary mechanics of soil behaviour: saturated remoulded soils. AA Balkema
28. Zhu Q-Y, Yin Z-Y, Hicher P-Y, Shen S-L (2016) Nonlinearity of one-dimensional creep characteristics of soft clays. Acta Geotech 11(4):887–900
29. Yao Y, Hou W, Zhou A (2009) UH model: three-dimensional unified hardening model for overconsolidated clays. Geotechnique 59(5):451–469
30. Yao YP, Sun DA (2000) Application of Lade's criterion to Cam-clay model. J Eng Mech 126(1):112–119
31. Yao Y-P, Kong L-M, Zhou A-N, Yin J-H (2014) Time-dependent unified hardening model: three-dimensional elastoviscoplastic constitutive model for clays. J Eng Mech 141(6):04014162
32. Roscoe KH, Burland J (1968) On the generalized stress-strain behaviour of wet clay. In: Engineering plasticity. Cambridge University Press, Cambridge, UK, pp 535–609
33. Deb K, Pratap A, Agarwal S, Meyarivan T (2002) A fast and elitist multiobjective genetic algorithm: NSGA-II. IEEE Trans Evol Comput 6(2):182–197
34. Deep K, Thakur M (2007) A new mutation operator for real coded genetic algorithms. Appl Math Comput 193(1):211–230
35. Jin Y-F, Yin Z-Y, Shen S-L, Hicher P-Y (2016) Selection of sand models and identification of parameters using an enhanced genetic algorithm. Int J Numer Anal Methods Geomech 40(8):1219–1240
36. Jin Y-F, Yin Z-Y, Shen S-L, Hicher P-Y (2016) Investigation into MOGA for identifying parameters of a critical-state-based sand model and parameters correlation by factor analysis. Acta Geotech 11(5):1131–1145
37. Zeng L (2010) Deformation mechanism and compression model of natural clays. Southeast University, Nanjing, China
38. Wang L-Z, Yin Z-Y (2012) Stress dilatancy of natural soft clay under an undrained creep condition. Int J Geomech 15(5):A4014002
39. Yin Z-Y, Jin Y-F, Shen S-L, Huang H-W (2017) An efficient optimization method for identifying parameters of soft structured clay by an enhanced genetic algorithm and elastic–viscoplastic model. Acta Geotech 12(4):849–867

Chapter 7
Parameter Identification for Granular Materials

7.1 Introduction

The chapter proposes an efficient optimization procedure for identifying parameters of easily crushable sand, which is then applied to the pile driving simulation. The Nelder–Mead Simplex genetic algorithm (NMGA) is first proposed and a newly enhanced elastoplastic breakage model is adopted. Then, the performance of NMGA is validated by identifying parameters from synthetic tests, and further verified by triaxial tests on limestone grains, based on which the necessary number of objective tests is also suggested. The role of grain breakage in bearing capacity of driven pile is also discussed. All comparisons demonstrate that the proposed NMGA with breakage-model-based parameter identification procedure is efficient and effective for easily crushable sand.

7.2 Overview

The impact of grain breakage on the mechanical behavior of granular materials has been highlighted in the past decades [1–4]. In order to describe the effect of grain breakage, numerous constitutive methods considering the grain breakage have been developed [5–15]. Especially in advanced breakage models incorporating the critical state concept, additional parameters are introduced which are difficult to be determined by conventional laboratory or field tests. Especially in advanced breakage models incorporating the critical state concept, additional parameters are introduced which are difficult to be determined by conventional laboratory or field tests. For instance, in critical-state-based breakage models, the critical state line kinematically moves with the amount of grain crushed. Thus, the critical-state-related parameters, the hardening parameters, and the dilatancy parameters become much more difficult to be calibrated in breakage model than in conventional critical-state-based models, such as the critical-state-line-related parameters (e.g., υ_f and λ_f) in the model of

Z. Yin and Y. Jin, *Practice of Optimisation Theory in Geotechnical Engineering*,
https://doi.org/10.1007/978-981-13-3408-5_7

Russell and Khalili [9]; the critical state parameters (e_{ref0} and λ) in the model of Hu et al. [8]; the critical state parameters ($e_{\Gamma i}$, ϕ_{cr}, and λ), the dilatancy parameters (k_d and A) and the hardening parameters (ξ_{L0}, ξ_L, ξ_U, and γ) in the model of Chen et al. [5]; the critical-state-related parameters (λ, e_{B0}, k_e, and χ_B) and two dilatancy constants (d_0 and k_d) in the model of Xiao and Liu [15]. This difficulty is much more pronounced for very easily crushable sand, and poses a challenge in terms of engineering practice. As a consequence, the breakage models were seldom used in the engineering practice. To improve the determination of parameters, more laboratory tests are becoming necessary, which results in a relatively high experimental cost. Therefore, an efficient approach to identify model parameters by a cost-saving way is useful and will be attractive for engineering practice.

Recently, the optimization method based on genetic algorithm (GA) has been applied successfully in the geotechnical engineering since its application can reduce the cost of laboratory testing or in situ monitoring. For example, the parameter identifications of various constitutive models for different soils have been carried out using optimizations with GA [16–22]. Along this way, it will be attractive if this powerful tool can be applied to identify parameters of breakage models for easily crushable sand.

Therefore, this chapter aims to propose an efficient optimization procedure for identifying parameters of easily crushable sand based on conventional laboratory tests which is then applied to the pile driving simulation. For this purpose, a recently enhanced elastoplastic breakage model is adopted for simulations. A novel Nelder–Mead Simplex enhanced genetic algorithm (NMGA) is developed to minimize the error function during the optimization process. Then, the parameter identification using three synthetic drained triaxial tests generated by the adopted model as objective tests is conducted to evaluate the performance of NMGA. After that, the model parameters for limestone grains are identified using the proposed optimization procedure with discussions on the necessary number of objective tests. Finally, the proposed procedure is applied to identify parameters of Dog's bay sand using four triaxial tests and validated by three other triaxial tests. The Coupled Eulerian-Lagrangian-based large deformation finite element analysis combined with the breakage model is carried out for pile driving in the same sand for which the finite element implementation of the model is also presented. Furthermore, the role of grain breakage in bearing capacity of driven pile is numerically discussed.

7.3 Adopted ElastoPlastic Grain Breakage Model

A novel constitutive model accounting for the effect of grain breakage developed by Yin et al. [10] was adopted to simulate the experiment in the optimization process. The total strain rate is conventionally composed of the elastic and plastic strain rates:

$$\dot{\varepsilon}_{ij} = \dot{\varepsilon}_{ij}^{e} + \dot{\varepsilon}_{ij}^{p} \tag{7.1}$$

The elastic behavior is assumed to be isotropic with the bulk modulus K adopting the same form of the shear modulus proposed by Richart et al. [23],

$$\dot{\varepsilon}_{ij}^{e} = \frac{1+\upsilon}{3K(1-2\upsilon)}\sigma_{ij}' - \frac{\upsilon}{3K(1-2\upsilon)}\sigma_{kk}'\delta_{ij} \tag{7.2}$$

$$K = K_0 \cdot p_{at}\frac{(2.97-e)^2}{(1+e)}\left(\frac{p'}{p_{at}}\right)^{\zeta} \tag{7.3}$$

where K_0 and ζ are elastic parameters; υ is Poisson's ratio; p' is the mean effective stress; p_{at} is the atmospheric pressure used as reference pressure ($p_{at} = 101.3$ kPa); δ_{ij} is the Kronecker delta.

The adopted model contains two yield surfaces: one for shear sliding (f_1) and one for compression (f_2). The plastic strain increment can be expressed as follows:

$$\dot{\varepsilon}_{ij}^{p} = \dot{\varepsilon}_{ij}^{p1} + \dot{\varepsilon}_{ij}^{p2} = \mathrm{d}\lambda_1\frac{\partial g_1}{\partial \sigma_{ij}'} + \mathrm{d}\lambda_2\frac{\partial g_2}{\partial \sigma_{ij}'} \tag{7.4}$$

where the scripts 1 and 2 represent the shear sliding and compression components, respectively. For one component with $f < 0$, the corresponding $\mathrm{d}\lambda$ is equal to zero.

The shear sliding yield function, noted f_1 has the following expression,

$$f_1 = \frac{q}{p'} - \frac{M_p\varepsilon_d^p}{(M_p p')/(G_p G) + \varepsilon_d^p} = 0 \tag{7.5}$$

where the deviatoric plastic strain ε_d^p is the hardening variable, q is the deviatoric stress ($q=\sqrt{3J_2} = \sqrt{3/2}\sqrt{s_{ij}s_{ji}}$ with $s_{ij} = s_{ij} - p'\delta_{ij}$); p' is the mean effective stress; G_p is the relative plastic stiffness controlling the hardening rate; G is shear modulus which can be calculated using the K and υ; $M_p = 6\sin(\phi_p)/(3-\sin(\phi_p))$ is the stress ratio corresponding to the peak strength and determined by the peak friction angle ϕ_p.

The plastic potential function is given by:

$$\frac{\partial g_1}{\partial \sigma_{ij}'} = \frac{\partial g_1}{\partial p'}\frac{\partial p'}{\partial \sigma_{ij}'} + \frac{\partial g_1}{\partial s_{ij}}\frac{\partial s_{ij}}{\partial \sigma_{ij}'}$$

$$\text{with } \frac{\partial g_1}{\partial p'} = A_d\left(M_{pt} - \frac{q}{p'}\right);\ \frac{\partial g_1}{\partial s_{ij}} = \{1\ 1\ 1\ 1\ 1\ 1\} \tag{7.6}$$

where g_1 is different from f_1; therefore, the model is non-associated. A_d is the stress–dilatancy parameter; $M_{pt} = 6\sin(\phi_{pt})/(3-\sin(\phi_{pt}))$ where ϕ_{pt} phase transformation friction angle. The double indices ij are simplified to be $1\hat{=}11$, $2\hat{=}22$, $3\hat{=}33$, $4\hat{=}12$, $5\hat{=}23$, $6\hat{=}31$ according to the Voigt rule since stress or strain tensors are symmetric.

The peak friction angle ϕ_p and the phase transformation friction angle ϕ_{pt} are associated with the critical friction angle, ϕ_μ and the critical void ratio, e_c is as follows:

$$\tan\phi_p = \left(\frac{e_c}{e}\right)^{\beta} \tan\phi_u; \quad \tan\phi_{pt} = \left(\frac{e_c}{e}\right)^{-\beta} \tan\phi_\mu \tag{7.7}$$

$$e_c = e_{\text{ref}} - \lambda \ln\left(\frac{p'}{p_{ref}}\right) \tag{7.8}$$

where β is a material constant, commonly equal to 1 according to Yin et al. [10]. λ is the slope of the critical state line in the e-ln p' plane; the two parameters (e_{ref}, p_{ref}) correspond to a reference point on the critical state line (CSL). The Lode-angle-dependent strength and stress–dilatancy were introduced as described in Yin et al. [24], which can also be incorporated by using the transformed stress method by Yao et al. [25–28].

In order to describe the compression behavior of crushable sand, a second yield surface was added,

$$f_2 = \frac{1}{2}\left(\frac{q}{p'M_p}\right)^3 p' + p' - p_m \quad \text{for} \quad \frac{q}{p'} \leq M_p \tag{7.9}$$

where p_m is the hardening variable controlling the size of the yield surface. An associated flow rule is adopted herein. This yield surface expands with the plastic volumetric strain, as in the modified Cam-Clay model

$$p_m = p_{m0} \exp\left(\frac{1+e_0}{\lambda - \kappa}\varepsilon_v^p\right) \tag{7.10}$$

where $\kappa = (1+e_0)p'/K$ is used with K by Eq. (7.3); λ takes the same value as that of the CSL.

The effects of grain breakage on the mechanical behavior can be expressed in two aspects:

(1) Influence of grain breakage on the critical state line. The evolution of the critical state line due to grain breakage is described as following expression:

$$p_{\text{ref}} = p_{\text{ref0}} \exp\left(-bB_r^*\right) \tag{7.11}$$

where parameter b controls the rate of the CSL shifting due to grain breakage; B_r^* is the modified breakage index proposed by Einav [7], which can be considered as a function of the plastic work during loading,

$$B_r^* = \frac{w_p}{a + w_p} \tag{7.12}$$

where a is a material constant controlling the evolution of the breakage amount.

Grain breakage was classified into three modes according to Guyon and Troadec [29]: (1) fracture which means a grain breaks into smaller grains of similar sizes, (2) attrition which means a grain breaks into one grain of a slightly smaller size and several much smaller ones, (3) abrasion which means the result is that the grain-size distribution remains almost constant but with a production of fine particles. These three modes also associate with the applied stress level: from low pressure to high pressure, the grain breakage from the mode (3) to mode (1) dominates. Equation (7.12) follows this phenomenon. Note that there exists a threshold plastic work or stress level from a physical standpoint for which one more parameter needs to be introduced. In fact, this threshold value is questionable in terms of parameter calibration since the grain breakage of mode (3) cannot be clearly measured. Fortunately, Eq. (7.12) can capture very slight amount of grain breakage corresponding to mode (3) mainly due to interparticle shear sliding at low stress level.

(2) Influence of grain breakage on compression behavior. The evolution of p_m due to grain breakage is added in Eq. (7.13) as follow,

$$p_m = p_{m0} \exp\left(\frac{1 + e_0}{\lambda - \kappa} \varepsilon_v^p\right) \exp\left(-bB_r^*\right) \tag{7.13}$$

Note that the same constant b is used in Eqs. (7.13) and (7.11) implying the same amount of shifting for isotropic compression line and critical state line, which was previously validated by Yin et al. [10]. Furthermore, Eq. (7.13) also assumed that the size of yield surface is controlled by two factors simultaneously: the volumetric plastic strain and the modified breakage index.

Overall, the parameters of the adopted breakage model can be divided into four groups: (1) elastic parameters (K_0, n, υ); (2) critical-state-related parameters (e_{ref}, λ, ϕ_u); (3) plastic parameters (G_p, A_d); and (4) grain-breakage-related parameters (p_{m0}, a, b).

7.4 Enhancement of RCGA

After formulating the error function, the selection of the search strategy is the key step of whether the optimized solution can be found or not. The solution to an optimization problem is a vector x_0 which satisfies the global minimum condition for any $x_l \ll x \ll x_u$,

$$F(x_0) \leq F(x) \tag{7.14}$$

For obtaining a more accurate solution, a highly efficient optimization method with the ability to search for a global minimum should be adopted. Since the GA is widely used in the geotechnical field as stated in the introduction, the adopted real-coded GA (RCGA) has a similar structure to those developed by Jin et al. [19] and Yin et al. [21]. To further accelerate the convergence speed and enhance the search

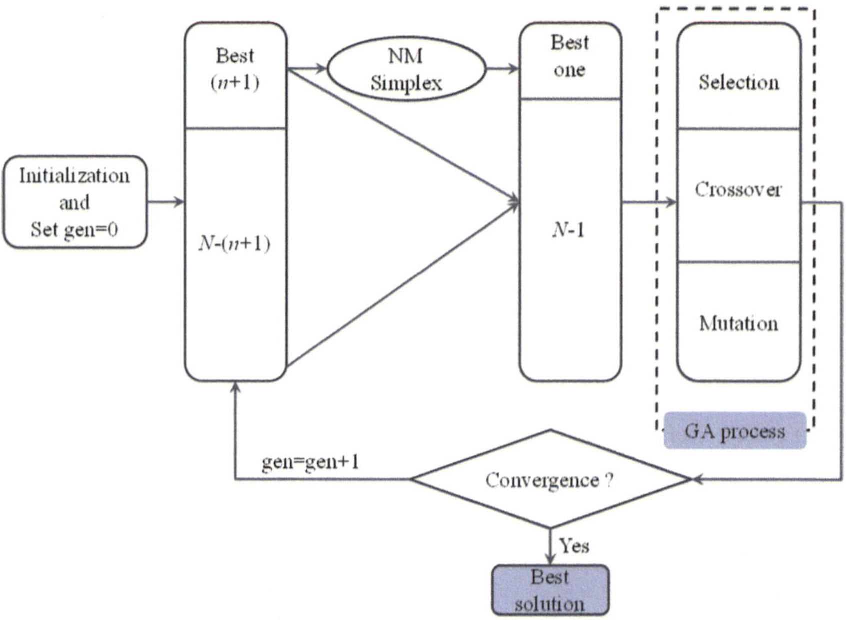

Fig. 7.1 Flowchart of the NMGA

ability, the adopted RCGA was enhanced by implementing a local search named "Nelder–Mead Simplex (NMS)" with excellent convergence speed. The proposed algorithm was named as "NMGA" for simplicity in the following sections.

Although the hybrid of Nelder–Mead optimization scheme and GA with its efficiency has been well recognized during past decades [30, 31], on the premise of keeping the intrinsic advantage of hybrid between GA and Simplex, adopting newly developed optimization schemes is necessary and useful for solving complex geotechnical problems with more parameters. In this study, the NMGA is different from previous works [30, 31] due to (1) the adoption of the newly developed RCGA with outstanding performance in identifying soil parameters and (2) a newly developed NMS operator for high-dimensional problems. Figure 7.1 shows the flowchart, which presents the simplest way of the enhancement. First, N initial individuals were generated by SOBOL and the generation number was set to zero. Then, all the individuals were sorted based on their fitness values, and the best $n + 1$ (n is the number of variables) individuals were selected to perform the NMS. For the NMS, a Nelder–Mead method with adaptive parameters for high-dimensional problems suggested by Gao and Han [32] was adopted, and a brief procedure of this method is presented in Appendix I. Based on the result of the NMS, the best individual was updated. Then, this best individual was recombined with the $n - 1$ remaining individuals to perform the GA global search. More details about the adopted RCGA can be found in Jin et al. [19]. In order to avoid a rapid loss of diversity, an elitism strategy was adopted in the RCGA. In the replacement process, 10% of individuals with the highest fitness were selected from the parents and children to survive to the next generation. The remainders were chosen by tournament selection from the mating pool composed of parents and children other than the 10% individuals. The iteration stopped when the convergence criterion or the maximum generation number was reached.

For the NMS, the tolerance for convergence was set to 10^{-4}, as recommended by Gao and Han [32]. For the adopted RCGA, the recommended settings can be found in Jin et al. [19].

During the optimization process, the NMS could accelerate the convergence speed, but could easily get trapped in the local minimum. On the other hand, the RCGA process can overcome the local minimum problem, however, the convergence speed is always slow and the approach is not applicable for the engineering applications [16, 21, 22]. Therefore, this hybrid strategy owns the advantages of both processes and results in better optimization performance.

7.5 Proposed Identification Procedure

In general, deformation and strength are two important indicators of the mechanical behavior of granular materials. Additionally, the grain-size distribution (GSD) is strongly associated with the grain-breakage effect during the loading process and hence influences the mechanical behaviors [33], especially for the crushable materials. Therefore, the GSD is another indicator for measuring the degree of gain breakage. For a laboratory triaxial test, the isotropic or anisotropic compression test is normally first conducted, followed by the shearing test. During the whole testing process, the measurements indicating the compression behavior, shearing behavior and grain breakage behavior are recorded. Some model parameters can be directly determined based on the measurements, e.g., the initial void ratio, Young's modulus [17, 34]. While some variables cannot be directly measured, e.g., the parameters related to grain breakage [5, 8]. Therefore, the parameter identification using optimization method is proposed. In this study, a mono-objective framework with four different criteria is considered:

$$\min[\mathrm{Error}(x)] = \min\{\mathrm{mean}[\mathrm{Error}_{\mathrm{com}}(e), \mathrm{Error}_{\mathrm{tri}}(q), \mathrm{Error}_{\mathrm{tri}}(e), \mathrm{Error}_{\mathrm{com\,+\,tri}}(\mathrm{GSD})]\} \tag{7.15}$$

where $\mathrm{Error}_{\mathrm{com}}(e)$ is the error of void ratio e between simulations and objectives for isotropic compression or anisotropic compression stage; $\mathrm{Error}_{\mathrm{tri}}(q)$ is the error in deviatoric stress q between simulations and objectives for the triaxial shearing stage; $\mathrm{Error}_{\mathrm{tri}}(e)$ is the error in void ratio e between simulations and objectives for triaxial shearing stage, and $\mathrm{Error}_{\mathrm{com+tri}}(\mathrm{GSD})$ is the error in GSD curve between simulations and objectives for both compression and triaxial shearing stages. All the errors mentioned above were calculated according to Eq. (7.15).

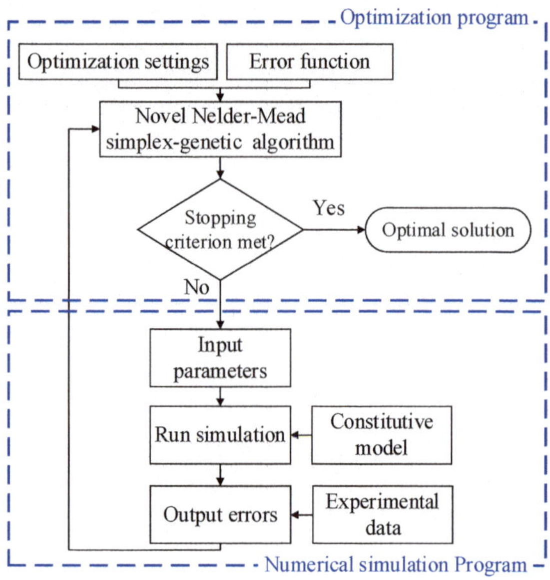

Fig. 7.2 Proposed mono-objective optimization procedure

Figure 7.2 shows the identification procedure, which aims to find values for the model parameters and provide the best attainable fit between model predictions and corresponding observations. In this study, the procedure is based on two different codes: the code for the integration of the constitutive model to simulate the objective tests and the code for the optimization process to find the minimum value of error function. For each optimization process, each individual in the initial population was generated using the SOBOL method [35].

7.6 Performance of NMGA on Identifying Model Parameters

For evaluating the performance of the proposed NMGA, a set of synthetic tests including three drained triaxial tests ($e_0 = 0.8$ with $p_0 = 100$, 200 and 400 kPa) and one isotropic compression test ($e_0 = 0.8$) was generated by the adopted breakage model [10] as the objectives in the optimization. In order to simulate the breakage behavior of easily crushable materials, a set of typical parameters: the elastic parameters ($K_0 = 70$, $n = 0.66$, $\upsilon = 0.25$), the critical-state-related parameters ($e_{\mathrm{ref}} = 1.0$, $\lambda = 0.05$, $\phi = 35$), the plastic parameters ($G_p = 5$, $A_d = 1$), and the grain-breakage-related parameters ($p_{m0} = 500$ kPa, $a = 150$ kPa, $b = 10$) were adopted to generate the synthetic tests, and the simulated results are shown in Fig. 7.3. Totally ten model parameters were optimized by using the proposed optimization procedure, as shown in Eq. (7.16).

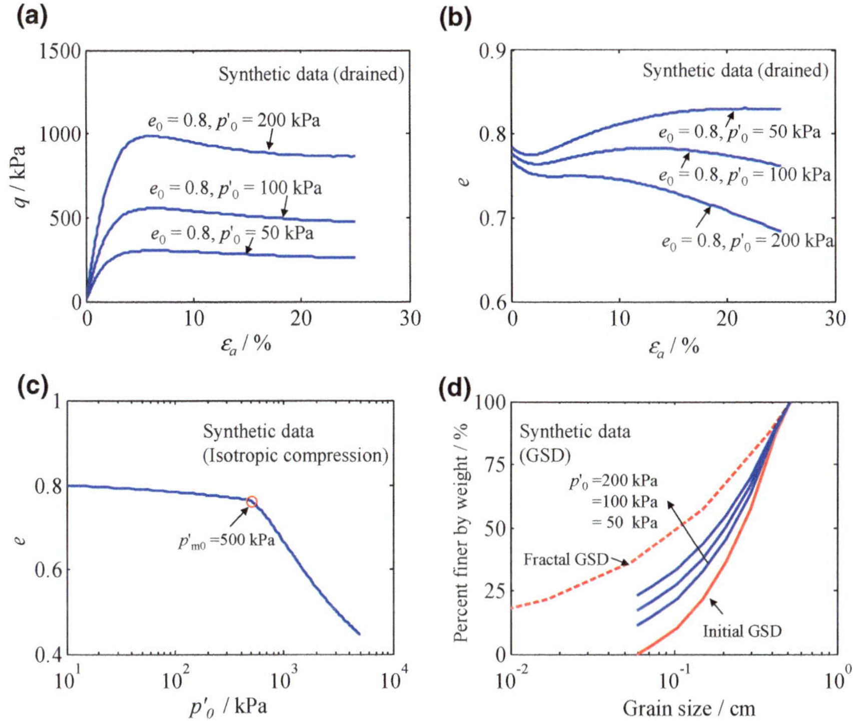

Fig. 7.3 Results of synthetic objective tests generated by grain-breakage model

$$\min f_e(x) = \min f_e\big(K_0,\ n,\ e_{\text{ref}},\ \lambda,\ \phi,\ G_p,\ A_d,\ p_{m0},\ a,\ b\big) \tag{7.16}$$

where f_e is the error function of the model parameters; x is the model parameters.

The initial population size was set to 100 and the maximum number of generations was set to 300. The interval of each parameter is given in Table 7.1, which covers a big range in terms of preset parameters. For comparison and verifying the performance of the proposed NMGA, the original RCGA without Nelder–Mead Simplex and another outstanding algorithm MOGA-II used by Papon et al. [36] were selected to conduct the same optimization. Optimal parameters obtained by NMGA, RCGA, and MOGA-II from synthetic data are summarized in Table 7.2. It can be seen that the preset parameters can be entirely detected within 300 generations by the proposed NMGA. However, the performance of RCGA and MOGA-II is worse satisfied comparing to that of the NMGA. Figure 7.4 shows the evolution of minimum objective error in each generation with the increasing number of generations. It can be seen that the proposed NMGA has the faster convergence speed than the MOGA-II. Therefore, all the comparisons demonstrate that the proposed NMGA has an excellent performance on identifying model parameters for easily crushable sands.

Table 7.1 Search domain and intervals of parameters for adopted grain-breakage model

Groups	Elastic		Critical-sate			Plastic		Gain breakage		
Parameters	K_0	n	e_{ref}	λ	ϕ_u	G_p	A_d	p_{m0}	a	b
Lower bound	10	0.2	0.5	0.01	20	1	0	50	50	0
Upper bound	400	0.8	2.0	0.5	50	50	5.0	10,000	10,000	20
Step size	10	0.01	0.001	0.001	1	0.1	0.1	10	10	1

Table 7.2 Optimal parameters obtained by NMGA, RCGA, and MOGA-II from synthetic data

Parameters	K_0	n	e_{ref}	λ	ϕ	G_p	A_d	p_{m0}	a	b	Objective error (%)
Preset	70	0.66	1.000	0.05	35	5.0	1.0	500	150	10	–
NMGA	70	0.66	1.000	0.05	35	5.0	1.0	500	150	10	0.00
RCGA	65	0.47	0.996	0.086	35	12.9	0.9	560	150	4	2.70
MOGA-II	60	0.53	1.019	0.065	34	10.2	0.8	620	150	8	2.22

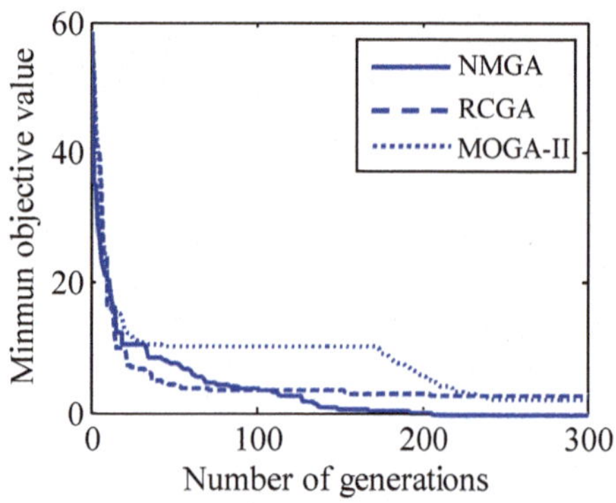

Fig. 7.4 Comparison of convergence speed for three optimization methods in identifying parameters for synthetic data

7.7 Verification by Limestone Grains

7.7.1 Brief Introduction of Laboratory Tests on Limestone Grains

In order to evaluate the effectiveness and efficiency of proposed optimization procedure, a series of conventional drained triaxial tests on limestone grains (a easily crushable sand, the Mohr's hardness equals to 3.5) by Lo and Roy [3] was selected as objective tests. Some main physical properties of this limestone grains are as follows: specific gravity $G_s = 2.71$, initial void ratio $e_0 = 0.81$, maximum void ratio $e_{\max} = 1.05$ and minimum void ratio $e_{\min} = 0.65$, mean particle size $d_{50} = 0.215$ mm, and

Table 7.3 Series of triaxial tests on limestone grains with initial void ratio ($e_0 = 0.81$)

Test number	①	②	③	④	⑤	⑥	⑦	⑧	⑨
e_0 before shearing	0.8	0.77	0.75	0.726	0.67	0.606	0.510	0.448	0.397
σ_c' (kPa)	172	345	517	690	1380	2760	5520	8275	11,030

uniformity coefficient $C_u = 2.85$. For simplicity, all the tests with different confining pressures were marked by the sequence number. The sequence number and the information for the corresponding test are summarized in Table 7.3. Before conducting the triaxial shear test, each sample was compressed from the initial void ratio 0.81 to the corresponding confining pressure. After shearing, the change of grain-size distribution for each sample was measured.

7.7.2 Optimization Results and Discussion

Following the proposed procedure, the model parameters for limestone grains were identified by the proposed NMGA. In addition, the RCGA and the MOGA-II were also adopted to conduct the same identification procedure for comparisons. For both adopted algorithms, the initial individuals and the maximum number of generations were set to 100. Other settings of RCGA recommended by Jin et al. [19] and MOGA-II recommended by Papon et al. [36] were kept in this case. To highlight the performance of the proposed procedure in identifying breakage-related parameters, the parameters related to compression behavior ($K_0 = 24$, $n = 0.5$ and $p_{m0} = 1000$ kPa) which can be directly measured from isotropic compression curve were not put into the optimization procedure. Poisson's ratio υ is assumed to be 0.25 which is a typical value for sand. Furthermore, the remaining parameters (e_{ref}, λ, ϕ, G_p, A_d, a, b) can in also be determined a direct way according to Yin et al. [10] for sand that has a high yielding stress. However, for easily crushable sand, the direct way to determine these parameters (e_{ref}, λ, ϕ, G_p, A_d, a, b) based on laboratory tests is difficult due to the grain breakage occurring even in a very low stress level. Therefore, apart from the parameters related to compression behavior ($K_0 = 24$, $n = 0.5$ and $p_{m0} = 1000$ kPa), the remaining parameters (e_{ref}, λ, ϕ, G_p, A_d, a,

Table 7.4 Optimal parameters with the optimal errors of testing on limestone grains

Method	Optimal parameters							Average error (%)
	e_{ref}	λ	ϕ	G_p	A_d	a	b	
NMGA	0.930	0.107	41	32.1	2.1	2300	3	5.83
RCGA	0.890	0.080	41	30	2.7	2350	7	6.82
MOGA-II	0.884	0.066	42	20.2	2.3	2300	10	7.44

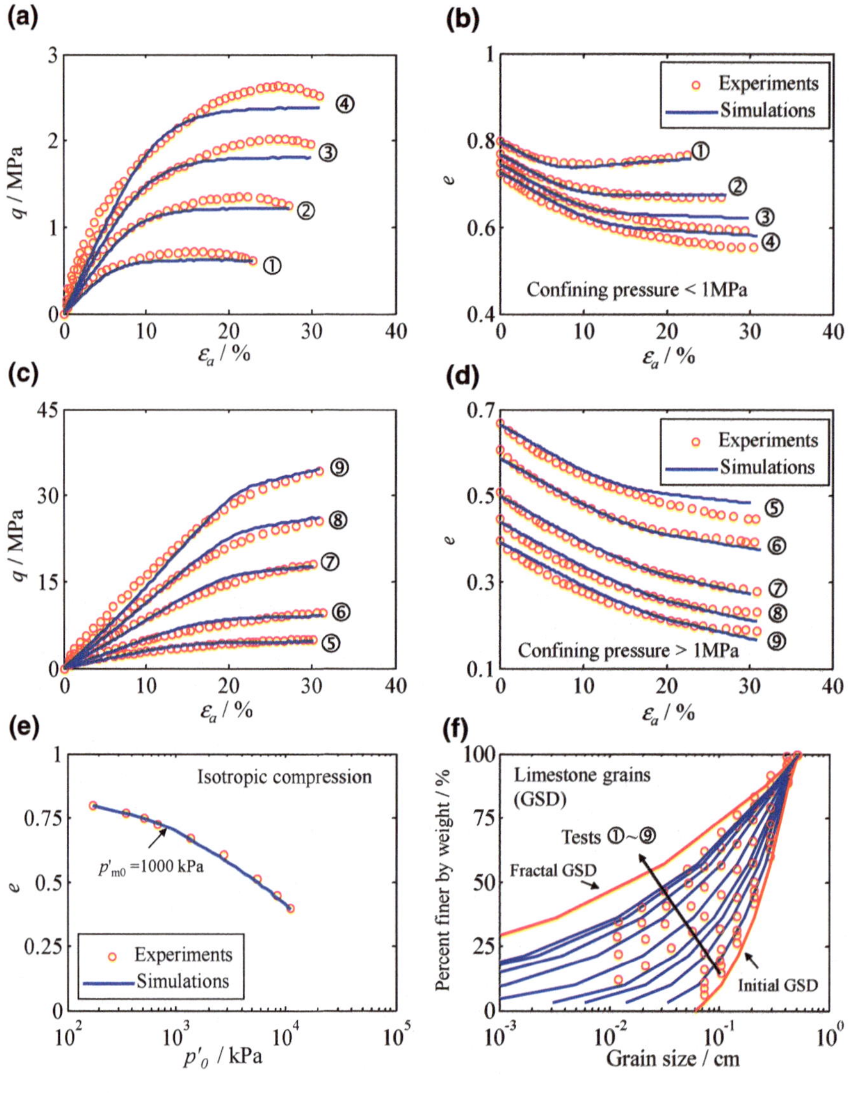

Fig. 7.5 Comparisons between simulation and experiments

b) were optimized using the proposed procedure. First, for obtaining a set of most accurate parameters, all the experimental data (in total nine tests) were used as the objective in the optimization process. The objective error is defined as the average of $\mathrm{Error}_{\mathrm{com}}(e)$, $\mathrm{Error}_{\mathrm{tri}}(q)$, $\mathrm{Error}_{\mathrm{tri}}(e)$, and $\mathrm{Error}_{\mathrm{com+tri}}(\mathrm{GSD})$. The interval of each parameter is also given in Table 7.1.

Table 7.4 shows the comparison of optimal parameters and the corresponding objective error obtained by NMGA, RCGA, and MOGA-II. It can be seen that the

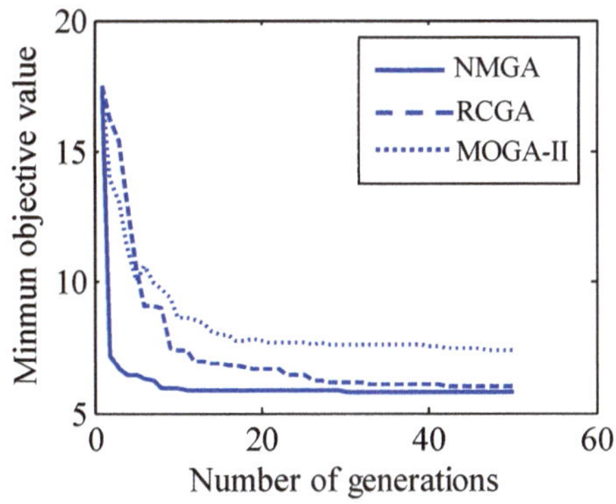

Fig. 7.6 Comparison of convergence speed for three optimization methods in identifying parameters for limestone

NMGA gives a set of optimal parameters with smallest objective error among them. Using the optimal parameters, the objective tests were simulated using the adopted breakage model. Figure 7.5 shows the comparisons of "$q - \varepsilon_a$" (deviatoric stress versus axial strain), isotropic compression curve and GSD curves between simulated and experimental results. A good agreement between the simulations and observations indicates that the optimal parameters are reasonable and reliable, which also indicates that NMGA performs well in identifying breakage model parameters for easily crushable sand.

The convergence rates of NMGA, RCGA, and MOGA-II are shown in Fig. 7.6. It can be seen that the convergence speed of NMGA is much faster than that of RCGA and MOGA-II, which indicates a higher efficiency of NMGA in the parameter identification. This should be attractive for engineering application.

Furthermore, for investigating the influence of the weight of GSD in the error function on optimized parameters, a scheme for evaluating the influence of the weight of GSD is presented in Table 7.5 based on the error function in Eq. (7.17).

$$\min[\mathrm{Error}(x)] = \min\{[\mathrm{mean}(A(\mathrm{Error}_{\mathrm{com}}(e), \mathrm{Error}_{\mathrm{tri}}(q), \mathrm{Error}_{\mathrm{tri}}(e)) \,, \\ (1 - A)\mathrm{Error}_{\mathrm{com+tri}}(\mathrm{GSD}))]\} \tag{7.17}$$

where the A is weight factor varying between 0 and 1.

Following the same optimization procedure, the optimal parameters and objective errors corresponding to different weights of GSD are finally summarized in Table 7.5. It can be seen that the weight of GSD ranging from 0.3 to 0.7 gives a good performance in identifying the parameters of the adopted breakage model. Therefore, for obtaining an equal contribution of each term in the error function to the optimal parameters, the weight of GSD is set to 0.5. The following optimizations are then conducted based on the error function in Eq. (7.17) with a value of 0.5 for the weight of GSD.

Table 7.5 Optimal parameters and errors for different weights of GSD

No.	Weight	Optimal parameters							Objective error (%)	Total average error (%)
	A	e_{ref}	λ	ϕ (°)	G_p	A_d	a	b		
1	1	0.99	0.110	42	21.6	1.6	350	2	7.35	12.96
2	0.7	0.95	0.115	41	31	1.9	1800	2	6.09	6.45
3	0.5	0.95	0.117	41	33.8	1.7	1950	2	5.26	6.49
4	0.3	0.95	0.116	41	33.7	1.8	1950	2	4.46	6.47
5	0	1.42	0.189	34	4.3	1.9	1800	11	0.34	154

7.7.3 Estimation of Minimum Number of Tests for Identifying Parameters

Traditionally, three triaxial tests are needed for determining mechanical parameters of soils according to standard criterion of laboratory test. However, for determining parameters of critical-state-based models, more laboratory tests including isotropic compression and shearing stages are appreciated [34]. For easily crushable sand, it is questionable that how many tests are needed to determine the model parameters associated with grain breakage. It is clear that the process of parameter identification based on large number of tests is time-consuming and cost-consuming, which may not be preferred by the engineers. In order to reduce the test cost in parameter identification, the minimum number of tests involved in the objective to obtain a set of accurate parameters needs to be estimated.

Since the amount of grain breakage significantly associates with the applied stress level, two necessary tests, i.e., one test performed under a lower confining stress and one test performed under a higher confining stress, should be involved in the objective. Due to this, the tests ① and ⑧ (see Table 7.3) were first selected as a basic component in all combinations (one combination is one objective for optimization), considering that the stress level of the test ⑧ is high enough. Then, combinations with three tests, or four tests, or five tests were constructed by adding one, or two, or three additional tests from database of tests ② to ⑦, based on which a total of 41 combinations were formed as shown in Table 7.6. All combinations were successively used for optimization to identify model parameters.

The examined combinations with the test number, the obtained optimal parameters, the objective error, and the total average error are summarized in Table 7.6. Based on the total number of tests in the objective, all the combinations were divided into four groups and marked as "three tests", "four tests", "five tests", and "eight tests". Figure 7.7 shows the minimum error and the average error in each group. It can be seen that both the average error and the minimum error generally decrease with the increasing number of tests in the objective. For some scatter points with big errors (as combinations 1, 2, 3, and 12), a possible reason is the inaccuracy of critical-state-related parameters caused by the unreasonable selection of objective tests [34]. It can be seen that the minimum error and average error of the group containing four tests with more possible high stress tests or five tests is close to that obtained from eight tests. In terms of the obtained optimal parameters, the average value of each optimal parameter for each group has been calculated and is summarized in Table 7.7. It can be found that with increasing the quantity of tests in the objective, the obtained optimal parameters gradually approach to the most accurate set of parameters from all eight tests. The difference of the obtained optimal parameters, the objective error, and the total average error between the group with five tests and the group with all eight tests is very small, which can be ignored in the engineering practice. Therefore, it can be concluded that a total of four tests including two basic tests and two additional high stress tests, or a total of five tests including two basic tests and three additional tests are sufficient for the parameter identification.

Table 7.6 Optimal parameters and errors of different combinations of tests

No.	Combination of tests	Optimal parameters							Objective error (%)	Total average error (%)
		e_{ref}	λ	ϕ (°)	G_p	A_d	a	b		
1	①②⑧	0.97	0.131	41	33.0	1.5	1700	1	7.53	8.74
2	①③⑧	0.97	0.128	41	29.4	1.3	2000	1	7.22	8.07
3	①④⑧	0.96	0.129	41	42.0	1.6	2050	1	7.30	8.78
4	①⑤⑧	0.96	0.129	40	34.5	1.6	2050	1	6.18	7.35
5	①⑥⑧	0.97	0.118	41	25.7	1.4	1900	2	5.71	7.04
6	①⑦⑧	0.96	0.115	41	28.1	1.6	1600	2	6.05	6.93
7	①②③⑧	0.95	0.120	41	40.3	1.9	1900	2	5.37	6.61
8	①②④⑧	0.96	0.132	41	48.5	1.8	2050	1	5.78	7.30
9	①②⑤⑧	0.93	0.109	41	41.8	2.1	2250	3	5.90	6.02
10	①②⑥⑧	0.94	0.108	41	32.6	1.9	2050	3	5.97	6.09
11	①②⑦⑧	0.95	0.123	40	34.5	1.9	1750	1	5.50	7.01
12	①③④⑧	0.97	0.140	41	49.3	2.0	2250	1	7.41	8.30
13	①③⑤⑧	0.94	0.118	40	43.2	2.2	2250	2	6.17	6.17
14	①③⑥⑧	0.94	0.108	41	36.3	1.9	2000	3	6.33	6.27
15	①③⑦⑧	0.95	0.118	41	37.8	1.7	1900	2	5.60	6.47
16	①④⑤⑧	0.94	0.111	41	36.4	2.1	2300	3	6.35	5.96
17	①④⑥⑧	0.94	0.114	40	34.3	2.0	2150	2	6.42	6.15
18	①④⑦⑧	0.95	0.124	40	38.0	1.8	1900	1	5.71	6.83
19	①⑤⑥⑧	0.94	0.114	40	33.3	2.1	2240	2	6.33	6.07
20	①⑤⑦⑧	0.95	0.117	40	28.9	1.7	2000	2	5.61	6.44

(continued)

Table 7.6 (continued)

No.	Combination of tests	Optimal parameters							Objective error (%)	Total average error (%)
		e_{ref}	λ	ϕ (°)	G_p	A_d	a	b		
21	①⑥⑦⑧	0.97	0.119	41	23.4	1.4	1800	2	5.38	6.75
22	①②③④⑧	0.98	0.143	41	40.8	1.8	2350	1	7.43	6.53
23	①②③⑤⑧	0.95	0.124	41	42.8	1.9	2500	2	5.71	6.49
24	①②③⑥⑧	0.93	0.106	41	33.2	2.0	2250	3	5.80	5.92
25	①②③⑦⑧	0.94	0.115	41	38.3	2.0	1800	2	5.37	6.38
26	①②④⑤⑧	0.94	0.114	41	39.9	2.1	2500	3	5.72	6.07
27	①②④⑥⑧	0.93	0.107	41	38.7	2.1	2300	3	5.93	5.96
28	①②④⑦⑧	0.95	0.117	41	36.1	1.8	1800	2	5.26	6.50
29	①②⑤⑥⑧	0.94	0.107	41	29.7	2.0	2200	3	5.94	5.97
30	①②⑤⑦⑧	0.93	0.107	41	33.5	2.0	2100	3	5.68	5.93
31	①②⑥⑦⑧	0.94	0.112	40	29.4	2.0	1900	2	5.09	6.27
32	①③④⑤⑧	0.93	0.110	41	43.4	2.4	2400	3	6.48	5.95
33	①③④⑥⑧	0.93	0.106	41	36.5	2.3	2350	3	6.78	5.93
34	①③④⑦⑧	0.96	0.121	41	32.5	1.7	2100	2	6.25	6.40
35	①③⑤⑥⑧	0.93	0.114	40	42.1	2.2	2400	2	6.48	5.98
36	①③⑤⑦⑧	0.94	0.115	40	35.2	2.2	2150	2	5.96	6.07
37	①③⑥⑦⑧	0.95	0.108	41	25.8	2.0	2050	3	5.58	6.10
38	①④⑤⑥⑧	0.93	0.106	40	34.1	2.5	2300	3	6.34	6.00
39	①④⑤⑦⑧	0.94	0.115	40	34.8	2.1	2200	2	6.06	6.06
40	①④⑥⑦⑧	0.94	0.114	40	34.7	1.9	1900	2	5.55	6.34
41	①⑤⑥⑦⑧	0.94	0.113	40	33.7	2.1	2000	2	5.63	6.21

Table 7.7 Average optimal parameters and errors for each group of different number of tests

Groups	e_{ref}	λ	ϕ	G_p	A_d	a	b	Range of objective error (%)	Total average error (%)
3 tests	0.965	0.125	40.8	32.1	1.50	1900	1.3	5.71–7.53	7.80
4 tests	0.948	0.118	40.6	37.2	1.90	2050	2.0	5.37–7.41	6.60
5 tests	0.941	0.114	40.7	35.8	2.05	2200	2.4	5.26–7.43	6.10
8 tests	0.940	0.107	41.0	32.1	2.10	2300	3.0	5.80	5.90

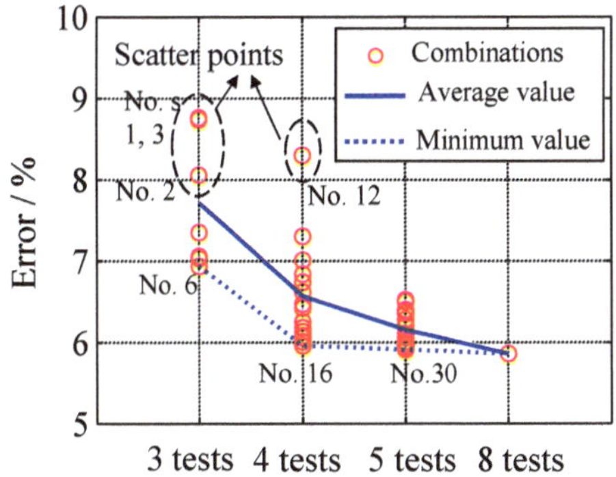

Fig. 7.7 Evolution of simulation error with increased number of tests for different combinations

7.8 Application to Identify Parameters of Carbonate Sand

This section presents the application of the proposed identification procedure to identify parameters of Dog's bay sand. Four drained constant p' triaxial tests with different initial confining pressures by Bandini and Coop [37] and one isotropic compression tests by Coop [1] performed on Dog's bay sand were selected as the objective in the optimization. Due to the limited test data, only two GSD curves were used in the optimization procedure. For the simplicity, all the selected constant p' tests were marked as "D-1, 2, 3, 4, 5, 6, and 7". Table 7.8 shows the initial void ratio and confining pressure for each selected test. Based on the necessary number of tests suggested previously, the tests of D-1, 2, 3, 4, and 5 were used as the objectives in the optimization, and the tests of D-6 and 7 were used for validation.

Following the proposed procedure, the parameters of the adopted breakage model for Dog's bay sand were identified by using NMGA. Table 7.9 summarizes the optimal parameters. Using the optimal parameters, the objective tests were first simulated. Figure 7.8a–d show the comparison of isotropic compression test, constant p' tests and GSD curves between the optimal simulations and experiments. Then, three additional drained constant p' triaxial tests were used for validation, as shown in Fig. 7.8e–f. It can be seen that the simulations generally well agree with the measurements, which demonstrates that the identified parameters of the adopted breakage model by the proposed procedure are reliable.

Table 7.8 Series of triaxial tests on Dog's bay sand (constant mean effective stress)

Tests	D-1	D-2	D-3	D-4	D-5	D-6	D-7
e_0	1.365	1.582	1.557	1.751	1.757	1.744	1.705
σ_c'(kPa)	500	500	1000	4000	500	1000	4000

Table 7.9 Optimal parameters of adopted breakage model with objective error for Dog's bay sand

Parameters	K_0	n	e_{ref}	λ	ϕ (°)	G_p	A_d	p_{m0} (kPa)	a	b	Error (%)
Value	160	0.56	1.79	0.0148	44	19	0.5	670	2500	10	6.62

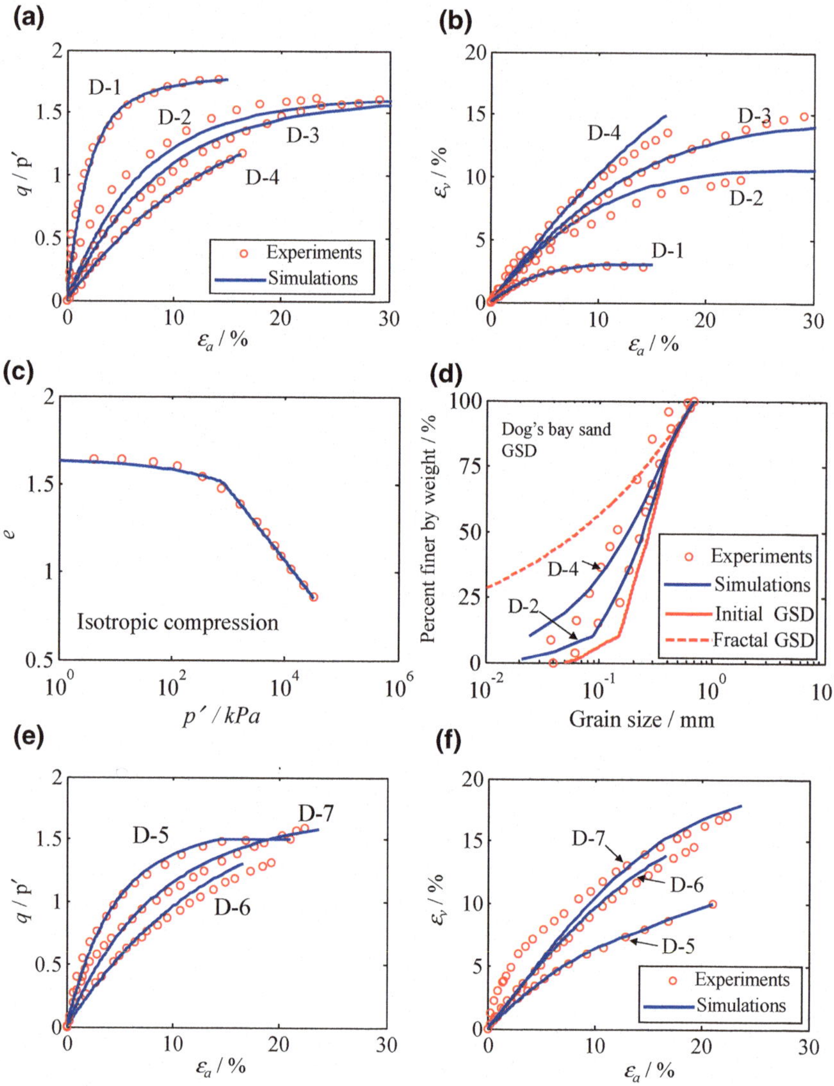

Fig. 7.8 Comparisons between simulation and experiments on Dog's bay sand

7.9 Summary

An efficient optimization procedure to identify the breakage model's parameters for easily crushable sand has been proposed. In the optimization procedure, a newly developed elastoplastic constitutive model accounting for grain breakage was adopted to simulate the objective tests for measuring the error between the simulations and objective tests. To improve the performance of identification, a new optimization algorithm NMGA was proposed by implementing the Nelder–Mead Simplex into a recently developed real-coded genetic algorithm. First, the computational effectiveness and efficiency of the proposed NMGA were evaluated by comparing with the RCGA and the MOGA-II algorithms. All the three optimization methods were applied to conduct the same parameter identification. In the optimization, the objective was three synthetic drained triaxial tests generated by the adopted breakage model with a set of parameters representing an easily crushable sand. The results show that the performance of proposed NMGA on search ability and convergence speed is better than the RCGA and MOGA-II.

Then, the proposed optimization procedure was examined by identifying the model parameters of limestone grains adopting all the experimental tests as the objective. The results show that the obtained optimal parameters are accurate and reasonable. To reduce the cost of parameters determination, the minimum number of tests involved in the objective was investigated. The results indicate that five tests including one test at low confining pressure and one test at high confining pressure are sufficient for the parameter identification, which can be adopted in testing code for practice. Finally, the adopted breakage model's parameters of a typically carbonate sand (Dog's bay sand) were optimized using the proposed procedure.

Overall, it can be concluded that the proposed optimization procedure is efficient and reliable on identifying the model parameters for easily crushable sand, and the developed numerical platform is applicable for estimating the capacity of pile driving in this kind of sand [38, 39].

References

1. Coop M (1990) The mechanics of uncemented carbonate sands. Geotechnique 40(4):607–626
2. Coop M, Sorensen K, Freitas TB, Georgoutsos G (2004) Particle breakage during shearing of a carbonate sand. Geotechnique 54(3):157–163
3. Lo KY, RoY M (1973) Response of particulate materials at high pressures. Soils Found 13(1):61–76
4. Miao G, Airey D (2013) Breakage and ultimate states for a carbonate sand. Géotechnique 63(14):1221–1229
5. Chen Q, Indraratna B, Carter JP, Nimbalkar S (2015) Isotropic–kinematic hardening model for coarse granular soils capturing particle breakage and cyclic loading under triaxial stress space. Can Geotech J 53(4):646–658
6. Daouadji A, Hicher P-Y, Rahma A (2001) An elastoplastic model for granular materials taking into account grain breakage. Eur J Mech-A/Solids 20(1):113–137
7. Einav I (2007) Breakage mechanics—part I: theory. J Mech Phys Solids 55(6):1274–1297

8. Hu W, Yin ZY, Dano C, Hicher PY (2011) A constitutive model for granular materials considering grain breakage. Sci China-Tech Sci 54(8):2188–2196
9. Russell AR, Khalili N (2004) A bounding surface plasticity model for sands exhibiting particle crushing. Can Geotech J 41(6):1179–1192
10. Yin Z-Y, Hicher P-Y, Dano C, Jin Y-F (2016) Modeling mechanical behavior of very coarse granular materials. J Eng Mech C4016006
11. Yao Y-P, Yamamoto H, Wang N-D (2008) Constitutive model considering sand crushing. Soils Found 48(4):603–608
12. Cecconi M, Desimone A, Tamagnini C, Viggiani GMB (2002) A constitutive model for granular materials with grain crushing and its application to a pyroclastic soil. Int J Numer Anal Meth Geomech 26(15):1531–1560
13. Kong Y, Xu M, Song E (2017) An elastic-viscoplastic double-yield-surface model for coarse-grained soils considering particle breakage. Comput Geotech 85:59–70
14. Liu M, Gao Y (2016) Constitutive modeling of coarse-grained materials incorporating the effect of particle breakage on critical state behavior in a framework of generalized plasticity. Int J Geomech 17(5):04016113
15. Xiao Y, Liu H (2016) Elastoplastic constitutive model for rockfill materials considering particle breakage. Int J Geomech 17(1):04016041
16. Jin Y-F, Yin Z-Y, Shen S-L, Hicher P-Y (2016) Selection of sand models and identification of parameters using an enhanced genetic algorithm. Int J Numer Anal Meth Geomech 40(8):1219–1240
17. Jin Y-F, Yin Z-Y, Shen S-L, Hicher P-Y (2016) Investigation into MOGA for identifying parameters of a critical-state-based sand model and parameters correlation by factor analysis. Acta Geotech 11(5):1131–1145
18. Jin Y-F, Wu Z-X, Yin Z-Y, Shen JS (2017) Estimation of critical state-related formula in advanced constitutive modeling of granular material. Acta Geotech 12(6):1329–1351
19. Jin Y-F, Yin Z-Y, Shen S-L, Zhang D-M (2017) A new hybrid real-coded genetic algorithm and its application to parameters identification of soils. Inverse Probl Sci Eng 25(9):1343–1366
20. Yin Z-Y, Jin Y-F, Shen JS, Hicher P-Y (2018) Optimization techniques for identifying soil parameters in geotechnical engineering: comparative study and enhancement. Int J Numer Anal Methods Geomech 42(1):70–94
21. Yin Z-Y, Jin Y-F, Shen S-L, Huang H-W (2017) An efficient optimization method for identifying parameters of soft structured clay by an enhanced genetic algorithm and elastic–viscoplastic model. Acta Geotech 12(4):849–867
22. Jin Y-F, Yin Z-Y, Riou Y, Hicher P-Y (2017) Identifying creep and destructuration related soil parameters by optimization methods. KSCE J Civ Eng 21(4):1123–1134
23. Richart F, Hall J, Woods R (1970) Vibrations of soils and foundations. International series in theoretical and applied mechanics. Prentice-Hall, Englewood Cliffs, NJ
24. Yin Z-Y, Xu Q, Hicher P-Y (2013) A simple critical-state-based double-yield-surface model for clay behavior under complex loading. Acta Geotech 8(5):509–523
25. Yao Y, Hou W, Zhou A (2009) UH model: three-dimensional unified hardening model for overconsolidated clays. Geotechnique 59(5):451–469
26. Yao Y, Lu D, Zhou A, Zou B (2004) Generalized non-linear strength theory and transformed stress space. Sci China Ser E Technol Sci 47(6):691–709
27. Yao Y, Sun D, Luo T (2004) A critical state model for sands dependent on stress and density. Int J Numer Anal Meth Geomech 28(4):323–337
28. Yao Y, Sun D, Matsuoka H (2008) A unified constitutive model for both clay and sand with hardening parameter independent on stress path. Comput Geotech 35(2):210–222
29. Guyon É, Troadec J-P (1994) Du sac de billes au tas de sable: Odile Jacob
30. Durand N, Alliot J-M (1999) A combined Nelder-Mead simplex and genetic algorithm. In: Proceedings of the genetic and evolutionary computation conference GECCO1999, pp 1–7
31. Chelouah R, Siarry P (2003) Genetic and Nelder-Mead algorithms hybridized for a more accurate global optimization of continuous multiminima functions. Eur J Oper Res 148(2):335–348

32. Gao F, Han L (2012) Implementing the Nelder-Mead simplex algorithm with adaptive parameters. Comput Optim Appl 51(1):259–277
33. Liu Y-J, Li G, Yin Z-Y, Dano C, Hicher P-Y, Xia X-H et al (2014) Influence of grading on the undrained behavior of granular materials. CR Mec 342(2):85–95
34. Jin YF, Yin ZY, Shen SL, Hicher PY (2016) Selection of sand models and identification of parameters using an enhanced genetic algorithm. Int J Numer Anal Methods Geomech 40:1219–1240
35. Sobol IM (1967) On the distribution of points in a cube and the approximate evaluation of integrals. Zhurnal Vychislitel'noi Matematiki i Matematicheskoi Fiziki 7(4):784–802
36. Papon A, Riou Y, Dano C, Hicher PY (2012) Single-and multi-objective genetic algorithm optimization for identifying soil parameters. Int J Numer Anal Meth Geomech 36(5):597–618
37. Bandini V, Coop MR (2011) The influence of particle breakage on the location of the critical state line of sands. Soils Found 51(4):591–600
38. Jin Y-F, Yin Z-Y, Wu Z-X, Zhou W-H (2018) Identifying parameters of easily crushable sand and application to offshore pile driving. Ocean Eng 154:416–429
39. Jin Y-F, Yin Z-Y, Wu Z-X, Daouadji A (2018) Numerical modeling of pile penetration in silica sands considering the effect of grain breakage. Finite Elem Anal Des 144:15–29

Chapter 8
Optimization-Based Selection of Sand Models

8.1 Introduction

This chapter first aims to discuss the selection of sand models and the identification of their parameters with a genetic algorithm. Conventional triaxial tests on Hostun sand have been chosen as the objectives. Four relative simple constitutive models with gradually increasing numbers of features, referred to as MC (Mohr–Coulomb model), NLMC (nonlinear Mohr–Coulomb model), CS-NLMC (critical-state-based nonlinear Mohr–Coulomb model), and CS-TS (critical-state-based two-surface model), have been selected for optimization. For each model, the optimized parameters were used to simulate other tests on the same sand to evaluate the model's predictive ability. Once the appropriate model with its associated features was determined, then the selection of the type of tests (e.g., drained and/or undrained tests) as the objectives to identify the model parameters was evaluated. The number of tests in the objective is then examined to obtain the relative accuracy and reliability of the parameters. Finally, the strain levels of objective tests for identifying parameters are estimated. Further, this paper also discusses the selection of these critical-state-related formulas and parameter identification. Three formulas of critical state line together with two formulas of critical-state-dependent interlocking effect are combined to propose six elastoplastic models. Drained and undrained triaxial tests on four different granular materials are selected for simulations. In order to eliminate artificial errors, a new hybrid genetic algorithm-based intelligent method is proposed and used to identify parameters and estimate simulations with minimum errors for each granular material and each model. Then, the performance of each CSL and each state parameter is evaluated using two information criteria. Furthermore, the performance was evaluated by simulating three footing tests using finite element analysis in which the models are implemented. All comparisons demonstrate that the incorporating of nonlinear critical state line combined with the state parameter e/e_c in constitutive modeling can result in relatively more satisfied simulated results.

Z. Yin and Y. Jin, *Practice of Optimisation Theory in Geotechnical Engineering*,
https://doi.org/10.1007/978-981-13-3408-5_8

8.2 Genetic Algorithm-Based Optimization

In this section, the genetic algorithm-based optimization is introduced. Before conducting the optimization, three key points need to be clearly introduced: (1) the error function, to measure the difference between model predictions and corresponding observations; (2) the initialization method, to generate the initial population for the optimization; and (3) the optimization algorithm, to control the optimization process.

8.2.1 Error Function

The discrepancy between the measured and the modeled behavior has been expressed by a scalar error function, "Error", in the sense of the least square method introduced by Levasseur et al. [1]. Note that different cost function made of error functions or the error function with different weights for different variables can result in different results, as discussed in Levasseur et al. [1]. In our case, the scale effects on the fitness between the experimental and the simulated results can be eliminated by this normalized formula of Eq. (8.1), and the same weight for different variables is adopted for ensuring the whole performance. Additionally, the objective error calculated by this function is a dimensionless variable; thus, any difference in error can be avoided for different objectives with different variables.

8.2.2 Adopted Hybrid Real-Coded Genetic Algorithm and Initialization Method

The newly developed RCGA in Chap. 3 was employed to conduct this optimization process. The initial population for a genetic algorithm is usually generated by Sobol sampling method. The aim is to obtain a uniform sampling of the design space. It has been reported as being suitable for problems with up to twenty variables.

8.2.3 Optimization Procedure

The aim of the inverse modeling procedure (see Fig. 8.1) is to find values for the model parameters that provide the best attainable fit between model predictions and corresponding observations. In this study, a mono-objective framework with two criteria was considered:

$$\min[\mathrm{Error}(x)] = \min\left[\frac{\mathrm{Error}(q) + \mathrm{Error}(e)}{2}\right] \tag{8.1}$$

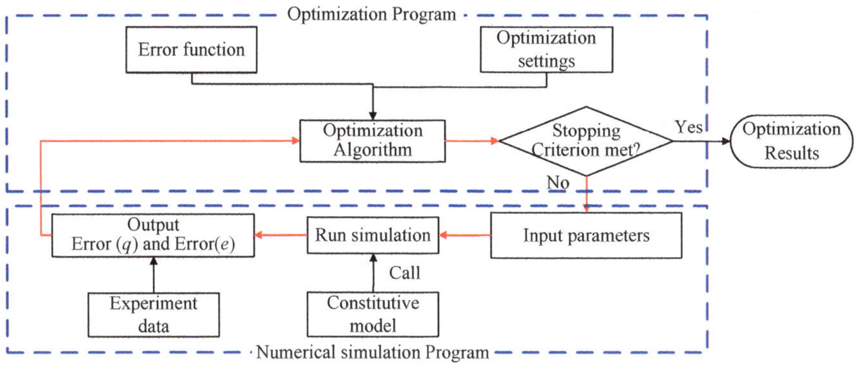

Fig. 8.1 Identification procedure

where Error(q) is the average difference in deviatoric stress between simulations and objectives; Error(e) is the average difference in void ratio between simulations and objectives.

In order to focus on the key parameters which are not easily determined or where their determination could involve more experimental costs, parameters such as Young's modulus and Poisson's ratio which can be directly measured from experiment are not considered in the optimization. Thus, the number of parameters and their physical attributes for each model are not always the same, and they depend on which model is selected. Therefore, the intervals of parameters for every selected model are given in the following sections according to the sand models selected in this study.

For the optimization algorithm, each population was generated using the initialization algorithm Sobol. The initial population was set to 100 individuals and the size of the population kept constant during the optimization process. The number of generations was set to 50 and was tested sufficiently to obtain the optimal results for this study.

8.3 Selection of Features of Sand Necessary for Constitutive Modeling

This section is based on laboratory tests on sand and the use of the genetic optimization method to identify which features of sand have to be taken into account in constitutive modeling. The purpose is to find a constitutive model which can adequately describe the sand behavior with only a limited number of parameters to be identified. Note that the study is based on industry-demand conventional triaxial tests, and thus high-level features (such as anisotropy, non-coaxial behavior, and cyclic behavior) are not considered in this study.

8.3.1 Brief Introduction of Selected Tests

The tests selected for this study are drained triaxial tests performed on Hostun sand by Liu et al. [2] and Li et al. [3]. Hostun sand has a high siliceous content.

Table 8.1 shows the main physical properties of Hostun sand. In the optimization, three drained triaxial tests ($p'_0 = 100$ kPa, $e_0 = 0.66$; $p'_0 = 200$ kPa, $e_0 = 0.83$; and $p'_0 = 400$ kPa, $e_0 = 0.82$) were selected as the objective to obtain the critical-state-related parameters. All the tests were isotropically consolidated to the corresponding consolidation pressure before shearing. The experimental results for the three drained triaxial tests are shown in Fig. 8.2.

In order to identify the sand features which should be considered in constitutive modeling, four sand models with a gradually increasing number of features and different numbers of parameters were chosen to simulate the objective tests, which are (1) an elastic-perfectly plastic Mohr–Coulomb model (MC), (2) a nonlinear Mohr–Coulomb model (NLMC), (3) a critical-state-based nonlinear Mohr–Coulomb model (CS-NLMC), and (4) a critical-state-based two-surface model (CS-TS). The comparisons of constitutive laws for the four sand models are shown in Table 8.2. For MC, Young's modulus is constant. For NLMC, CS-NLMC, and CS-TS, Young's modulus is expressed as follows, according to Richard et al. [4],

$$E = E_0 \cdot p_{\text{at}} \frac{(2.97 - e)^2}{(1 + e)} \left(\frac{p'}{p_{\text{at}}} \right)^{\zeta} \tag{8.2}$$

Table 8.1 Index properties of Hostun sand

Particle shape	SiO_2	G_s	d_{50}/mm	e_{max}	e_{min}	C_u
Angular to subangular	>99.24%	2.6	0.35	0.881	0.577	1.4

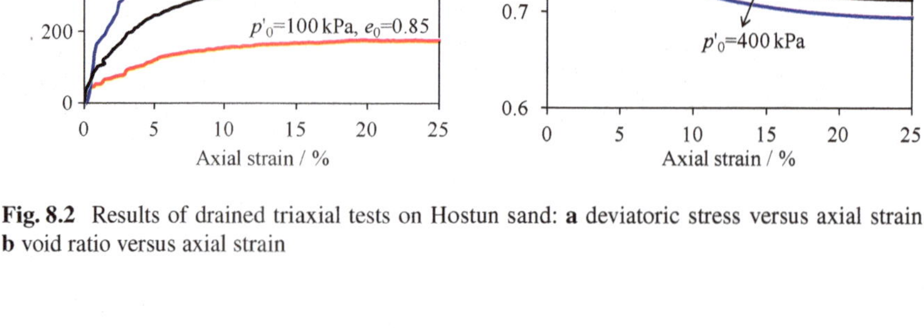

Fig. 8.2 Results of drained triaxial tests on Hostun sand: **a** deviatoric stress versus axial strain; **b** void ratio versus axial strain

Table 8.2 Typical constitutive relations of four selected sand models

Constitutive models	MC	NLMC	CS-NLMC	CS-TS
Elastic behavior	$\dot{\varepsilon}_{ij}^{e} = \frac{1+\upsilon}{E}\sigma'_{ij} - \frac{\upsilon}{E}\sigma'_{kk}\delta_{ij}$			
Yield function	$f = \frac{\sigma_1-\sigma_3}{2} - \frac{\sigma_1+\sigma_3}{2}\sin\phi$	$f = \frac{q}{p'} - H = 0$	$f = \frac{q}{p'} - H = 0$	$f = (q - p\alpha)^2 - m^2p^2 = 0$
Potential function	$g = \frac{\sigma_1-\sigma_3}{2} - \frac{\sigma_1+\sigma_3}{2}\sin\psi$	$\frac{\partial g}{\partial p'} = M_{\text{pt}} - \frac{q}{p'}$ with $\frac{\partial g}{\partial q} = 1$ $M_{\text{pt}} = \frac{6\sin\phi_{\text{pt}}}{3-\sin\phi_{\text{pt}}}$ with $\phi_{\text{pt}} = \phi_\mu - \psi$	$\frac{\partial g}{\partial p'} = A_d\left(M_{\text{pt}} - \frac{q}{p'}\right)$ with $\frac{\partial g}{\partial q} = 1$ $M_{\text{pt}} = \frac{6\sin\phi_{\text{pt}}}{3-\sin\phi_{\text{pt}}}$	$D = A_d(M_{\text{pt}} - \alpha)$ $M_{\text{pt}} = \frac{6\sin\phi_{\text{pt}}}{3-\sin\phi_{\text{pt}}}$
Hardening law	–	$H = \frac{M_{\text{p}}\varepsilon_{\text{d}}^{p}}{k_{\text{p}}+\varepsilon_{\text{d}}^{p}}$ with $M_{\text{p}}=\frac{6\sin\phi_{\text{p}}}{3-\sin\phi_{\text{p}}}$	$H = \frac{M_{\text{p}}\varepsilon_{\text{d}}^{p}}{k_{\text{p}}+\varepsilon_{\text{d}}^{p}}$ with $M_{\text{p}}=\frac{6\sin\phi_{\text{p}}}{3-\sin\phi_{\text{p}}}$	$h = k_p\frac{\lvert\mathbf{b}:\mathbf{n}\rvert}{b_{\text{ref}}-\lvert\mathbf{b}:\mathbf{n}\rvert}$
Critical state	–	–	$e_{\text{c}} = e_{\text{ref}} - \lambda\ln\left(\frac{p'}{p_{\text{at}}}\right)$	$e_{\text{c}} = e_{\text{ref}} - \lambda\ln\left(\frac{p'}{p_{\text{at}}}\right)$
Interlocking	–	–	$\tan\phi_{\text{p}} = \left(\frac{e_c}{e}\right)^{n_p}\tan\phi_\mu$; $\tan\phi_{\text{pt}} = \left(\frac{e_c}{e}\right)^{-n_d}\tan\phi_\mu$	$\tan\phi_{\text{p}} = \left(\frac{e_c}{e}\right)^{n_p}\tan\phi_\mu$; $\tan\phi_{\text{pt}} = \left(\frac{e_c}{e}\right)^{-n_d}\tan\phi_\mu$
Number of parameters	4	5	10	10

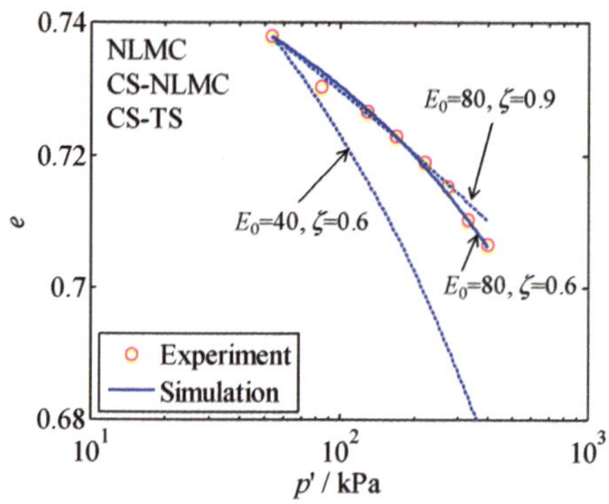

Fig. 8.3 Calibration of elasticity parameters by using isotropic compression test on Hostun sand

where E_0 is the reference value of Young's modulus; e is the void ratio; p' is the mean effective stress; p_{at} is the atmospheric pressure used as reference pressure (p_{at} = 101.3 kPa); and ζ is a constant.

The parameters of each selected model can be divided into: (1) elastic parameters; (2) plastic shear hardening-related parameters; (3) stress–dilatancy-related parameters, and (4) critical-state-related parameters for critical-state-based models. The two elastic parameters, E_0 and ζ, were easily obtained from isotropic compression tests as shown in Fig. 8.3. A typical value of Poisson's ratio $\upsilon = 0.2$ was assumed for the sand. All the other parameters were identified by the optimization method. Note that for MC, the elastic parameter was selected for optimization because the overall deformation before the maximum shear strength is entirely controlled by the elastic stiffness. For the optimization, the intervals of the parameters are given in Table 8.3, which cover their typical values for sand (Table 8.2).

8.3.2 Performance of the Enhanced RCGA

The parameter identification is performed by using the CS-NLMC model and one test result ($p'_0 = 200$ kPa, $e_0 = 0.83$) as example, the computational effectiveness and efficiency of the enhanced RCGA was assessed. In order to highlight the advantages of the new RCGA, the multi-objective genetic algorithm (MOGA-II, a binary-coded genetic algorithm) presented by Poles et al. [5] with high searchability [6] was chosen as a comparative objective to conduct the same optimization. The parameters of two GAs are shown in Table 8.4. The optimal parameters are presented in Table 8.5. It can be seen that two sets of parameters are almost the same. It demonstrates that the new RCGA has also an outstanding searchability for tackling the problem of parameter identification.

Moreover, the efficiency is important for assessing an algorithm. Figure 8.4 shows the evolution of the minimum objective in each generation with the increase of the number of generations. It can be seen that the convergence speed is lower during

Table 8.3 Search domain for different parameters of constitutive models

Model	MC			NLMC			CS-NLMC and CS-TS							
Parameters	E_0	ϕ_u	ψ	ϕ_μ	ψ	k_p	e_{ref}	λ	ϕ_μ	k_p	k_p (CS-TS)	A_d	n_p	n_d
Lower bound	0	10	0	10	0	0	0.5	0	10	0	0	0	0	0
Upper bound	50,000	50	20	50	20	0.1	1	0.1	50	0.1	100	5	10	10
Step	1000	0.5	0.5	0.5	0.5	10^{-4}	0.001	0.0001	0.5	10^{-4}	1	0.1	0.1	0.1

Table 8.4 Parameters of selected algorithms

Algorithm	PopSize	NumGens	Selection	p_C	p_D	p_M	Elitism
New RCGA	100	50	Tournament	0.9	0.5	0.05	Yes
MOGA-II	100	50	Tournament	0.9	0.5	0.05	Yes

Table 8.5 Optimal parameters with the optimal errors of testing for two selected GAs

Initialization method	Optimal parameters							Average error/%
	e_{ref}	λ	ϕ_μ	k_p	A_d	n_p	n_d	
RCGA	0.745	0.030	28.9	0.0039	1.1	2.8	1.9	3.88
MOGA-II	0.743	0.029	28.9	0.0037	1.0	2.8	2.4	3.82

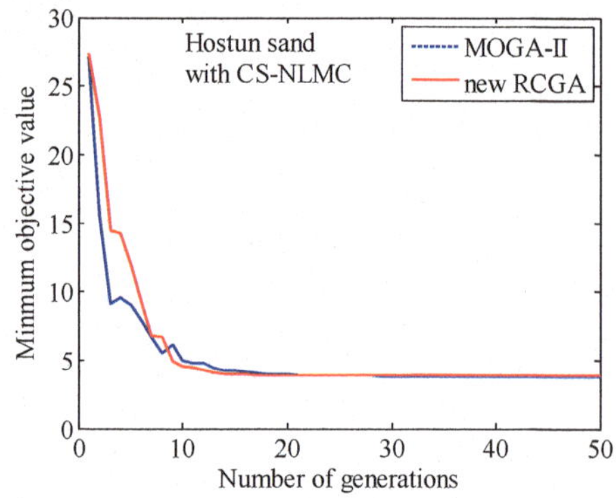

Fig. 8.4 Evolution of minimum objective error in each generation with increasing the number of generations

small number of generations and higher in a high number of generations for the new RCGA, compared to MOGA. This is due to the DRM used in the new RCGA. The DRM is a self-adaptive mutation, which provides a greater chance of population variation by producing a relatively large allowable step size for the mutation at every beginning evolution period. This can result in a higher probability for escaping from the local traps. When the population gradually converges to the optimum solution, a small mutation region produced by DRM is likely to enhance the precision of the obtained solution. The number of generations corresponding to convergence is 26 for the new RCGA and 33 for the MOGA-II, which indicates that the new RCGA shows a faster convergence speed than MOGA-II. This is a key point for GA optimization in identifying parameters from tests.

Overall, the proposed enhanced RCGA performs well in searching the optimal solution and has a faster convergence speed than MOGA-II. Furthermore, judging from the continuity of the geotechnical problem, the new RCGA is more suitable than other classical binary GAs due to its advantages in encoding. Therefore, only the new RCGA is used to conduct the optimization procedure in the following sections.

8.3.3 Optimization Results and Discussion

The optimization using the MC model was conducted first, followed by NLMC, CS-NLMC, and CS-TS in sequence. Since this problem is mono-objective, the set of parameters with the lowest error was selected and was considered as the optimal set of results. The optimization results with objective error are shown in Table 8.6.

The comparisons between the optimal simulations and the objective tests are shown in Fig. 8.5. The errors between the optimal simulations and the objective tests of the four selected models are shown in Fig. 8.6. It can be seen that the worst performance of the simulations is found in MC, followed by NLMC. Both CS-NLMC and CS-TS perform well and are much better in stress–strain behavior than the MC and NLMC models. The reason for this is that the four selected models have different features in describing the sand behavior. First, since the MC model is an elastic-perfectly plastic model, the stress–strain nonlinearity cannot be described. In contrast to MC, a nonlinear plastic stress–strain behavior is incorporated into NLMC, which results in a better performance than that given by MC. In other words, the incorporation of nonlinear elastic and plastic stress–strain features is essential for all sand models. In terms of CS-NLMC and CS-TS, a better agreement between the simulations and the experiments is obtained than when using NLMC. This indicates that it is necessary to incorporate the critical state concept in sand models for simulation. Note that the comparison of predictions is not surprising based on studies on critical states of sand during the last few decades, and this section also serves to show the performance of GA optimization as a basis for the following sections.

Additionally, the slight difference between CS-NLMC and CS-TS is related to the plastic hardening law. In contrast to CS-NLMC, which incorporates a hyperbolic plastic hardening law, the CS-TS model incorporates the bounding surface concept based hardening law with an small elastic domain proposed by Manzari and Dafalias [7], which is slightly more accurate in the simulation.

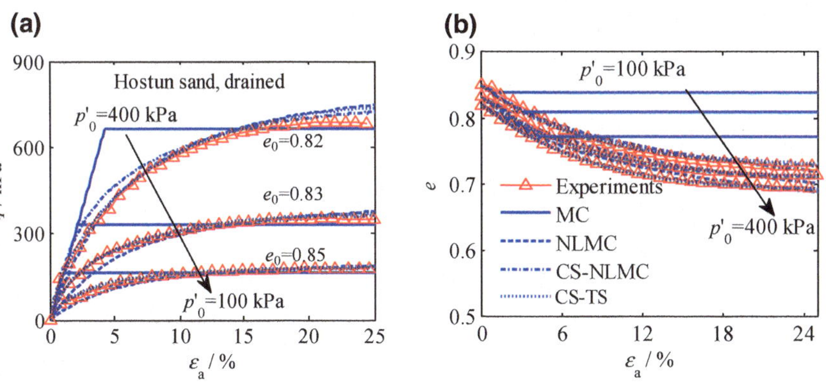

Fig. 8.5 Comparisons between the simulations and the objective tests for four selected models

Table 8.6 Optimal parameters and error for four sand models

Model	MC			NLMC			CS-NLMC (CS-TS)						
Parameters	E_0	ϕ_u	ψ	ϕ_μ	ψ	k_p	e_{ref}	λ	ϕ_μ	k_p	A_d	n_p	n_d
Values	15,500	27.0	0.0	31.5	0.0	0.022	0.739 (0.739)	0.0253 (0.0253)	29.0 (29.0)	0.0061 (29)	0.8 (0.7)	1.9 (1.7)	4.3 (5.4)
Error/%	36.43			5.31			2.91 (2.81)						

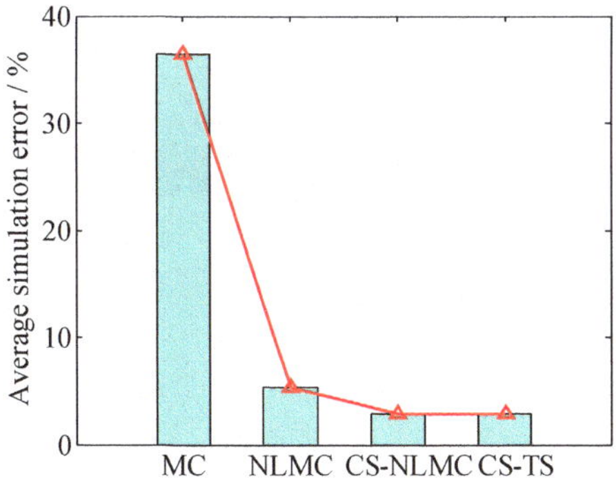

Fig. 8.6 Average errors between optimal simulations and objective tests of four selected models

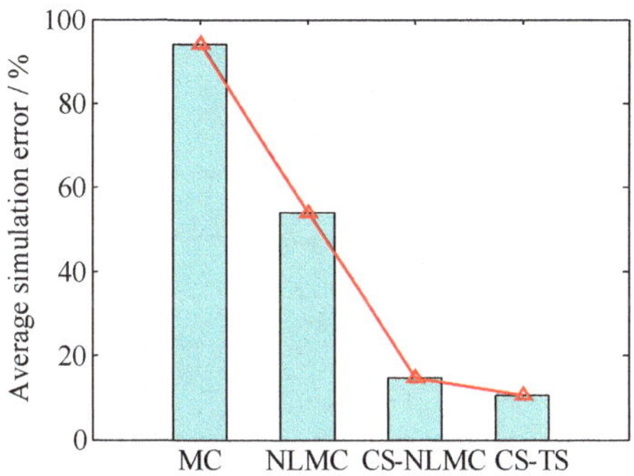

Fig. 8.7 Average simulation errors of MC, NLMC, C-SNLMC, and CS-TS

Overall, the features of sand necessary in constitutive modeling are nonlinear plastic hardening behavior, and the critical state concept with an interlocking effect. On the basis of these feathers, both the CS-NLMC and the CS-TS models are recommended to simulate sand behavior.

In order to further validate the ability of the selected models to describe the sand behavior, other triaxial tests performed on the same Hostun sand were simulated by selected models using the optimized parameters. The error between simulations and experiments was calculated simultaneously. Figure 8.7 shows the average simulation error of the four models for all the tests. Again, the CS-TS model results in the best performance in the simulation, followed by CS-NLMC, NLMC, and MC.

Figure 8.8 shows the comparisons between the simulations and the experiments for CS-NLMC and CS-TS, and based on these, the CS-NLMC and CS-TS models are still recommended. When two numerical models perform equally well in predicting test phenomena, additional criteria need to be selected to judge the merit of the models. One useful guideline is to evaluate the complexity of the formulae adopted in the model and the type and number of parameters. According to this criterion, the CS-NLMC model is more suitable due to its relatively simple formulae compared to those in CS-TS. Numerical convergence is easier to obtain when the simple formulae are used to deal with complex geotechnical problems. Since the bounding surface

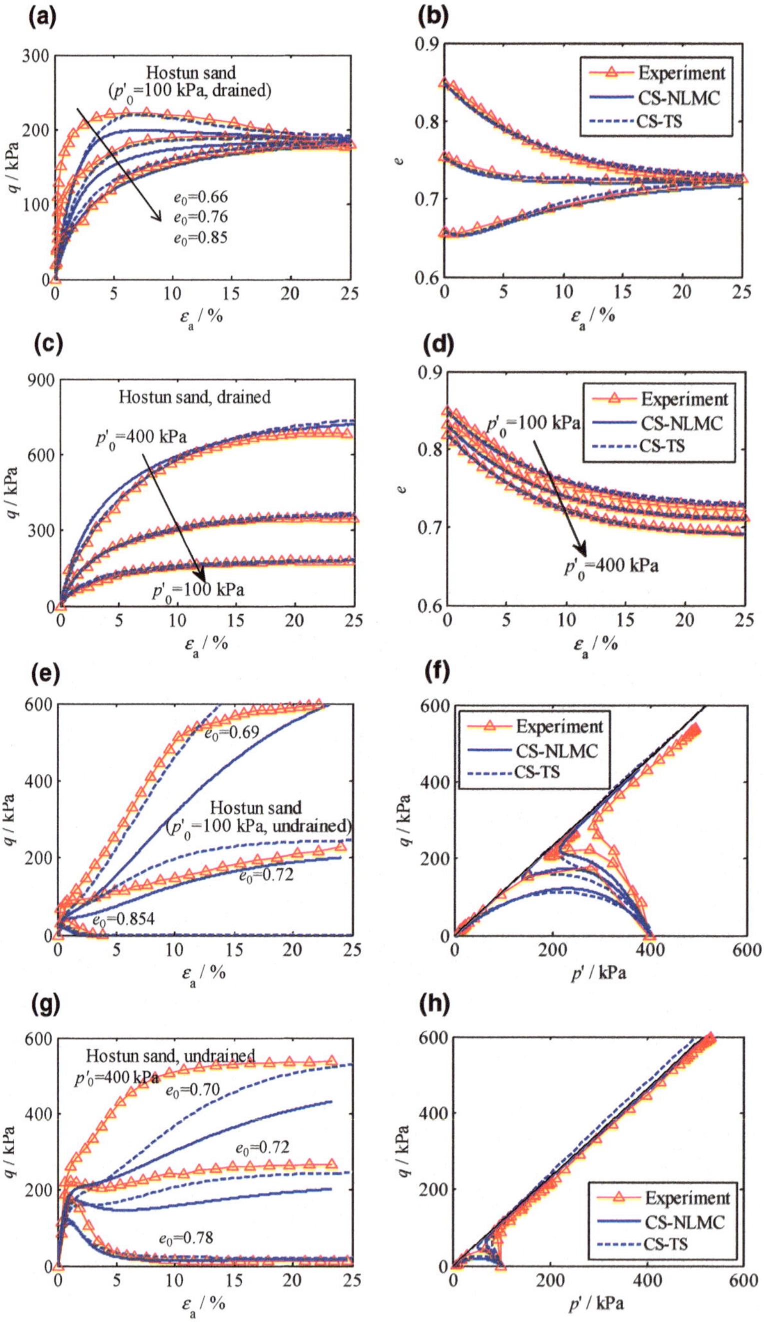

Fig. 8.8 Comparisons between simulation and experiments for CS-NLMC and CS-TS

concept is not necessary for describing monotonic behavior, the CS-NLMC model was chosen as an appropriate sand model for the following sections.

8.4 Selection of Test Type for Identification of Parameters

Besides drained triaxial tests, undrained triaxial tests can also be conducted for estimating soil properties. To better identify the parameters, the performance of different combinations of drained and undrained triaxial tests as objective tests needs to be examined. For this purpose, three drained and three undrained triaxial tests performed on Hostun sand were selected for possible combinations of the GA objective. The results of the selected tests are shown in Fig. 8.9 and are marked by the sequence number. The sequence number and the information for the corresponding test are presented in Table 8.7. Three tests were selected randomly as a combination from the total of six tests. Thus, twenty different combinations in total are examined in this section and are summarized in Table 8.8.

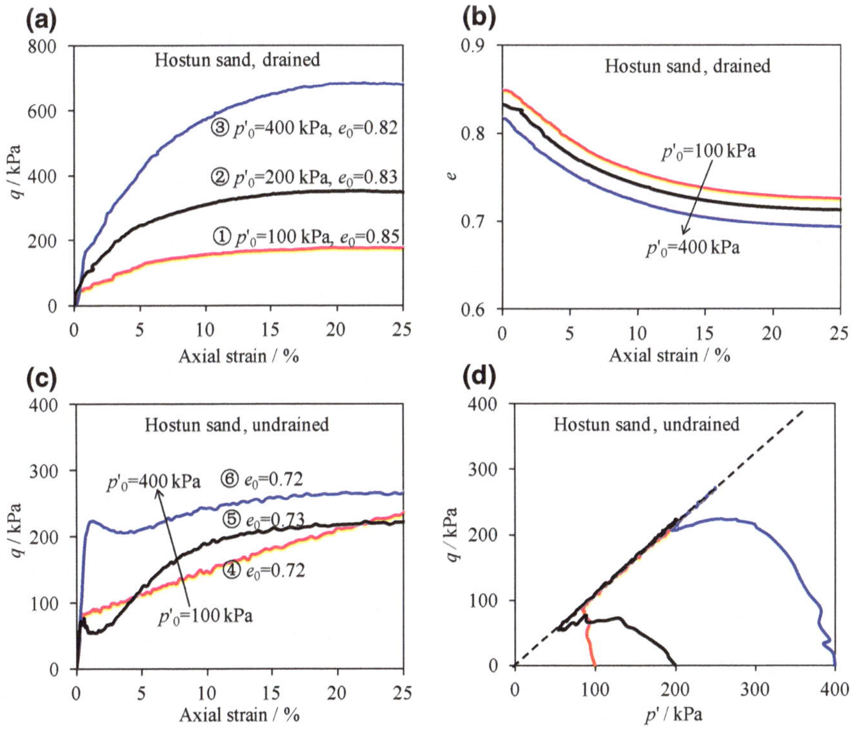

Fig. 8.9 Results of drained and undrained triaxial tests of Hostun sand

Table 8.7 Number of optimum objectives

Number of tests	Initial void ratio e_0	Confining pressure σ_3/kPa	Drainage conditions
①	0.85	100	CD
②	0.83	200	CD
③	0.82	400	CD
④	0.72	100	UD
⑤	0.73	200	UD
⑥	0.72	400	UD

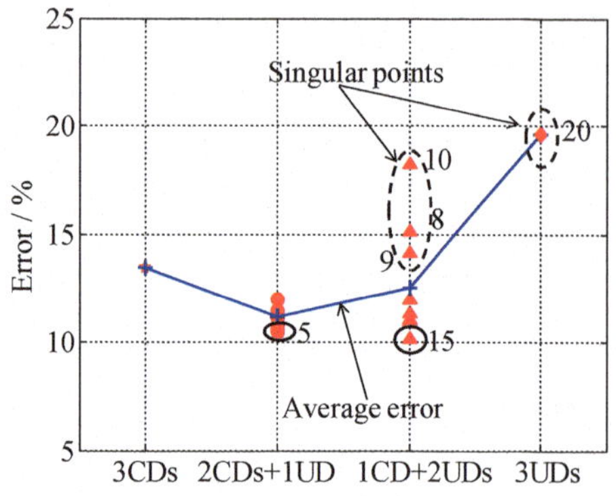

Fig. 8.10 Simulation errors based on optimal parameters of different combinations

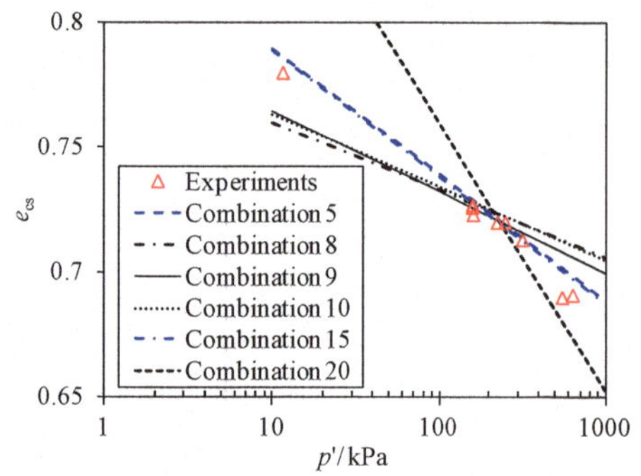

Fig. 8.11 Critical state lines of different combinations

The same optimization procedure was carried out for all combinations. The optimal parameters and the corresponding objective errors for the different combinations are listed in Table 8.8. In order to evaluate the performance of each combination, the optimal set of parameters was applied to simulate five drained tests and six undrained tests with different confining pressures and void ratios on the same Hostun sand, as performed by Liu et al. [2] and Li et al. [8]. Simulation errors were also calculated, as shown in Table 8.8.

In order to analyze the effect of the test type on the identification of parameters, all the combinations in Table 8.8 were divided into four groups according to the number

Table 8.8 Optimal parameters and errors of different combinations

Number	Combinations	Optimal parameters							Objective error/%	Average error/%
		e_{ref}	λ	ϕ_μ	k_p	A_d	n_p	n_d		
1	①②③	0.739	0.0253	28.5	0.0038	1.1	2.4	2.6	3.46	13.43
2	①②④	0.735	0.0188	29.0	0.0025	0.7	3.1	5.0	5.16	11.11
3	①②⑤	0.739	0.0212	29.0	0.0013	1.7	4.1	0.2	8.98	14.28
4	①②⑥	0.735	0.0181	29.0	0.0023	0.9	3.3	3.6	5.04	12.00
5	①③④	0.739	0.0260	28.5	0.0037	0.8	2.7	4.6	5.11	10.46
6	①③⑤	0.743	0.0281	29.0	0.0017	1.7	4.2	0.0	7.71	11.47
7	①③⑥	0.740	0.0262	29.0	0.0023	0.9	3.9	3.4	4.83	11.29
8	①④⑤	0.733	0.0117	28.0	0.0017	0.7	1.7	5.0	15.21	15.09
9	①④⑥	0.732	0.0142	28.0	0.0018	1.0	2.5	3.9	6.91	14.09
10	①⑤⑥	0.734	0.0127	28.5	0.0017	1.6	3.5	1.4	12.76	18.16
11	②③④	0.744	0.0286	29.0	0.0058	0.7	2.4	5.0	5.42	10.77
12	②③⑤	0.753	0.0340	29.0	0.0026	1.7	3.7	0.0	5.04	11.18
13	②③⑥	0.749	0.0314	29.0	0.0031	1.0	3.5	2.6	3.03	10.57
14	②④⑤	0.750	0.0334	29.5	0.0057	0.8	2.1	3.7	11.89	10.89
15	②④⑥	0.738	0.0219	29.0	0.0035	0.7	2.6	5.0	6.93	10.09
16	②⑤⑥	0.755	0.0374	28.5	0.0018	1.9	3.10	0.0	5.97	11.26
17	③④⑤	0.749	0.0317	29.0	0.0056	0.9	2.8	3.0	11.87	11.01
18	③④⑥	0.745	0.0294	29.5	0.0054	0.6	2.9	5.0	7.79	10.15
19	③⑤⑥	0.752	0.0333	28.0	0.0023	1.6	3.7	0.5	5.64	11.93
20	④⑤⑥	0.760	0.0467	28.5	0.0028	0.8	0.5	2.3	11.64	19.61

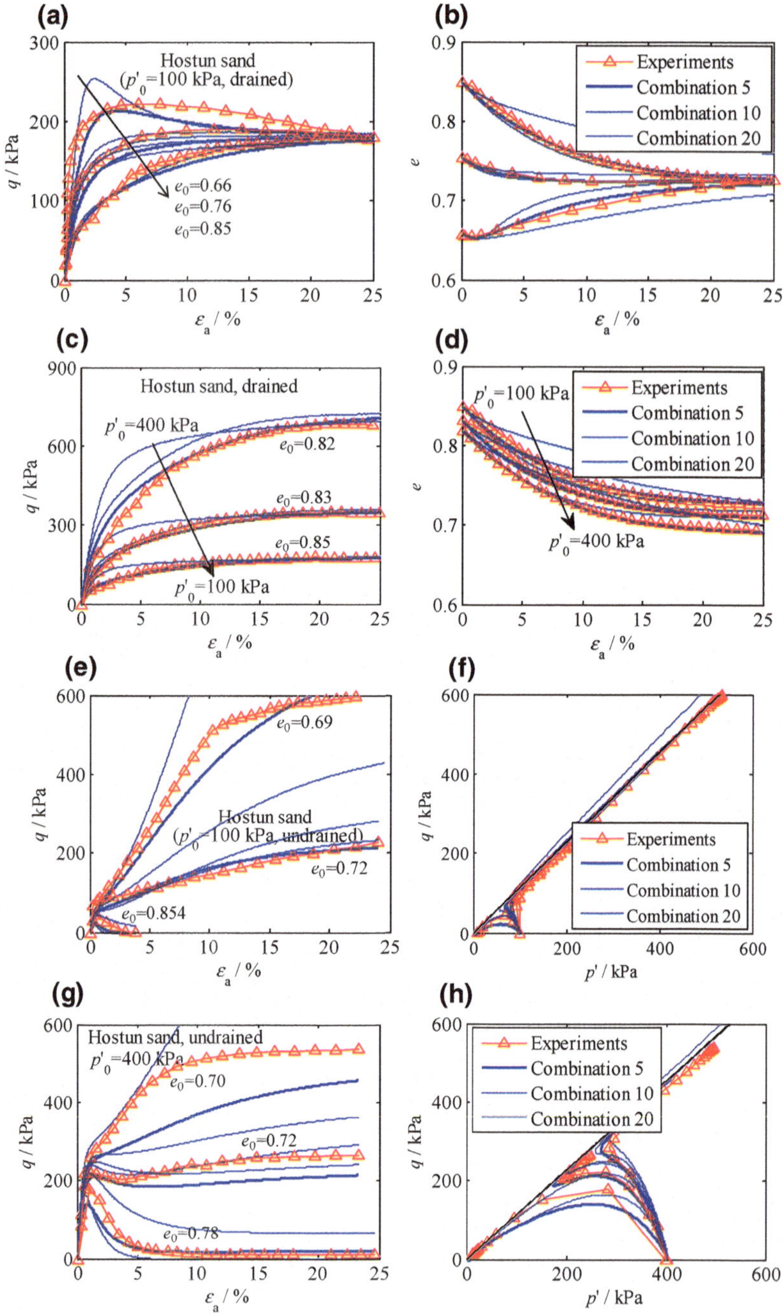

Fig. 8.12 Results of simulation based on different combinations

of undrained tests in the objective and were marked as 3CDs, 2CDs + UD, CD + 2UDs, and 3UDs (CD and UD representing drained and undrained tests, respectively). Four groups with simulation errors are plotted in Fig. 8.10. It can be found that the average error first decreases and then increases with the increasing number of undrained tests in the objective. However, there are scatter points with large simulation errors among all the combinations. A possible reason for the poor simulations is the determination of CSL parameters, as shown in Fig. 8.11, which shows a comparison of the critical state line between predictions and experiments for different combinations. Note that the experimental critical states in the figure are apparent points corresponding to a strain level of 25%. It can be seen that the combinations with close final states of e, p' in the e-log p' space could lead to an incorrect CSL, as found in combinations 8, 10, and 20. These incorrect CSLs may lead to poor simulated results. In contrast, the combinations with dissimilar final states of e, p' may give a generally accurate critical state line and result in a good simulation performance, such as combinations 5 and 15. Figure 8.12 shows the comparison of results between experiment and simulation for three typical CSLs.

Overall, the performance of parameter identification can be further improved by using the combinations which contain both undrained and drained tests as objectives, apart from those combinations with close final states of e, p' which cause incorrect CSLs.

8.5 Estimation of Minimum Number of Tests for Identification of Parameters

As previously mentioned, the objective with one undrained test could result in a generally better performance. At the same time, the inaccurate CSL determined using selected tests could result in unsatisfactory parameters and simulations, which was highlighted previously. One possible way to avoid this problem is to add more tests to the objective in the optimization. Traditionally, three triaxial tests have been proposed for estimating strength parameters (e.g., cohesion, c, and friction angle, ϕ). However, for critical-state-based modeling, more tests should be used. Thus, this section aims to estimate the minimum number of tests required for modeling based on critical state.

In this case, in order to focus on the effect of the number of tests for the identification of parameters, there are two possibilities for adding more tests to the standard set of three drained tests. These are: (1) adding drained tests and (2) adding undrained tests. For adding drained tests, one or two more tests (marked as 3 + 1 or 3 + 2) were examined. For adding undrained tests, one to four more tests (marked as 3 + 1, 3 + 2, 3 + 3, and 3 + 4) were examined, based on the available tests carried out by Liu et al. [2] and Li et al. [3]. The test which is easy to carry out in the laboratory at low cost should be selected first. Following this rule, the test on dense sand with relatively low

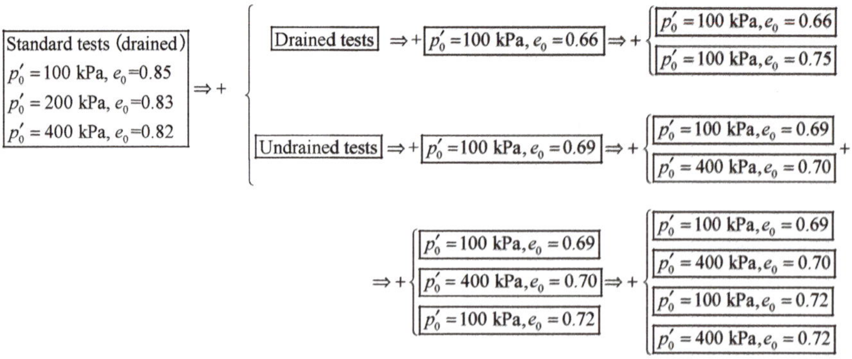

Fig. 8.13 Program for selecting the effective number of tests

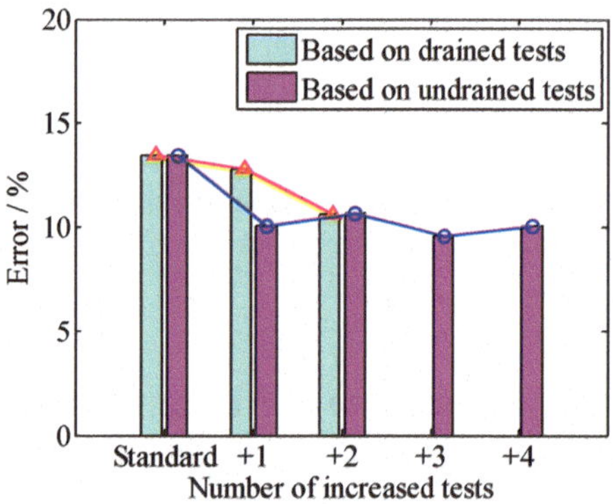

Fig. 8.14 Variation tendency of errors with the increase of the number of drained or undrained tests

confining pressure was first selected, and then, the test with high confining pressure was subsequently added. The program for choosing tests is presented in Fig. 8.13.

The same optimization procedure was conducted for objectives with different numbers of tests. The optimal parameters are summarized in Table 8.9. In order to estimate the number of tests, other tests in addition to the objectives were simulated by CS-NLMC using each set of optimal parameters. Meanwhile, the differences between simulations and experiments were also computed, and the values of simulation errors are summarized in Table 8.9. The variation of errors with the increasing number of drained or undrained tests is plotted in Fig. 8.14.

It can be found that adding two tests to the basic standard combination is sufficient to obtain accurate parameters. By using the optimal parameters obtained by adding two tests to the basic standard combination, the comparisons between experimental and simulated results are shown in Fig. 8.15. Moreover, the results suggest also that model parameters identified by using three tests in practice are not reliable for critical-state-based constitutive models. Therefore, the minimum recommended number of tests for critical-state-based modeling is five.

Table 8.9 Optimization parameters and error based on critical state sand model

Total quantity	Additional tests		Optimal parameters							Average error/%
	Drained	Undrained	e_{ref}	λ	ϕ_μ	k_p	A_d	n_p	n_d	
3	0	0	0.739	0.0253	28.5	0.0038	1.1	2.4	2.6	13.43
3 + 1	+1		0.736	0.0273	28.5	0.0035	0.7	2.8	4.6	12.75
		+1	0.740	0.0275	29.0	0.0023	0.8	3.4	4.3	10.04
3 + 2	+2		0.737	0.0241	29.5	0.0033	0.8	2.9	3.8	10.60
		+2	0.740	0.0268	29.0	0.0019	0.8	3.4	4.3	10.66
3 + 3		+3	0.741	0.0272	29.0	0.0031	0.8	3.1	4.3	9.60
3 + 4		+4	0.742	0.0279	29.0	0.0022	1.0	3.4	2.7	10.04

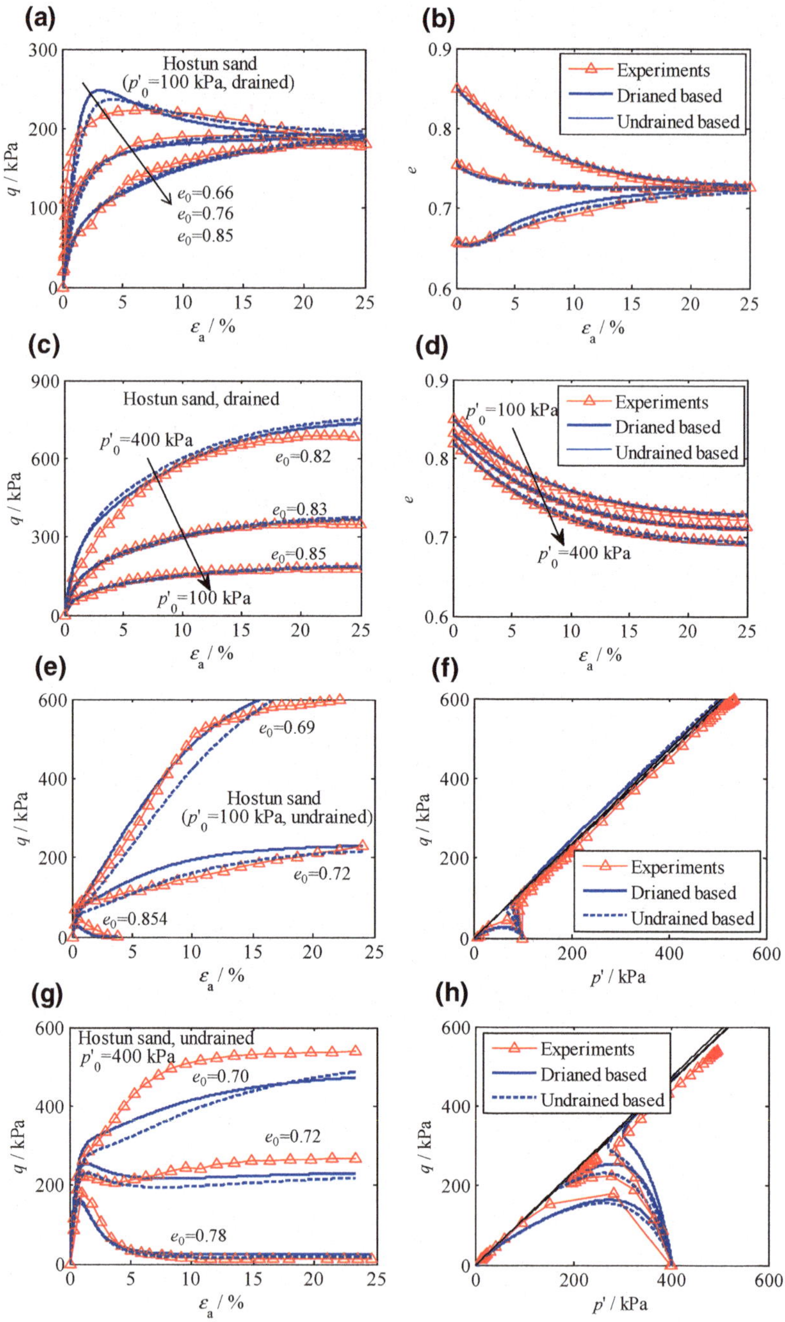

Fig. 8.15 Simulation results of Hostun sand based on the optimal parameters

8.6 Estimation of Strain Level of Tests for Identification of Parameters

It is well known that the critical state cannot be accurately reached during conventional triaxial tests on sand. The reason is that the sample becomes inhomogeneous with the increase of the strain level due to localizations or instabilities. In reality, therefore, the critical state parameters cannot be directly measured from triaxial tests. In this case, the optimization method should be applied to the tests at limited strain levels with samples being still more or less homogenous. Therefore, it is necessary to confirm the smaller suitable strain level of tests for the identification of parameters by the optimization method.

According to the conclusions from previous section, two groups with five tests (3CDs + 2CDs, 3CDs + 2UDs) were selected as the objective to examine the smaller suitable strain level of tests for the identification of parameters. The optimization procedure was conducted based on the objective tests with strain levels of 5, 10,

Table 8.10 Optimal parameters of Hostun sand for different strain levels

Strain levels/%	3CDs + 2CDs (3CDs + 2UDs)						
	e_{ref}	λ	ϕ_μ	k_p	A_d	n_p	n_d
5	0.750 (0.765)	0.0565 (0.038)	28.3 (28.4)	0.0021 (0.0048)	0.6 (0.7)	3.0 (3.2)	5.3(5.0)
10	0.735 (0.780)	0.0345 (0.0445)	29.0 (29.0)	0.0020 (0.0076)	0.8 (0.8)	2.8 (2.5)	3.5(4.8)
15	0.740 (0.760)	0.0335 (0.036)	29.5 (29.0)	0.0029 (0.0046)	0.8 (0.8)	2.9 (2.9)	3.4(4.7)
20	0.736 (0.78)	0.0273 (0.0505)	29.5 (29.0)	0.0035 (0.0048)	0.8 (1.1)	2.8 (3.2)	4.3(2.3)
25	0.737 (0.74)	0.0241 (0.0268)	29.5 (29.0)	0.0033 (0.0019)	0.8 (0.8)	2.9 (3.4)	3.8(4.3)

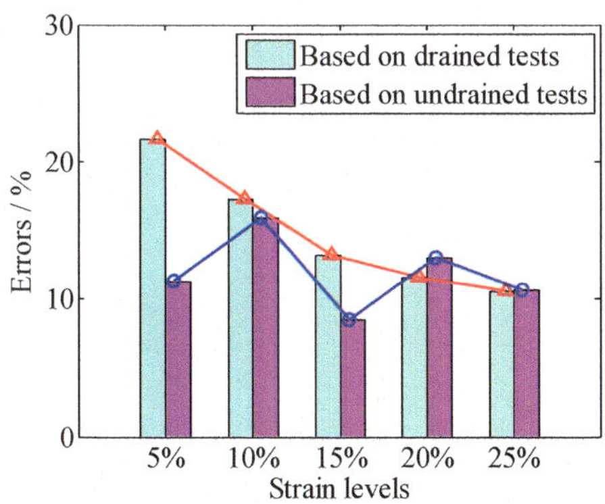

Fig. 8.16 Evolution of average simulation errors with the strain levels for Hostun sand

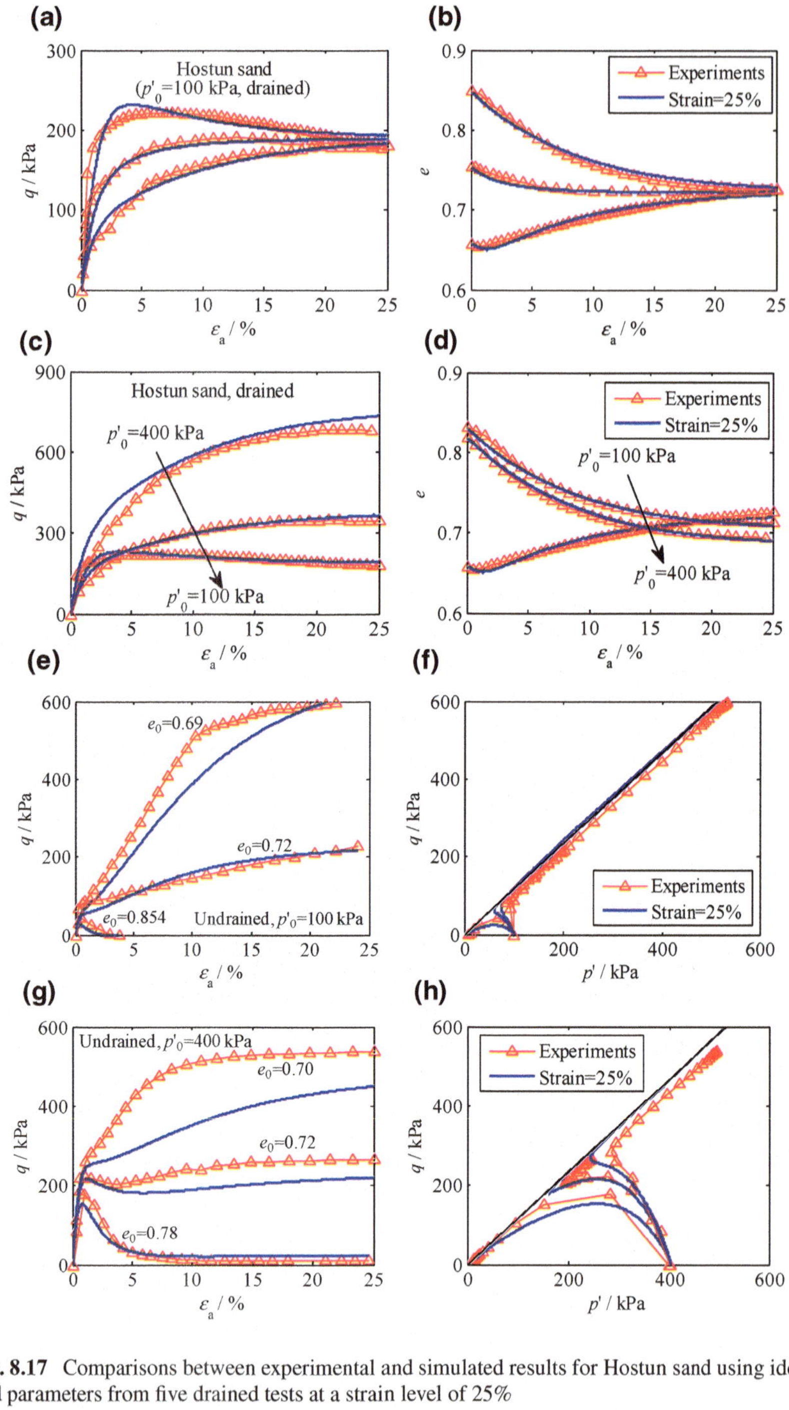

Fig. 8.17 Comparisons between experimental and simulated results for Hostun sand using identified parameters from five drained tests at a strain level of 25%

15, 20, and 25% successively. The optimization results are shown in Table 8.10. In order to evaluate the performance of the optimal parameters by GA optimization, other drained and undrained tests on the same Hostun sand were simulated again by using the optimal parameters. The errors were then taken average based on all test simulations. The variation of errors with the increasing strain level for all tests is plotted in Fig. 8.16. It can be found that the parameter identification based on all drained tests becomes acceptable when the strain level of tests becomes bigger than 20%, and based on drained combined with undrained tests that is not stable due to high nonlinear undrained stress–strain curves, as found in Fig. 8.14. Therefore, the minimum strain level is recommended as 20% when all five drained tests are adopted and as 25% when three drained tests with two undrained tests are adopted.

Comparisons between experimental and simulated results using parameters identified from five drained tests at a strain level of 25%, as shown in Fig. 8.17, demonstrate a good agreement. Overall, the objective tests up to an axial strain of 25% can give the relatively reliable and reasonable parameters by optimization.

Note that the all tests used as the objective were performed on the loose and medium Hostun sand; therefore, the effect of localization on the selecting minimum strain level is too slight. And in the undrained tests, no localization occurs. The purpose to conduct the estimation is to find the minimum strain level to avoid the localization effect.

8.7 Selection of Critical-State-Related Formulas for Advanced Sand Models

8.7.1 Current Critical State Line Formulas

8.7.1.1 Critical State Line

Several typical critical state lines have been proposed to describe the evolution of critical state in the stress–strain space. The most typical one is the linear formula, which was assumed to be a straight line in *e*-log *p*' plane. The relationship has traditionally been written as follows:

$$e_c = e_{ref} - \lambda \ln\left(\frac{p'}{p_{ref}}\right) \tag{8.3}$$

where e_c is void ratio at this critical state line at a mean effective stress p'; e_{ref} is a reference void ratio corresponding to a reference mean effective stress p_{ref} (for the convenience, $p_{ref} = p_{at} = 101.325$ in this study); λ is the slope of CSL in *e*-log p' plane. Then, two parameters (e_{ref} and λ) are required for this CSL.

Another critical state line is a nonlinear formula proposed by Li and Wang [9], which is an extension of linear formula in e-log p' plane with one more parameter ξ, and can be expressed as follows,

$$e_c = e_{ref} - \lambda\left(\frac{p'}{p_{ref}}\right)^{\xi} \tag{8.4}$$

Moreover, there is a third CSL proposed by Gudehus [10] from the physical view of material, which is also a nonlinear formula but with a "s" form by considering a limit of critical void ratio at high stress level. It can be expressed as follows,

$$e_c = e_{cu} + (e_{ref} - e_{cu})\left(-\left(\frac{p'}{p_{ref} \cdot \lambda}\right)^{\xi}\right) \tag{8.5}$$

where e_{cu} is critical void ratio when $p' =$ infinity, which can be assumed equal to minimum void ratio of granular material e_{min}.

For the convenience, three critical state lines are marked as CSL[1], CSL[2], and CSL[3], respectively. Note that this study focuses on the normal stress level of geotechnical structures without considering the grain crushing effect into critical state during the loading, since the grain crushing changes the grain size distribution of material and thus changes the material itself.

8.7.1.2 Formulas of Interlocking Effect

Generally, the interlocking effects are always expressed by the change of material density. For clay, different OCRs can represent different density of samples, and the clay with different OCRs has different behaviors [11–14]. Like clay, the granular material also has the similar concept of density. In order to describe the density effect on the behavior of granular material, the state parameter to measure the distance of void ratio between current state point and corresponding critical state point at the same mean effective stress in e-log p' plane has been proposed. Two typical definitions have been proposed. One was proposed by Biarez and Hicher [15] and defined as the ratio of current void ratio to the critical state void ratio, expressed as follows,

$$\psi = \frac{e}{e_c} \tag{8.6}$$

where e is current void ratio. This state parameter was used for estimating the peak friction angle ϕ_p and the friction angle at phase transformation state ϕ_{pt} as follows,

$$\begin{cases} \phi_p = \mathrm{atan}\left(\psi^{-n_p} \tan\phi_\mu\right) \\ \phi_{pt} = \mathrm{atan}\left(\psi^{n_d} \tan\phi_\mu\right) \end{cases} \tag{8.7}$$

where n_p and n_p are material constants controlling the interlocking effect; ϕ_μ is critical friction angle. Then, the slope of peak stress state line M_p and the slope of phase transformation line M_{pt} in p'-q plane can be obtained by their friction angles in the same way as for the slope of critical state line M.

The other state parameter was originally proposed by Been and Jefferies [16], which is a measure of distance and expressed as follows,

$$\psi = e - e_c \tag{8.8}$$

Similar to Eq. (8.7) but in a more direct way, the slope of peak stress state line M_p and the slope of phase transformation line M_{pt} in p'-q plane were estimated as follows,

$$\begin{cases} M_p = M \exp(-n_p \psi) \\ M_{pt} = M \exp(n_d \psi) \end{cases} \tag{8.9}$$

For triaxial compression, the $M = 6 \sin \phi_\mu / (3 - \sin \phi_\mu)$ can be obtained.

The effects of state parameter on mechanical behavior of granular material can be expressed by describing the contraction and dilation properties of granular material. For a loose structure with $e > e_c$ or $e - e_c > 0$, the contractive is allowed during deviatoric loading. In a dense structure with $e < e_c$ or $e - e_c < 0$, it allows the dense structure to be first contractive and then dilative during deviatoric loading. For both loose and dense structures, when the stress state reaches the critical state line, the void ratio e becomes equal to the critical void ratio e_c, and then no dilation or contraction takes place. Thus, the constitutive equations guarantee that stresses and void ratio reach simultaneously the critical state in the p'-q-e space.

8.7.2 *Simple Critical-State-Based Models*

In order to conduct the optimization procedure, six critical-state-based nonlinear elastoplastic models with above critical-state-related formulas were proposed in this paper. The constitutive relations are introduced as follows:

The total strain rate is conventionally composed of the elastic and plastic strain rates:

$$\dot{\varepsilon}_{ij} = \dot{\varepsilon}_{ij}^e + \dot{\varepsilon}_{ij}^p \tag{8.10}$$

The elastic behavior is assumed to be isotropic with the bulk modulus K adopting the same form of the shear modulus proposed by Richart et al. [4],

$$\dot{\varepsilon}_{ij}^e = \frac{1 + \upsilon}{3K(1 - 2\upsilon)} \sigma'_{ij} - \frac{\upsilon}{3K(1 - 2\upsilon)} \sigma'_{kk} \delta_{ij} \tag{8.11}$$

$$K = K_0 \cdot p_{\text{at}} \frac{(2.97 - e)^2}{(1 + e)} \left(\frac{p'}{p_{\text{at}}} \right)^{\zeta} \tag{8.12}$$

where K_0 and ζ are elastic parameters; υ is Poisson's ratio.

The plastic strain rate is based on the shear sliding:

$$\dot{\varepsilon}_{ij}^{p} = d\lambda \frac{\partial g}{\partial \sigma'_{ij}} \tag{8.13}$$

The yield surface for shear sliding can be expressed in a similar way to that proposed by many previous researchers (such as [17, 18–20]):

$$f = \frac{q}{p'} - \frac{M_{\text{p}} \varepsilon_{\text{d}}^{p}}{k_{\text{p}} + \varepsilon_{\text{d}}^{p}} = 0 \tag{8.14}$$

where q is the deviatoric stress; k_{p} controls the plastic shear modulus; M_{p} is the stress ratio corresponding to the peak strength and determined by the peak friction angle $\phi_{\text{p}}(M_{\text{p}} = 6\sin(\phi_{\text{p}})/(3 - \sin(\phi_{\text{p}})))$ by Eq. (8.7) or directly by Eq. (8.9); $\varepsilon_{\text{d}}^{p}$ is the deviatoric plastic strain.

The potential surface for stress–dilatancy can be implied as:

$$\frac{\partial g}{\partial \sigma'_{ij}} = \frac{\partial g}{\partial p'} \frac{\partial p'}{\partial \sigma'_{ij}} + \frac{\partial g}{\partial s_{ij}} \frac{\partial s_{ij}}{\partial \sigma'_{ij}} \quad \text{with} \quad \frac{\partial g}{\partial p'} = A_{\text{d}} \left(M_{\text{pt}} - \frac{q}{p'} \right); \frac{\partial g}{\partial s_{ij}} = \{1\ 1\ 1\ 1\ 1\ 1\} \tag{8.15}$$

where A_{d} is the stress–dilatancy parameter; $M_{\text{pt}} = 6\sin(\phi_{\text{pt}})/(3 - \sin(\phi_{\text{pt}}))$ can be calculated from the phase transformation friction angle ϕ_{pt} by Eq. (8.7) or directly by Eq. (8.9); the double indices ij is simplified to be $1 \triangleq 11, 2 \triangleq 22, 3 \triangleq 33, 4 \triangleq 12, 5 \triangleq 23, 6 \triangleq 31$. The Lode-angle-dependent strength and stress–dilatancy were introduced as described in Yin et al. [21], which can also be incorporated by using the transformed stress method by Yao et al. [22–25].

Six combinations by using three types of critical state lines and two interlocking laws presented in Sect. 8.2 were adopted to formulate six different models.

The plastic multiplier dλ can be calculated in a conventional way according to plasticity:

$$\text{d}\lambda = \frac{[\partial f/\partial \sigma_{ij}]^{\text{T}} [D_{ijkl} \cdot d\varepsilon_{kl}]}{[\partial f/\partial \sigma_{ij}]^{T} [D_{ijkl} \cdot \partial g/\partial \sigma_{kl}] - \partial f/\partial \varepsilon_{\text{d}}^{p} \cdot \partial g/\partial q} \tag{8.16}$$

Combining Eqs. (8.10) to (8.16) and critical-state-related equations, the stress–strain relationship can be solved for test simulations.

The proposed model requires a calibration of ten or eleven parameters, which can be divided into four groups: (1) elasticity-related parameters: K_0, ζ, and υ; (2) shear-sliding-related parameters: ϕ_u and k_p; (3) critical-state-line-related parameters: e_{ref}

Table 8.11 Search domain for critical state nonlinear soil model

Parameters	e_{ref}	λ			ξ	ϕ_μ	A_d	k_p	n_d	n_p
		CSL[1]	CSL[2]	CSL[3]						
Lower bound	0.1	0	0	0	0	10	0	0	0	0
Upper bound	1.0	0.1	1.0	100	1.0	50	5	0.1	10	10
Step	0.001	0.0001	0.0001	0.01	0.001	0.1	0.1	0.0001	0.1	0.1

and λ for CSL[1], e_{ref}, λ and ξ for CSL[2] and CSL[3]; (4) interlocking-effect-related parameters: A_d, n_d, and n_p.

The bulk modulus, K_0, with, ζ, can be easily obtained based on isotropic compression curves (see [26, 27]), and a typical value for Poisson's ratio, $\upsilon = 0.2$, can be assumed. This study focuses on the identification of plasticity-related parameters (groups 2–4) using optimization methods. The intervals of parameters given in Table 8.11 are much larger than the ones corresponding to typical values.

8.7.3 *Estimation of CSL Formulation*

Four granular materials were selected as examples to discuss the problem in the selection of critical-state-related formulas in sand models. The particle shape of Hostun sand, Toyoura sand, and glass ball are round, flat, and perfect ball, respectively. Table 8.12 shows the physical properties of four selected materials (Hostun sand by Li et al. [8] and Liu et al. [2], Toyoura sand by Verdugo and Ishihara [28] glass ball by Li et al. [8], and Alaskan sand by Jefferies and Bean [29]). The elasticity-related parameters for four selected materials were determined based on isotropic compressions, shown in Fig. 8.18. Meanwhile, in this part the state parameter e/e_c with its interlocking formula was kept with three different formulas of critical state line in models.

In order to determine the critical state line, four conventional triaxial tests were selected as objective. The selected tests with uniformly distributed position and wide

Table 8.12 Physical properties of four experimental materials

Materials	Shape of particle	G_s	e_{max}	e_{min}	D_{50} (mm)	C_u
Hostun sand	Round	2.60	0.881	0.577	0.9	1.4
Toyoura sand	Elliptic	2.65	0.977	0.597	0.17	1.7
Glass ball	Perfect ball	2.60	0.432	0.160	0.9	20
Alaskan sand	–	2.70	0.856	0.565	–	–

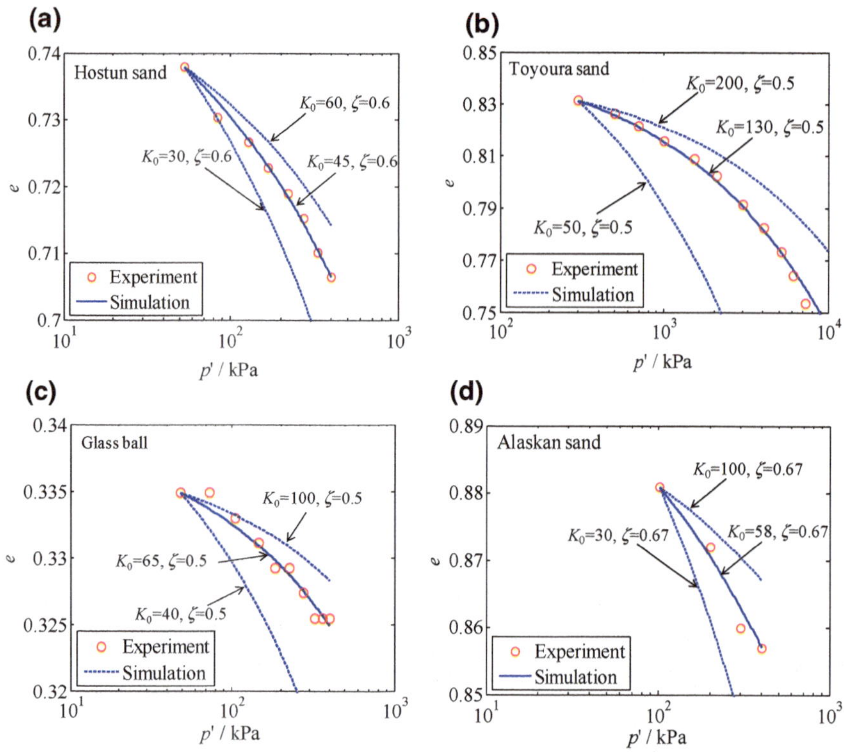

Fig. 8.18 Determination of elasticity-related parameters for four selected materials

stress range of final points (e, p') in e-log p' plane are necessary for getting a relatively accurate CSL. It is pointed out that the mean effective stress levels of all the selected tests are between 10 and 1000 kPa. According to the previous experimental works [30], the behaviors of granular materials can involve particle crushing at high stresses (in shearing tests with $p'_0 > 400$ kPa on Quartz sand). Particle crushing makes the change of materials no longer the same as before, and thus the position of critical state line in e-log p' plane is no longer valid (see Hu et al. [31]; Yin et al. [32]). Therefore, in order to avoid the effect of grain breakage on the selection of CSL, the difference between three critical state lines at high pressure will not be discussed in this study. The critical state lines based on four selected tests of four materials are shown in Fig. 8.19.

After optimization, the optimal parameters with objective errors for four selected materials are summarized in Table 8.13. There is a slight difference for parameters (except 2–3 CSL parameters) obtained by GA between three models. To further evaluate the optimal parameters for three different CSLs, other drained and undrained tests on same materials were simulated: For Hostun sand, five drained tests and five undrained tests were simulated; for Toyoura sand, four drained tests and four

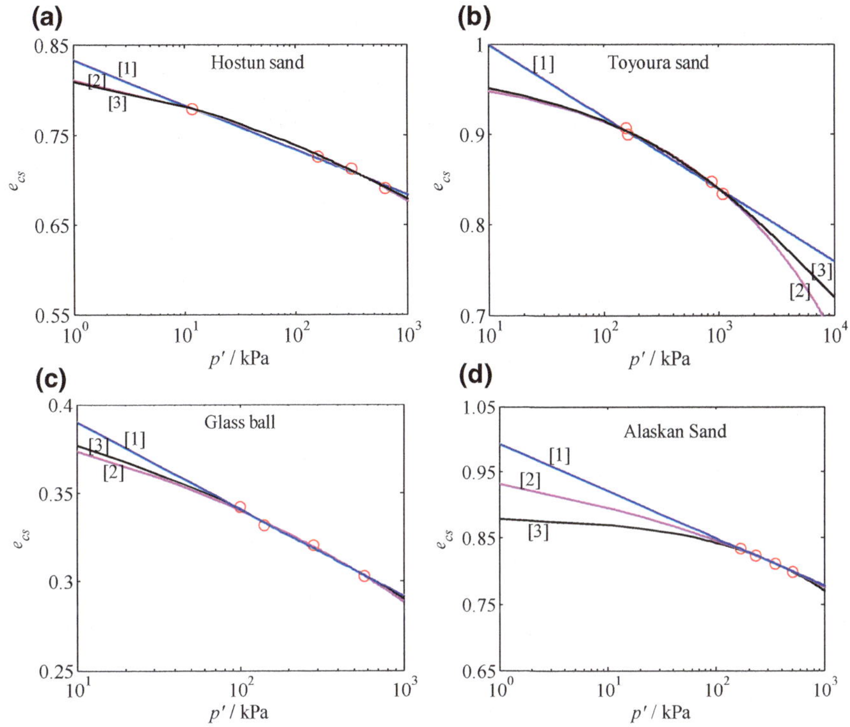

Fig. 8.19 Different critical state lines for four selected materials

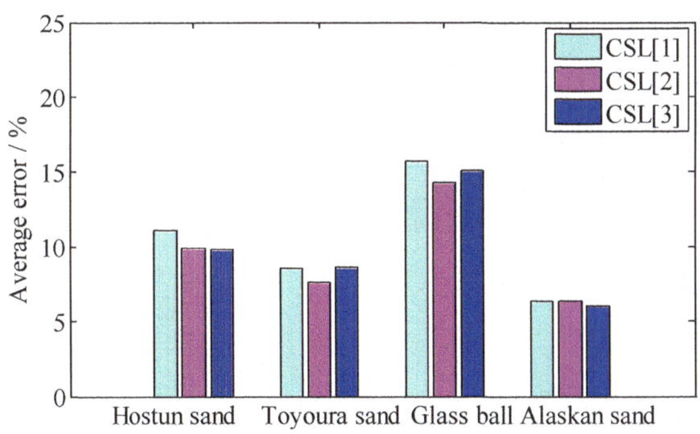

Fig. 8.20 Comparisons of simulation errors between three different CSLs for four selected materials

Table 8.13 Optimal parameters and error for four materials with e/e_c

Materials	CSL Type	CSL		Optimal parameters						Objective error/%	Simulation error/%
		e_{ref}	λ	ξ	ϕ_μ	k_p	A_d	n_p	n_d		
Hostun sand	[1]	0.734	0.0215	–	29.5	0.0031	0.7	3.2	4.7	4.33	11.09
	[2]	0.881	0.143	0.156	29.5	0.0033	0.7	3.2	4.7	4.06	9.93
	[3]	0.844	11.53	0.279	29.5	0.0029	0.8	3.4	3.8	4.17	9.79
Toyoura sand	[1]	0.923	0.0363	–	31.5	0.0049	0.7	2.6	3.6	7.39	8.52
	[2]	0.977	0.0596	0.365	31.5	0.0044	0.7	2.4	2.9	6.33	7.64
	[3]	0.977	64.19	0.428	31.5	0.0045	0.8	2.3	2.7	6.66	8.65
Glass ball	[1]	0.341	0.021	–	24.0	0.002	1.1	5.0	3.4	12.06	15.74
	[2]	0.432	0.0919	0.194	24.0	0.0018	1.1	2.1	4.3	12.06	14.32
	[3]	0.432	32.69	0.255	24.0	0.0018	1.0	3.4	5.2	12.11	15.10
Alaskan sand	[1]	0.85	0.0311	–	31.4	0.0013	0.6	1.7	2.6	6.35	34.88
	[2]	0.964	0.118	0.205	31.4	0.0017	0.3	1.9	3.5	6.33	31.34
	[3]	0.883	50.6	0.51	30.9	0.0013	0.5	1.9	3.1	6.14	30.53

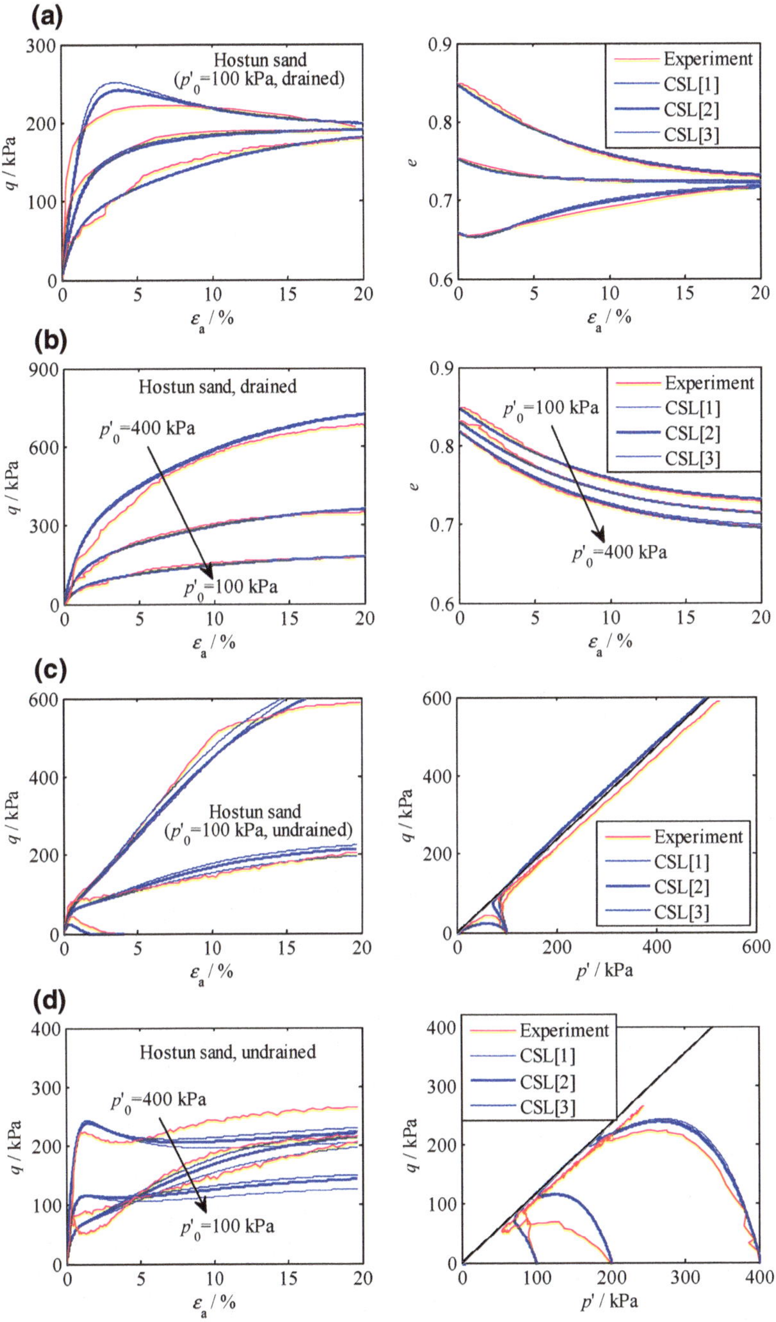

Fig. 8.21 Comparisons of experimental data and simulations for triaxial compression tests on Hostun sand

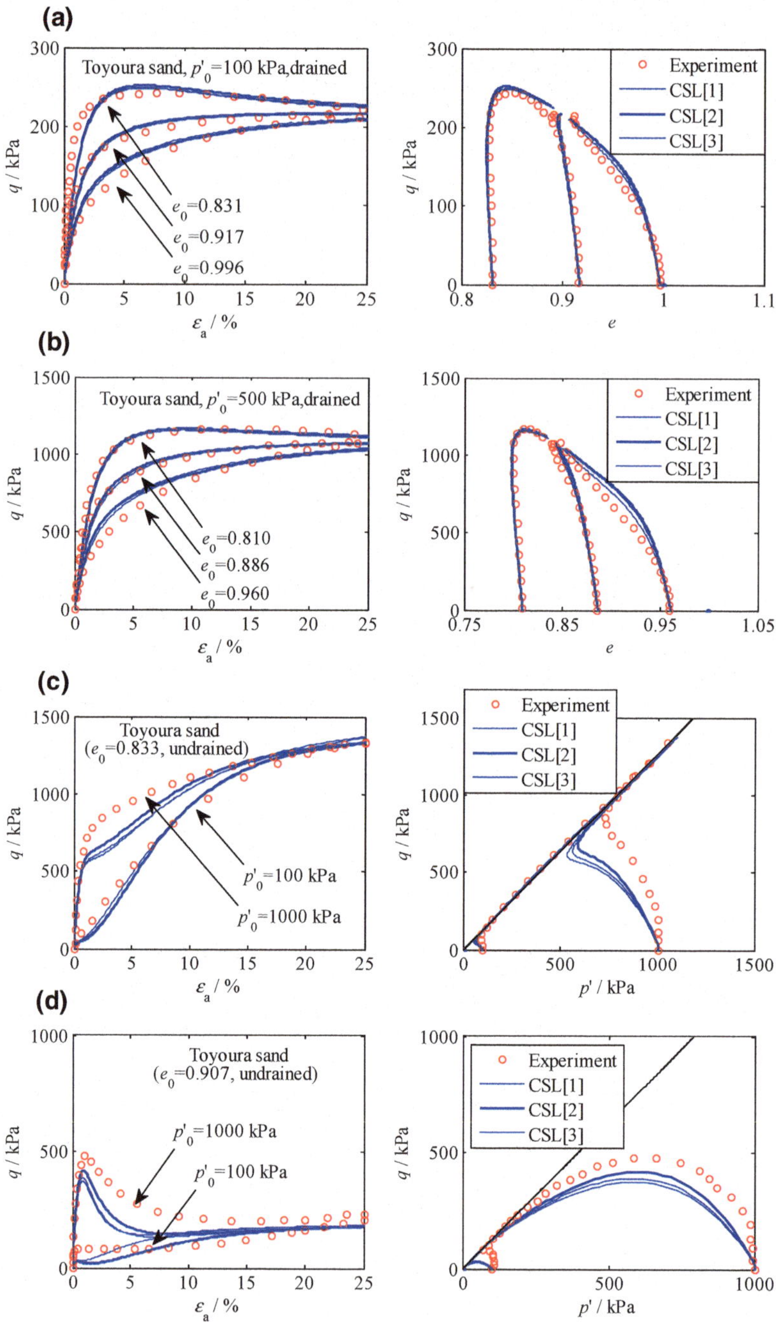

Fig. 8.22 Comparisons of experimental data and simulations for triaxial compression tests on Toyoura sand

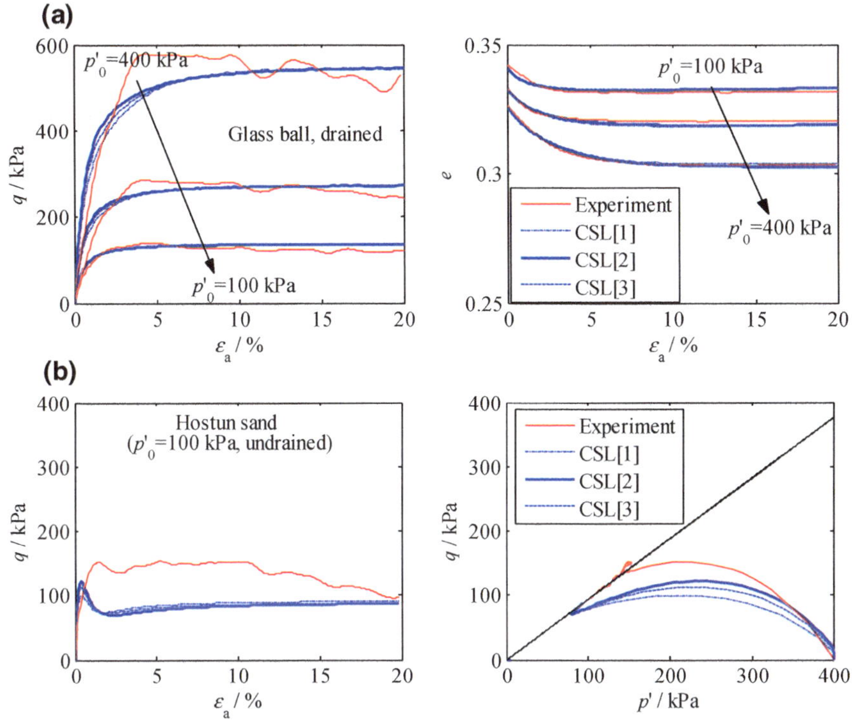

Fig. 8.23 Comparisons of experimental data and simulations for triaxial compression tests on glass ball

undrained tests were simulated; for glass ball, only the objective tests were simulated owing to no available additional test data; for Alaskan sand, four drained tests and five undrained tests were simulated. The difference between the experiments and simulations was calculated during simulation process noted as "simulation error", and shown in Table 8.13. For four selected materials, the comparisons of simulation errors for all three CSLs are presented in Fig. 8.20. It can be seen that the simulation performance by optimal parameters using the model with CSL[2] is the best for glass ball and Toyoura sand, and the model with CSL[3] is the best for Hostun sand and Alaskan sand based on objective tests. However, the difference of average error between CSL[2] and CSL[3] is slight for Hostun sand and Alaskan sand. This demonstrates that the formula of CSL[2] could more accurately describe the behavior of critical state than the other two CSLs under the stress level before particle crushing. Figures 8.21, 8.22, 8.23 and 8.24 show the comparisons between simulated and experimental results on all other additional tests on Hostun sand, Toyoura sand, glass ball, and Alaskan sand, respectively. It is obvious that the difference of simulated results for different CSLs will be enlarged when the soil undergoes a very low stress level, as shown in undrained tests on various very loose samples of Alaskan sand in Fig. 8.24. The comparisons of objective and simulation errors for three models

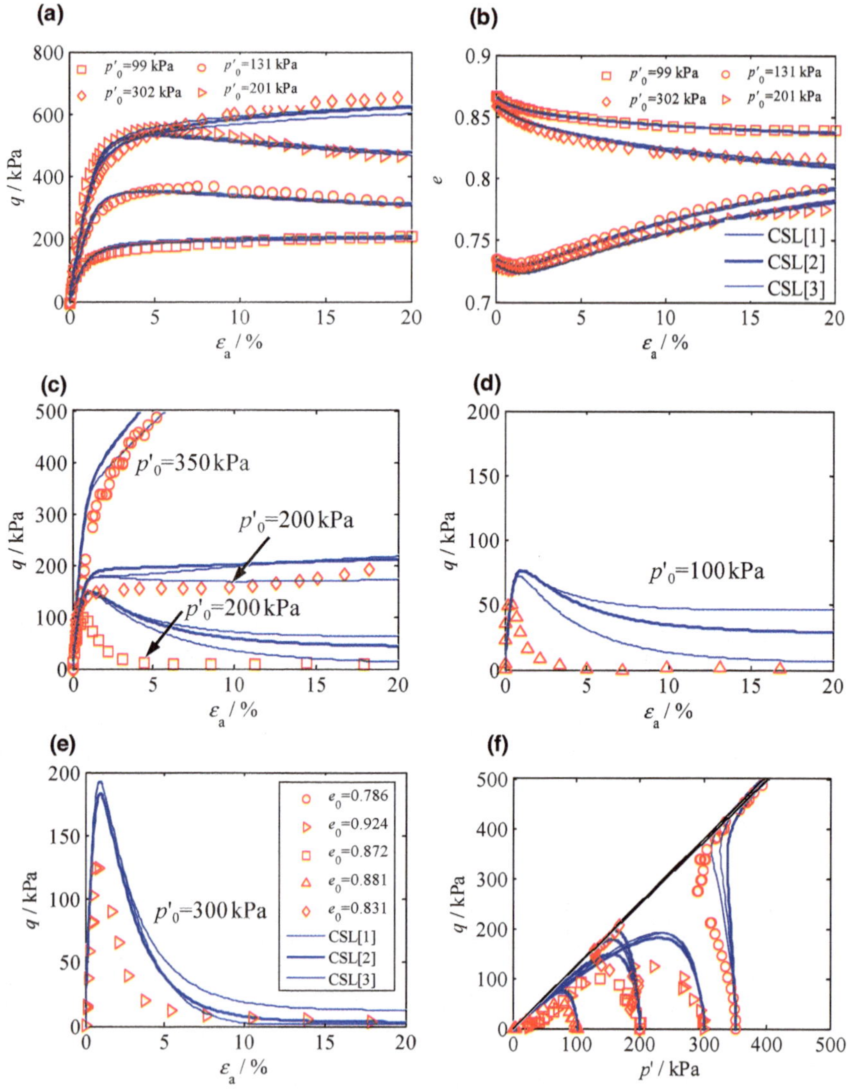

Fig. 8.24 Comparisons of experimental data and simulations for triaxial compression tests on Alaskan sand

with three different CSLs are shown in Fig. 8.25 marked by "e/e_c". It reveals the best performance obtained by the model with CSL[2].

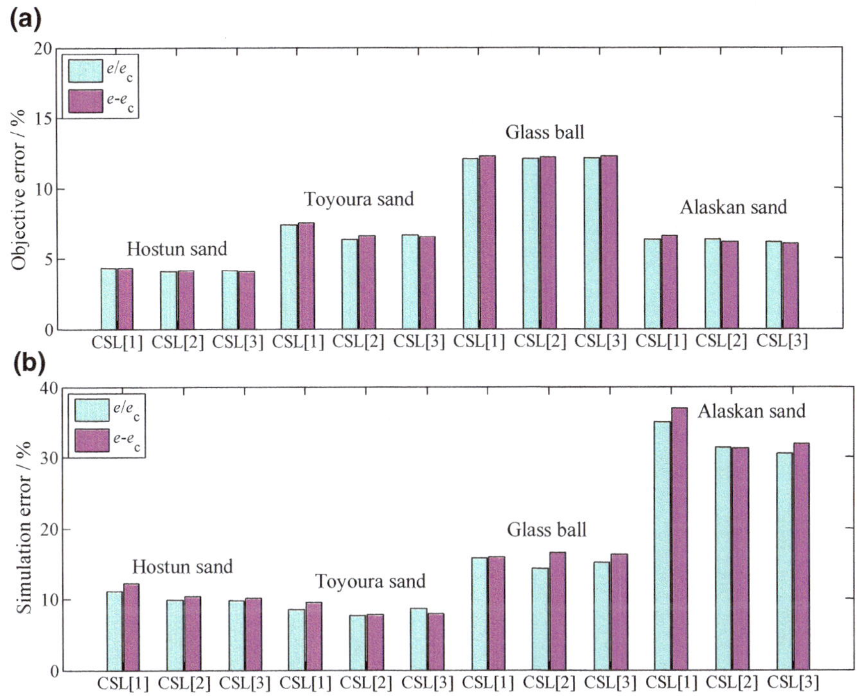

Fig. 8.25 Comparisons of errors between e/e_c and $e - e_c$ for different CSLs **a** objective error; **b** average simulation error

8.7.4 Estimation of State Parameter and Interlocking Formulation

Except considering the selection of CSL in the constitutive modeling, the interlocking effects between different state parameters are also worth to be estimated.

To comprehensively compare the two different state parameters, each state parameter with its interlocking formulation combining with three CSLs introduced previously was examined. In the previous section, the state parameters e/e_c have been examined along with evaluating the performance of three different CSLs. Thus, only the simulations by using the state parameter $e - e_c$ with three different CSLs were performed and compared to those by e/e_c. Same materials and same objective tests were again used and simulated by three models. The same optimization procedure was conducted.

The optimal parameters with objective errors are summarized in Table 8.14. For the state parameter $e - e_c$, the difference between optimal parameters for different types of CSLs is slight. In order to further assess the second state parameter, same additional tests of four materials used in the previous section were again used and simulated by using the optimal parameters in Table 8.14. The difference between

Table 8.14 Optimization parameters and error based on critical-state-based model with $e - e_c$

Materials	CSL Type	CSL			Optimal parameters					Objective error/%	Simulation error/%
		e_{ref}	λ	ξ	ϕ_μ	k_p	A_d	n_p	n_d		
Hostun sand	[1]	0.734	0.0215	–	29.5	0.0035	0.5	3.9	7.9	4.32	12.20
	[2]	0.881	0.143	0.156	29.5	0.0029	0.7	4.3	5.3	4.12	10.33
	[3]	0.844	11.53	0.279	29.5	0.0032	0.6	4.2	6.7	4.07	10.09
Toyoura sand	[1]	0.923	0.0363	–	31.5	0.0044	0.9	2.4	2.5	7.53	9.56
	[2]	0.977	0.0596	0.365	31.5	0.0041	0.7	2.2	2.9	6.59	7.75
	[3]	0.977	64.19	0.428	31.5	0.0046	0.7	2.4	3.2	6.54	7.97
Glass ball	[1]	0.341	0.021	–	24.0	0.0018	1.2	10.0	9.9	12.22	15.94
	[2]	0.432	0.0919	0.194	23.5	0.0014	1.4	10.0	8.2	12.20	16.53
	[3]	0.432	32.69	0.255	24.0	0.0016	1.2	10.0	10.0	12.26	16.28
Alaskan sand	[1]	0.85	0.0311	–	31.4	0.0022	0.4	2.0	4.8	6.59	36.88
	[2]	0.964	0.118	0.205	31.3	0.0015	0.5	1.7	3.5	6.19	31.28
	[3]	0.883	50.6	0.51	31.2	0.0014	0.4	1.8	5.1	6.04	31.83

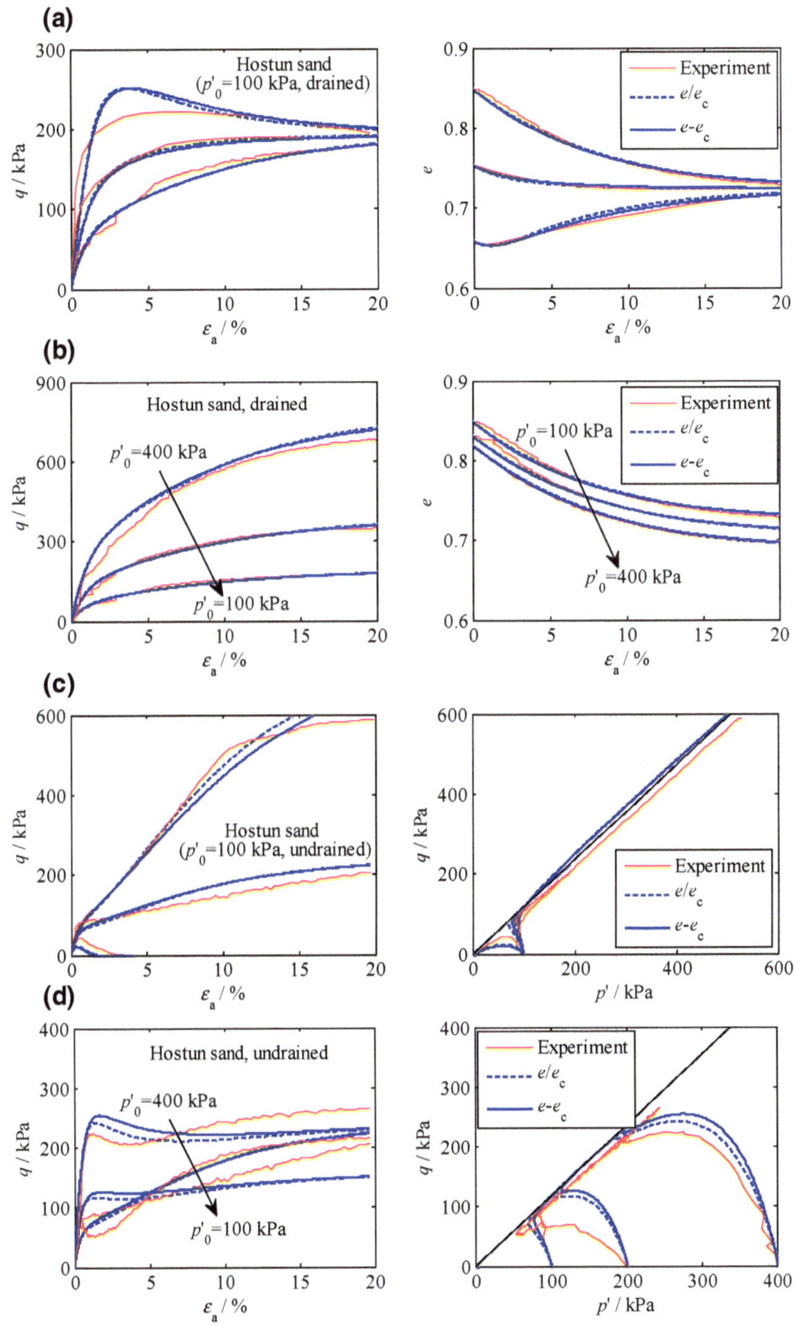

Fig. 8.26 Comparisons of experimental data and simulations for triaxial compression tests on Hostun sand

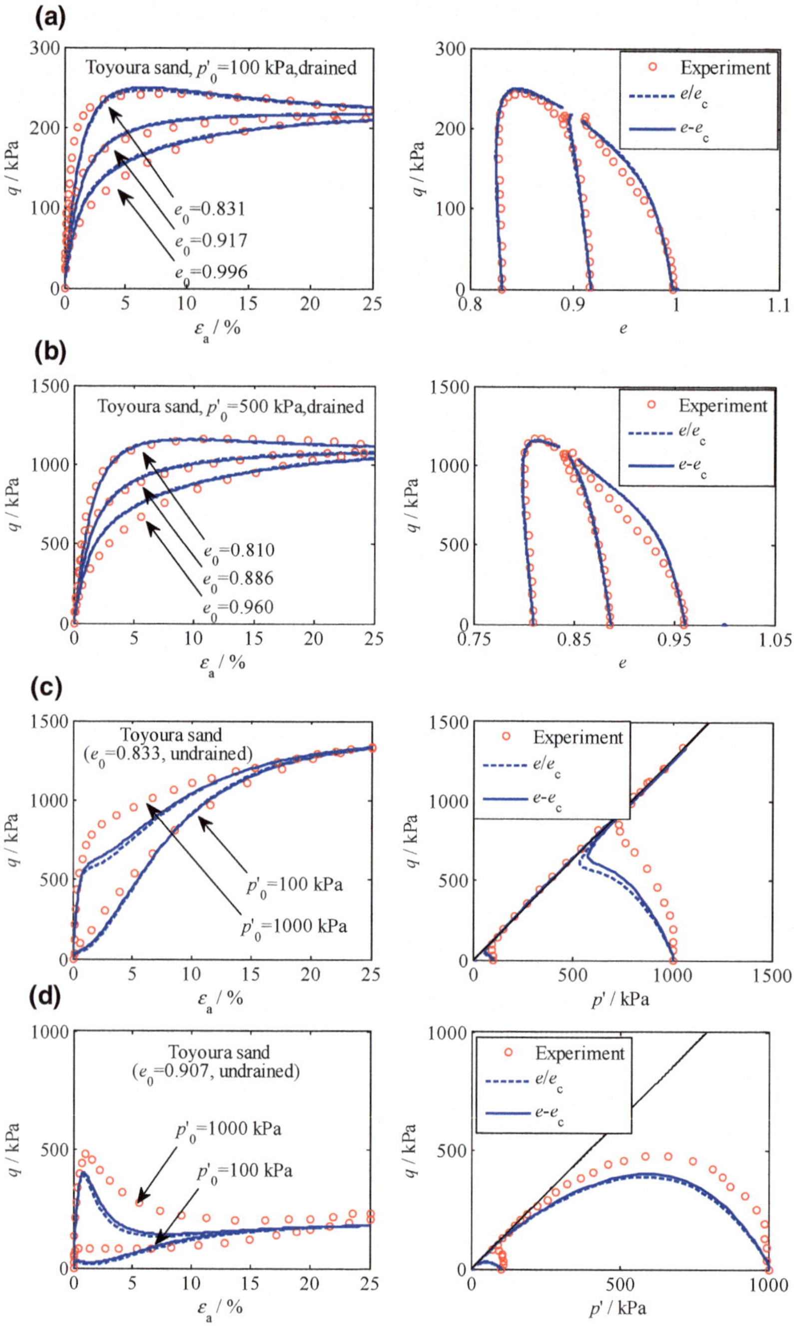

Fig. 8.27 Comparisons of experimental data and simulations for triaxial compression tests on Toyoura sand

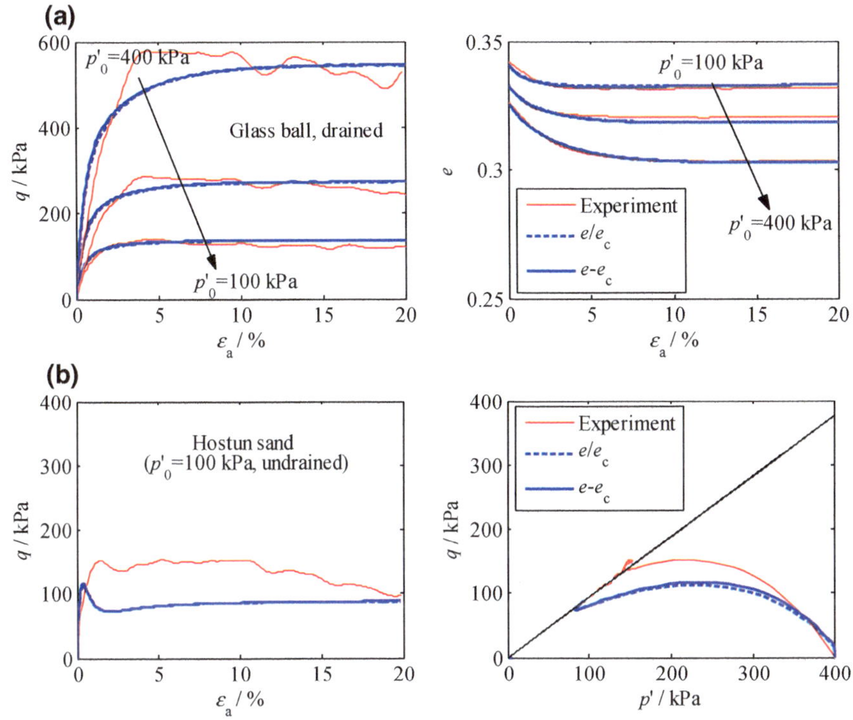

Fig. 8.28 Comparisons of experimental data and simulations for triaxial compression tests on glass ball

three models was measured by the error function, and the values are also summarized in Table 8.14. The comparisons of objective and simulation errors for three different CSLs are shown in Fig. 8.25. It can be seen that same performance of simulations can be achieved by different models with e/e_c or $e - e_c$ based on only the objective error. However, based on the simulation error from all additional tests, a better performance of simulations by adopting e/e_c is obtained for four materials. It demonstrates that the state parameter e/e_c with its formulation has a better ability in expressing the interlocking effect during the constitutive modeling.

Then, the generally best combination was estimated as the model using CSL[2] and "e/e_c". Figures 8.26, 8.27, 8.28, and 8.29 show all comparisons between experiments and simulations by using the best estimated model (with CSL[2] and "e/e_c") for Hostun sand, Toyoura sand, glass ball, and Alaskan sand, respectively, which is in fact slightly better than the model with CSL[2] and "$e - e_c$" also shown in these figures.

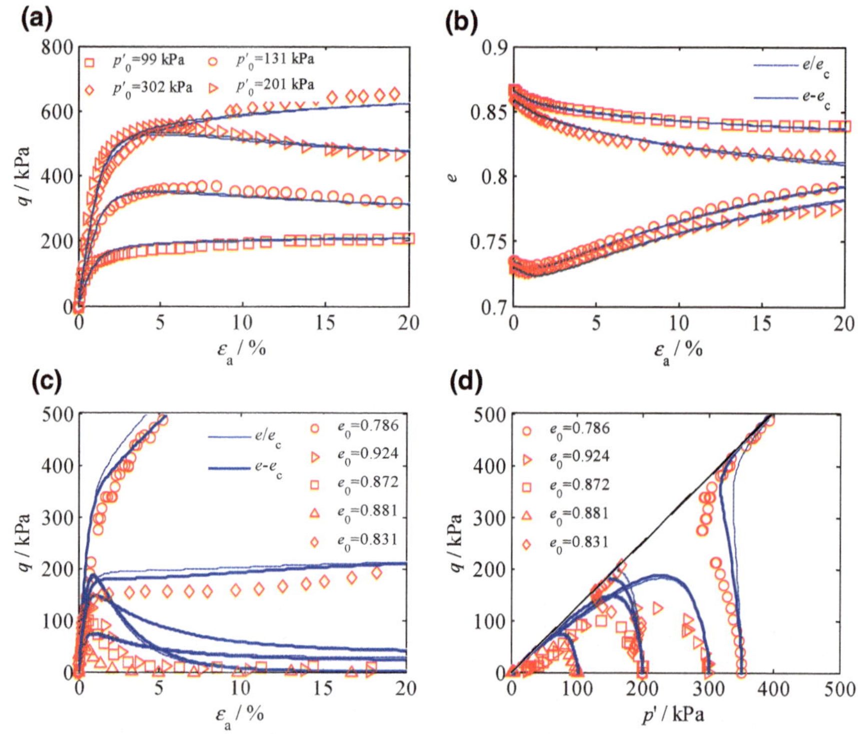

Fig. 8.29 Comparisons of experimental data and simulations for triaxial compression tests on Alaskan sand

8.8 Evaluation of Model's Performance

8.8.1 Evaluation by Information Criteria

It should be noticed that each CSL has different number of parameters. The evaluation of CSL should consider the number of parameters besides model predictive performance. There are two widely used criteria which can be adopted: (a) Akaike's information criterion and (b) the Schwartz's Bayesian information criterion to assess the performance of a model with respect to how well it explains the data.

Although these two terms address model selection, they are not the same. One can come across any difference between the two approaches of model selection. The AIC can be termed as a measure of the goodness of fit of any estimated statistical model. The BIC is a type of model selection among a class of parametric models with different numbers of parameters. For two criteria, the big difference is the penalty for additional parameters. Unlike the AIC, the BIC penalizes free parameters more strongly. Additionally, the AIC will present the danger that it would outfit and the BIC will present the danger that it would underfit. Therefore, two criteria were adopted together for selecting the best CSL.

Akaike's information criterion (AIC) by Akaike [33] provides a measure of model quality obtained by simulating the situation where the model is tested on a different data sets, which can be expressed as,

$$\mathrm{AIC} = -2\log L(\hat{\theta}) + 2k \tag{8.17}$$

where θ is set (vector) of model parameters; $L(\hat{\theta})$ is likelihood of candidate model given the data when evaluated at the maximum likelihood estimate of θ; k is number of estimated parameters in the candidate model.

The AIC in isolation is meaningless. Rather, this value is calculated for every candidate model and the "best" model is the candidate model with the smallest AIC. In this case, the AIC can be expressed equivalently,

$$\mathrm{AIC} = n \cdot \log(\mathrm{RSS}/n) + 2k \tag{8.18}$$

where RSS is residual sum of squares, $\mathrm{RSS} = \sum_{i=1}^{n}\left(U_{\exp}^{i} - U_{\mathrm{num}}^{i}\right)^2$; n is the number of values in the estimation data set.

Similar to AIC, the BIC is computed as follows according to Schwarz [34],

$$\mathrm{BIC} = n \cdot \log(\mathrm{RSS}/n) + k \cdot \log(n) \tag{8.19}$$

Then, the best model is the one that provides the minimum values of AIC and BIC. For further selecting the most "appropriate" CSL, the values of AIC and BIC corresponding to each validation tests (both drained and undrained) for each CSL were calculated. Note that for each test curve, the AIC and BIC were, respectively, computed on both deviatoric stress–axial strain curve (marked as "AIC_*q*" and "BIC_*q*") and void ratio–axial strain curve of drained test (marked as "AIC_*e*" and "BIC_*e*") or excess pore pressure vs. axial strain curve of undrained test (marked as "AIC_*u*" and "BIC_*u*"). All the computed results are summarized in Table 8.15. For each calculation, a relatively better CSL with smaller AIC or BIC was selected and saved using the CSL number (CSL[1], CSL[2], and CSL[3]) for four materials. Based on this, the most "appropriate" CSL is CSL[1] due to less input parameters, followed by CSL[2], although the difference of simulations is slight among three CSLs in terms of laboratory test simulations.

Similarly, the AIC and BIC of each validation test on different materials for three CSLs with "e/e_c" and "$e - e_c$" were computed. All the results are also summarized in Table 8.15. The calculation performance of each model was compared, and the better one corresponding to a smaller value of AIC or BIC is recorded using the CSL number. For CSL[1], the recorded times of "better one" for "e/e_c" is 13 and for "$e - e_c$" is 11 among 32 calculations, which demonstrates the model with "e/e_c" is slightly better than the model with "$e - e_c$". For CSL[2], the recorded time of "better one"

Table 8.15 Values of AIC and BIC for three critical-state-based models with e/e_c or $e - e_c$ from laboratory tests

Test types		e/e_c				$e - e_c$			
		CSL[1]	CSL[2]	CSL[3]	Best CSL	CSL[1]	CSL[2]	CSL[3]	Best CSL
		Hostun sand							
Drained	AIC_*q*	2694.1	2674.7	2694.5	[2]	2685.6	2689.1	2682.3	[3]
	AIC_*e*	−1520.5	−1524.4	−1521.4	[2]	−1437.5	−1533.6	−1547.0	[3]
	BIC_*q*	1363.7	1344.4	1364.2	[2]	1351.7	1358.7	1351.9	[1]
	BIC_*e*	−2850.8	−2854.7	−2851.8	[2]	−2771.3	−2864.0	- 2877.4	[3]
Undrained	AIC_*q*	2759.1	2708.2	2731.0	[2]	2732.4	2703.4	2708.3	[2]
	AIC_*u*	2714.1	2609.9	2608.8	[3]	2721.1	2606.4	2604.0	[3]
	BIC_*q*	1605.8	1558.1	1581.0	[2]	1579.1	1553.4	1558.2	[2]
	BIC_*u*	1560.8	1459.8	1458.7	[3]	1567.8	1456.3	1453.9	[3]
Toyoura sand									
Drained	AIC_*q*	2813.1	2817.0	2854.5	[1]	2871.9	2830.1	2823.7	[3]
	AIC_*e*	−1152.5	−1040.0	−1136.8	[1]	−1273.4	−1042.9	−1088.8	[1]
	BIC_*q*	1495.1	1502.5	1540.0	[1]	1553.9	1515.7	1509.2	[3]
	BIC_*e*	−2470.5	−2354.5	−2451.3	[1]	−2591.4	−2357.4	−2403.3	[1]
Undrained	AIC_*q*	3488.4	3487.2	3572.6	[2]	3583.6	3478.2	3491.9	[2]
	AIC_*u*	3290.0	3243.9	3326.2	[2]	3356.0	3212.7	3247.6	[2]
	BIC_*q*	2140.3	2142.6	2228.1	[1]	2235.5	2133.7	2147.3	[2]

(continued)

Table 8.15 (continued)

Test types		e/e_c				$e - e_c$			
		CSL[1]	CSL[2]	CSL[3]	Best CSL	CSL[1]	CSL[2]	CSL[3]	Best CSL
		Hostun sand							
	BIC_*u*	1941.9	1899.3	1981.6	[2]	2007.9	1868.2	1903.0	[2]
Glass ball									
Drained	AIC_*q*	3138.3	3328.9	3172.2	[1]	3160.4	3490.8	3147.9	[3]
	AIC_*e*	−1836.6	−1758.6	−1862.7	[2]	−1815.8	−1641.9	−1799.8	[1]
	BIC_*q*	1822.4	2016.6	1859.8	[1]	1844.6	2178.5	1835.6	[3]
	BIC_*e*	−3152.5	−3070.9	−3175.0	[3]	−3131.6	−2954.3	−3112.2	[1]
Undrained	AIC_*q*	2637.9	2674.8	2659.2	[1]	2637.1	2675.3	2656.8	[1]
	AIC_*u*	2582.6	2600.5	2593.6	[1]	2587.0	2602.5	2596.8	[1]
	BIC_*q*	1620.8	1660.9	1645.4	[1]	1620.0	1661.4	1643.0	[1]
	BIC_*u*	1565.5	1586.7	1579.7	[1]	1569.8	1588.6	1582.9	[1]
Alaskan sand									
Drained	AIC_*q*	−1252.3	−1296.3	−1241.6	[2]	−1268.9	−1242.5	−1249.8	[1]
	AIC_*e*	−1239.9	−1225.2	−1178.9	[1]	−1185.0	−1229.7	−1270.9	[2]
	BIC_*q*	−1219.2	−1259.9	−1201.8	[2]	−1235.8	−1206.1	−1210.1	[1]
	BIC_*e*	−1206.8	−1188.8	−1139.2	[1]	−1151.9	−1193.3	−1231.2	[3]
Undrained	AIC_*q*	1203.8	1184.6	1130.3	[3]	1220.3	1150.9	1143.0	[3]
	AIC_*u*	891.4	935.7	876.2	[3]	936.6	871.7	894.1	[2]
	BIC_*q*	1233.6	1217.3	1166.0	[3]	1250.0	1183.6	1178.7	[3]
	BIC_*u*	921.1	968.4	911.9	[3]	966.4	904.4	929.8	[2]
Summary		13*[1], 12*[2], 7*[3]				11*[1], 9*[2], 12*[3]			

for "e/e_c" is 12 and for "$e - e_c$" is 9 among 32 calculations, which demonstrates the model with "e/e_c" has same performance comparing to the model with "$e - e_c$". For CSL[3], the recorded times of "better one" for "e/e_c" is 7 and for "$e - e_c$" is 12 among 32 calculations, demonstrating that the model with "e/e_c" is worse than the model with "$e - e_c$".

Overall, for all calculations with "e/e_c" and with "$e - e_c$", the total times of "better one" for CSL[1], CSL[2], and CSL[3] are 24, 21, and 19, respectively. All the summarized data demonstrate that the model implementing CSL[1] with less parameters is the most "appropriate" sand model, followed by the model with CSL[2]. However, the selection of CSL based on the performance of simulating laboratory tests is not enough to give a final decision. Therefore, more estimations of CSL on different tests are required to be conducted.

8.8.2 Evaluation by Modeling of Footings

For further evaluating the performance of different CSLs and different state parameters, a series of footing tests performed on Toyoura sand were simulated by different combined models with three CSLs and two state parameters. The calculated load–settlement ("p–s") curves of footings by different combined models were compared with measurements for evaluating the "appropriate" CSL and state parameter.

8.8.2.1 Description of Footing Tests

A series of circular footing model tests on Toyoura sand were performed by Tomita et al. [35]. The footing has a circular cross section with a diameter of $B = 20$ mm and a height of 80 mm. The soil tank has a size of 400 × 550 × 110 mm ($l = 5B$). Three footing tests were performed on Toyoura sand with three different relative densities ($e_0 = 0.85, 0.71, 0.67$).

8.8.2.2 Model Implementation and Finite Element Simulations

For simulating the footing tests, the ABAQUS/Explicit with employing arbitrary Lagrangian–Eulerian (ALE) method was adopted to deal with the localization problem with large deformation in edge of footing. The combined models were implemented into ABAQUS as a user-defined material model via VUMAT to model the Toyoura sand. The model implementation follows the way of Hibbitt et al. [36]. In ABAQUS/Explicit combining with VUMAT, the increment of displacement in node Δu and the increment of strain on element $\Delta\varepsilon$ at time t is solved by ABAQUS using the time-explicit integration method. Then, the increment of stress $\Delta\sigma$ was updated through VUMAT using the $\Delta\varepsilon$ solved by ABAQUS (described as stress integration in the next section). Subsequently, the updated stress $\sigma_1(t + \Delta t)$ will be transited

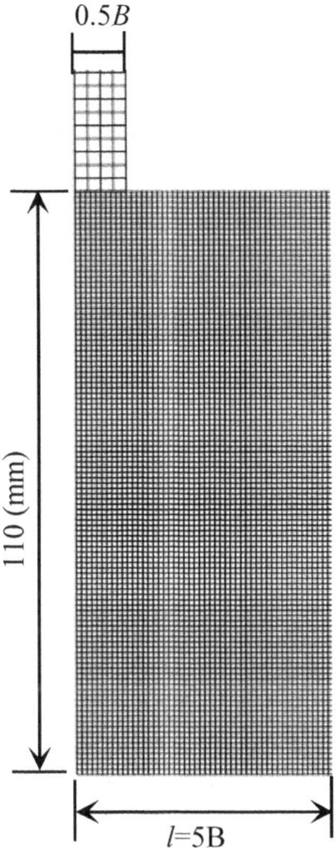

Fig. 8.30 Finite element model of footing tests in ABAQUS

to ABAQUS for calculating the increment of strain at next time step. The above procedure will be looped to the end of loading time.

The axisymmetric finite element model with 5628 elements was created, shown in Fig. 8.30. The dimension of the FE model is same as model tests by Tomita et al. [35]. The soil is modeled by using four-node axisymmetric element with one reduced integration point (named as CAX4R element in ABAQUS). Compared to soil, the deformation of footing structure is negligible. Therefore, the footing structure was modeled using rigid body.

According to Tomita et al. [35], the initial stress condition before loading is K_0 condition, and the value of K_0 was set to be 0.48 according to Jacky's formula. The contact between footing structure and soil is surface-to-surface with its interface described by the classical Coulomb friction law (the friction coefficient $\mu = \tan(\phi_\mu/2) = 0.28$). The parameters of Toyoura sand optimized previously were adopted here for all simulations.

8.8.2.3 Results and Comparisons

Figure 8.31 shows the comparisons of "*p*–*s*" curves between measurements and simulations for footing tests. The simulated mean effective stress field at the end of loading displacement by different combined models was selected because of its reflection to the bearing capacity and the critical state void ratio. Only simulations by models with e/e_c were plotted in Fig. 8.32, since choosing the two state parameters gives slight difference of simulations. For simulations with different CSLs and "e/e_c", the predictions of bearing capacity by the model with CSL[2] on different densities of Toyoura sand are in good agreement with the measurements. Meanwhile, the bearing capacity of footing is highly overestimated by the model with CSL[1] and slightly underestimated by the model with CSL[3]. In order to investigate why big difference is occurred among results with different CSLs, a representative Gauss point with the biggest mean effective stress under the footing was selected to show the evolution of stress development during vertical loading. The selected Gauss point for CSL[1] was "A", for CSL[2] was "B", and for CSL[3] was "C", shown in Fig. 8.32. The relationships of "p'-q", "p'-e", "p'-e_c" and "s-e_c/e" of these points for each simulation are plotted in Fig. 8.33. It can be seen that the ratio between current e and e_c corresponding to current stress for the CSL[1] is big when the mean effective stress is less than 100 kPa, which results in a big peak strength of soil and thus a higher simulated bearing capacity than others. However, for simulations with CSL[2] and CSL[3], the ratio e/e_c is relatively small, which leads to a small peak strength and a lower simulated bearing capacity. Note that the maximum mean effective stress in Toyoura sand for different footing tests was smaller than 300 kPa (see Fig. 8.32) which is far from the stress level of grain crushing for quartz sand. Thus, the difference of bearing capacity for different simulations is only caused by the difference of CSLs. The similar evolutions of p'-e and p'-e_c for CSL[2] and CSL[3] imply that the simulated results should be almost the same, which has been confirmed by the similar "*p*–*s*" curves for the simulations with different initial void ratios. Comparing to measurements, more reasonable predictions of "*p*–*s*" curve were achieved by CSL[2] and CSL[3] than CSL[1].

For simulations by different CSLs and "$e - e_c$", the performance of the model with CSL[3] is better than the model with CSL[1] and slightly better than the model with CSL[2]. For each CSL, all predictions by the models with "$e - e_c$" are slightly larger than that by the models with "e/e_c". The values of AIC and BIC for each simulation were also calculated for different models, summarized in Table 8.16. Overall, the model with CSL[2] and "e/e_c" can give a more precise accuracy and more reasonable predictions.

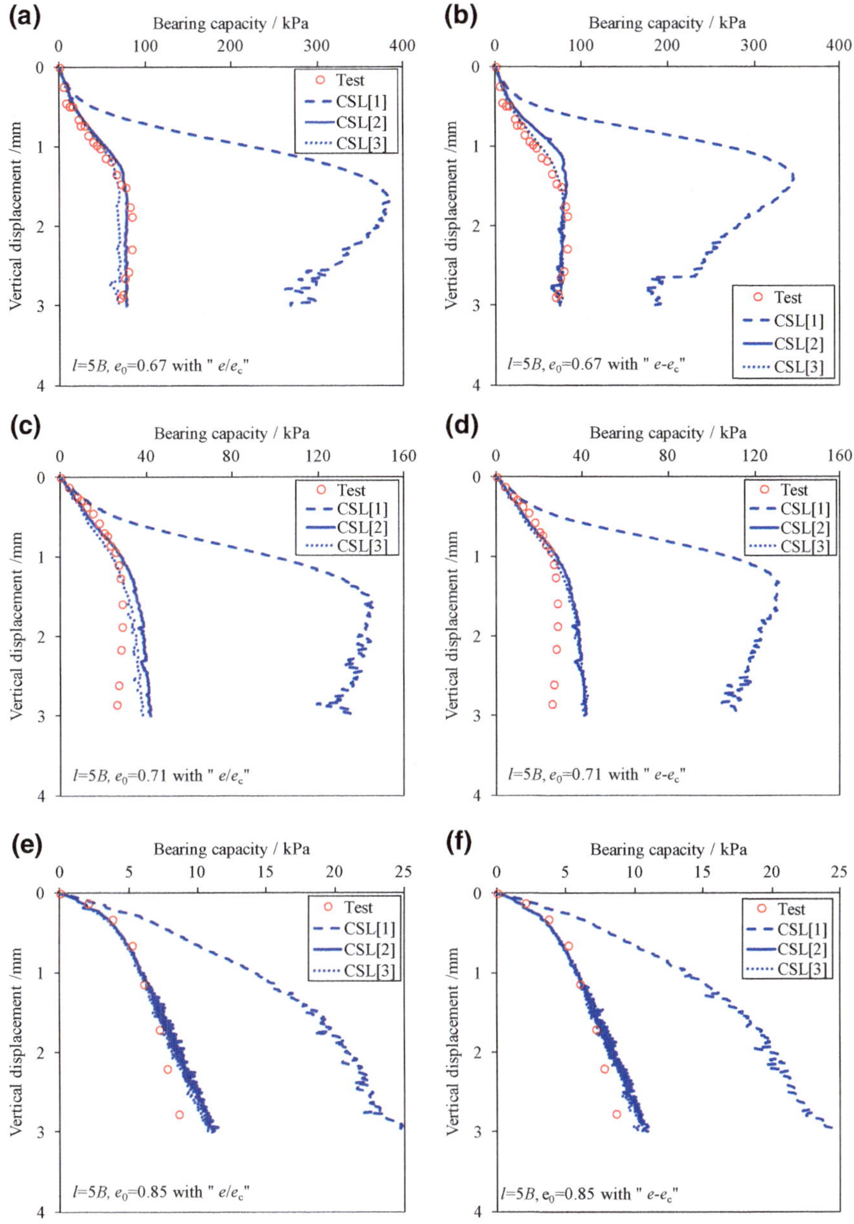

Fig. 8.31 Comparison between measurements and simulations for "*p*–*s*" curve of footing tests

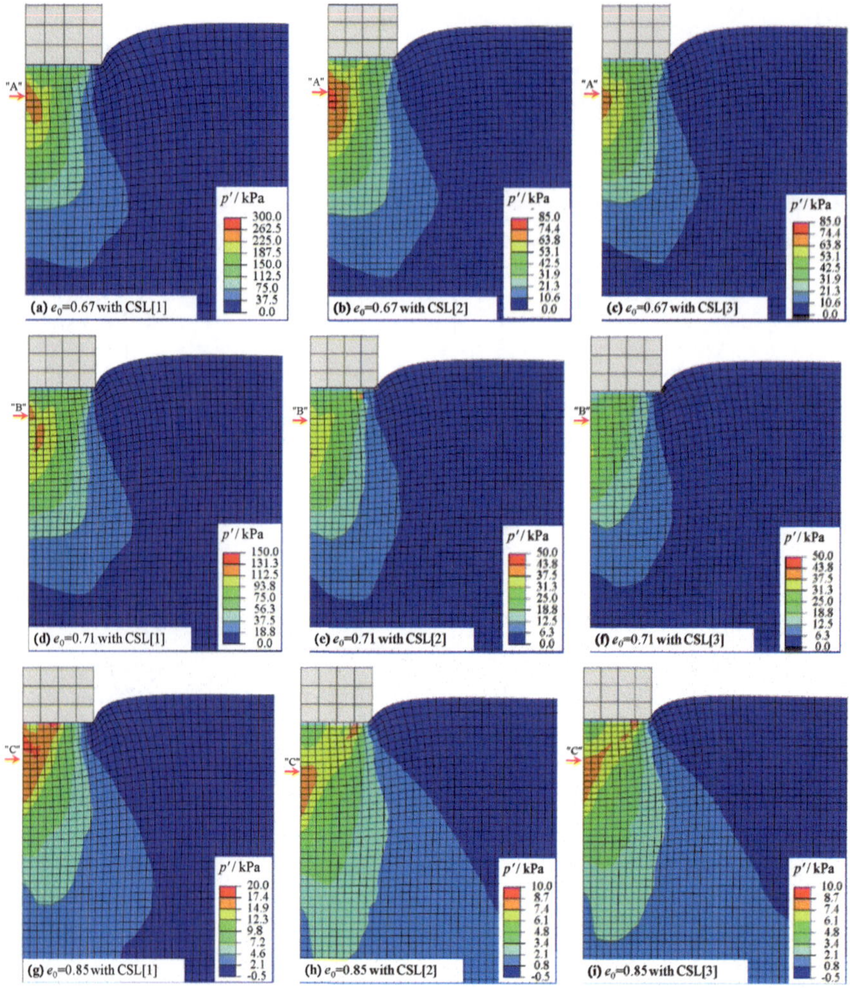

Fig. 8.32 Simulated mean effective stress field of footings by different models: **a–c** for $e_0 = 0.67$ with CSL[1], CSL[2], and CSL[3]; **d–f** for $e_0 = 0.71$ with CSL[1], CSL[2], and CSL[3]; **g–i** for $e_0 = 0.85$ with CSL[1], CSL[2], and CSL[3]

Above all, comparisons varying from laboratory tests to engineering practice demonstrate that the CSL[2] and "e/e_c" are the most "appropriate" combination elements in constitutive modeling of granular material.

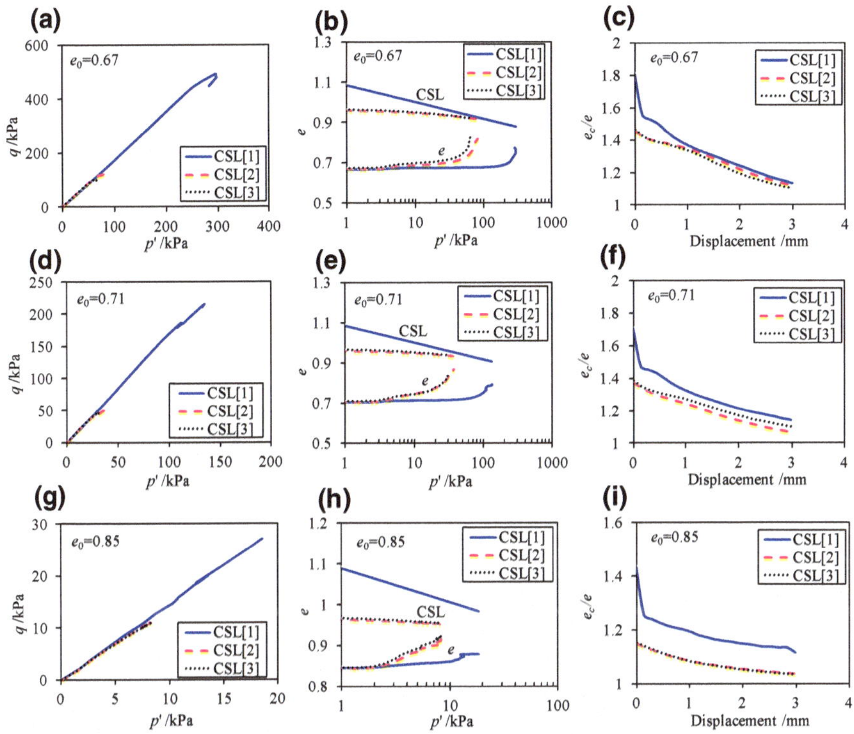

Fig. 8.33 Relationships of "p'-q", "p'-e", "p'-e_c", and "s-e_c/e" on representative Gauss point for simulations with different CSLs: **a–c** $e_0 = 0.67$; **d–f** $e_0 = 0.71$; **g–i** $e_0 = 0.85$

Table 8.16 Values of AIC and BIC for three critical-state-based models with e/e_c or $e - e_c$ from footing analyses

Footings			CSL[1]	CSL[2]	CSL[3]	Better CSL
$e_0 = 0.67$	AIC	e/e_c	163.12	94.67	99.26	[2]
		$e - e_c$	160.99	109.16	93.82	[3]
	BIC	e/e_c	123.79	54.73	58.69	[2]
		$e - e_c$	121.66	69.22	53.26	[3]
$e_0 = 0.71$	AIC	e/e_c	115.10	76.65	77.38	[2]
		$e - e_c$	113.38	79.71	81.01	[2]
	BIC	e/e_c	83.59	47.42	44.43	[3]
		$e - e_c$	81.87	47.48	48.06	[2]
$e_0 = 0.85$	AIC	e/e_c	47.39	33.89	34.16	[2]
		$e - e_c$	46.94	30.16	32.22	[2]
	BIC	e/e_c	28.35	13.05	13.02	[2]
		$e - e_c$	27.89	12.06	9.09	[3]

8.9 Conclusions

The selection of sand model and the parameter identification by genetic algorithm have been discussed in this chapter. Conventional triaxial tests on Hostun sand were selected as the objective in the optimization. Firstly, the determination of which features are required to be included in constitutive modeling of sand was discussed. Four models with gradually differing features were chosen from numerous sand models as examples for optimization. The results demonstrate that the appropriate sand model should incorporate nonlinear plastic stress–strain hardening, and the critical state concept with an interlocking effect. As a result, the critical-state-based models (CS-NLMC and CS-TS) were recommended. For the simplicity of modeling monotonic behavior, the CS-NLMC was selected for further study.

Then, the type of tests (drained and/or undrained) to be selected for parameter identification was discussed. It was found that the objective consisting of the drained test and the undrained test could result in relatively accurate optimal parameters in the optimization. Based on the criterion of least cost, two drained tests and one undrained tests were found to satisfy the requirement of obtaining the optimal parameters. In addition, the accuracy of optimal parameters would increase with the increasing number of tests in the objective. However, attention needs to be paid to the test combinations to avoid close final states of e, p' which may cause an incorrect determination of the CSL.

Thirdly, the minimum number of objective tests for identifying parameters was estimated. Optimizations based on two possibilities of adding tests were conducted. Comparisons between simulation and experiment demonstrate that five tests in the objective could give a good performance of parameter identification by genetic algorithm.

Finally, the smaller suitable strain level for identifying parameters was evaluated. Optimizations based on objectives with different strain levels were conducted. Comparisons between simulation and experiment suggest that five drained tests should be selected as the objective, and tests with a strain level of 20% can give relatively reliable and reasonable parameters by optimization.

The selection of critical state line and state parameter with its interlocking formulation by genetic algorithm has also been discussed. Six simple critical-state-based models with three different formulas of CSL and two different state parameters were formulated and adopted. Four different types of materials were selected as the typical example to evaluate the critical state line and state parameter with its interlocking formula. Conventional triaxial tests of four selected materials were selected as objective in the optimization.

Three different types of critical state line were first examined on four selected materials keeping the state parameter "e/e_c". The comparisons between simulations and experiments demonstrate that the nonlinear critical state line proposed by Li and Wang (1998) [9] can better describe the mechanical behavior of different materials. Then, the effect of different state parameters, e/e_c and $e - e_c$, on expressing the interlocking behavior was discussed. The results indicate that the state parameter

e/e_c has a better predictive ability in describing the interlocking effect. Finally, it is confirmed that the incorporating the nonlinear critical state line of Li and Wang combined with the interlocking law of e/e_c in the model can result in a relatively more satisfied simulated results.

Then, the performance of each CSL and each state parameter was first evaluated using AIC and BIC. The values of AIC and BIC were computed for all test simulations on four materials. The results indicated that all CSLs and state parameters are acceptable for constitutive modeling of laboratory tests. While the CSL[1] performed better among three CSLs due to its less number of parameters, the performance of "e/e_c" is better when using CSL[1], the performance of "e/e_c" and "$e - e_c$" is approximately the same when using CSL[2], and the performance of "$e - e_c$" is relatively better when using CSL[3].

The performance of each CSL and state parameter was further evaluated by modeling footing tests using different combined models in terms of engineering application. All comparisons between measurements and simulations demonstrate that the CSL[2] and "e/e_c" performed well among all combinations of CSLs and state parameters. Then, together with performance of models based on laboratory tests, it can be concluded that the CSL[2] and "e/e_c" can be considered as the most "appropriate" combination in constitutive modeling of granular material for both soil elementary behavior and engineering practice.

More details can be found in Jin et al. [26, 37].

References

1. Levasseur S, Malécot Y, Boulon M, Flavigny E (2008) Soil parameter identification using a genetic algorithm. Int J Numer Anal Methods Geomech 32(2):189–213
2. Liu Y-J, Li G, Yin Z-Y, Dano C, Hicher P-Y, Xia X-H et al (2014) Influence of grading on the undrained behavior of granular materials. CR Mec 342(2):85–95
3. Li G, Liu Y-J, Dano C, Hicher P-Y (2014) Grading-dependent behavior of granular materials: from discrete to continuous modeling. J Eng Mech
4. Richart F, Hall J, Woods R (1970) Vibrations of soils and foundations. In: International series in theoretical and applied mechanics. Prentice-Hall, Englewood Cliffs, NJ
5. Poles S, Rigoni E, Robic T (2004) MOGA-II performance on noisy optimization problems. In: International conference on bioinspired optimization methods and their applications BIOMA Ljubljana, Slovena (Oct 2004), pp 51–62
6. Papon A, Riou Y, Dano C, Hicher PY (2012) Single-and multi-objective genetic algorithm optimization for identifying soil parameters. Int J Numer Anal Methods Geomech 36(5):597–618
7. Manzari MT, Dafalias YF (1997) A critical state two-surface plasticity model for sands. Geotechnique 47(2):255–272
8. Li G, Liu Y, Dano C, Hicher P (2014) Grading-dependent behavior of granular materials: from discrete to continuous modeling. J Eng Mech 141(6):04014172
9. Li XS, Wang Y (1998) Linear representation of steady-state line for sand. J Geotech Geoenviron Eng 124(12):1215–1217
10. Gudehus G (1997) Attractors, percolation thresholds and phase limits of granular soils. Proc Powder Grains 97
11. Yin ZY, Chang CS (2009) Microstructural modelling of stress-dependent behaviour of clay. Int J Solids Struct 46(6):1373–1388

12. Yin Z-Y, Xu Q, Hicher P-Y (2013) A simple critical-state-based double-yield-surface model for clay behavior under complex loading. Acta Geotech 8(5):509–523
13. Yin ZY, Hattab M, Hicher PY (2011) Multiscale modeling of a sensitive marine clay. Int J Numer Anal Methods Geomech 35(15):1682–1702
14. Yin ZY, Xu Q, Chang CS (2013) Modeling cyclic behavior of clay by micromechanical approach. J Eng Mech 139(9):1305–1309
15. Biarez J, Hicher P-Y (1994) Elementary mechanics of soil behaviour: saturated remoulded soils. AA Balkema
16. Been K, Jefferies M (1985) A state parameter for sands. Geotechnique 35(2):99–112
17. Yin ZY, Chang CS, Hicher PY (2010) Micromechanical modelling for effect of inherent anisotropy on cyclic behaviour of sand. Int J Solids Struct 47(14–15):1933–1951
18. Vermeer P (1978) A double hardening model for sand. Geotechnique 28(4):413–433
19. Jefferies M (1993) Nor-sand: a simle critical state model for sand. Geotechnique 43(1):91–103
20. Gajo A, Wood M (1999) Severn-trent sand: a kinematic-hardening constitutive model: the q–p formulation. Geotechnique 49(5):595–614
21. Yin ZY, Chang CS (2013) Stress–dilatancy behavior for sand under loading and unloading conditions. Int J Numer Anal Methods Geomech 37(8):855–870
22. Yao Y, Hou W, Zhou A (2009) UH model: three-dimensional unified hardening model for overconsolidated clays. Geotechnique 59(5):451–469
23. Yao Y, Lu D, Zhou A, Zou B (2004) Generalized non-linear strength theory and transformed stress space. Sci China Ser E Technol Sci 47(6):691–709
24. Yao Y, Sun D, Luo T (2004) A critical state model for sands dependent on stress and density. Int J Numer Anal Methods Geomech 28(4):323–337
25. Yao Y, Sun D, Matsuoka H (2008) A unified constitutive model for both clay and sand with hardening parameter independent on stress path. Comput Geotech 35(2):210–222
26. Jin Y-F, Yin Z-Y, Shen S-L, Hicher P-Y (2016) Selection of sand models and identification of parameters using an enhanced genetic algorithm. Int J Numer Anal Methods Geomech 40(8):1219–1240
27. Jin Y-F, Yin Z-Y, Shen S-L, Hicher P-Y (2016) Investigation into MOGA for identifying parameters of a critical-state-based sand model and parameters correlation by factor analysis. Acta Geotech 11(5):1131–1145
28. Verdugo R, Ishihara K (1996) The steady state of sandy soils. Soils Found 36(2):81–91
29. Jefferies M, Been K (2006) Soil liquefaction: a critical state approach. CRC Press
30. Lade PV, Bopp PA (2005) Relative density effects on drained sand behavior at high pressures. Soils Found 45(1):1–13
31. Hu W, Yin ZY, Dano C, Hicher PY (2011) A constitutive model for granular materials considering grain breakage. Sci China-Technol Sci 54(8):2188–2196
32. Yin Z-Y, Hicher P-Y, Dano C, Jin Y-F (2016) Modeling mechanical behavior of very coarse granular materials. J Eng Mech C4016006
33. Akaike H (1998) Information theory and an extension of the maximum likelihood principle. Springer, Selected Papers of Hirotugu Akaike, pp 199–213
34. Schwarz G (1978) Estimating the dimension of a model. Ann Stat 6(2):461–464
35. Tomita Y, Nishigata T, Masui T, Yao S (2012) Load settlement relationships of circular footings considering dilatancy characteristics of sand. Int J GEOMATE Geotech Constr Mater Environ 2(1):148–153
36. Hibbitt, Karlsson, Sorensen (2001) ABAQUS/Explicit: User's Manual. Hibbitt, Karlsson and Sorenson
37. Jin Y-F, Wu Z-X, Yin Z-Y, Shen JS (2017) Estimation of critical state-related formula in advanced constitutive modeling of granular material. Acta Geotech 12(6):1329–1351

Chapter 9
Multi-objective Optimization-Based Updating of Predictions During Excavation

9.1 Introduction

In this chapter, an efficient multi-objective optimization (MOOP)-based updating framework is established, which involves (1) the development of an enhanced multi-objective differential evolution algorithm with good searching ability and high convergence speed, (2) the development of an enhanced anisotropic elastoplastic model considering small-strain stiffness with its implementation into a finite element code, and (3) the proposal of an identification procedure for parameters using field measurements followed by an updating procedure. The proposed updating framework is verified with a well-documented excavation case where the small-strain stiffness, the anisotropy of elasticity, the anisotropy of yield surface for natural clays, and the parameters of the supporting structures and diaphragm wall are consecutively updated during the staged excavation process. The advantages of the proposed updating framework compared to the Bayesian updating on the same case are also illustrated.

9.2 Overview of Optimizations Used in Excavation

Braced excavation plays an important role in construction in urban areas. However, this type of excavation can present complications in the interaction between the soil and the structure involving lateral wall deflection and ground movement (including lateral movement and surface settlement), which leads to damage to adjacent buildings and underground facilities [1–3]. To maintain safe construction conditions, the lateral wall deflection and ground movement in the later excavation stages should be estimated before next stages of the excavation. Overestimation will increase construction cost while underestimation may render the excavation unsafe, even resulting

Z. Yin and Y. Jin, *Practice of Optimisation Theory in Geotechnical Engineering*,
https://doi.org/10.1007/978-981-13-3408-5_9

in accidents. Therefore, an accurate and reliable prediction of lateral wall deflection and ground movement is crucial in good engineering practice. In addition to the interaction mechanism between the soil and the structure, many other factors including stratigraphy, soil properties, details of the support system, construction activities, contractual arrangements, construction quality (workmanship), and other environmental factors (such as temperature variation), can affect the ground movements associated with excavations, thus adding further challenges to accurate prediction. Therefore, the stage-updating prediction methodology, which is based on early field observations carried out stage by stage is useful in real practice.

In terms of staged-updating prediction methodologies, many approaches are available for back analysis of soil parameters and updating the wall and/or ground responses in the subsequent excavation stages with the updated soil parameters. Among them, the Bayesian probabilistic framework-based updating procedures, e.g., identifying soil parameters from excavation [4–7], modeling of tunnel excavation processes [8], structure inspection [9], and probabilistic finite element model updating of structural systems [10] are attractive and widely used. In a Bayesian updating procedure, only the maximum lateral displacement and the maximum ground surface settlement can be predicted with the updated soil parameters [5]. However, the specific positions corresponding to the maximum deformation and an accurate zone of influence pertaining to settlement induced by an excavation cannot be predicted. An accurate zone of influence for settlement is important for estimating the possibility of damage to adjacent buildings and utilities, especially in urban agglomerated regions. Furthermore, the accuracy of predictions mainly depends on the semi-empirical model used in the updating process [4, 5]. The cost for training a semi-empirical model with accuracy is commonly high. One possible way to obtain specific positions of maximum deformation and the zone of influence in settlement is to incorporate the finite element model as the calculation tool in the Bayesian updating process. However, when using a numerical model as the calculation tool for Bayesian updating, calculation time is prolonged as a result of more than 10,000 calculations that the Markov chain Monte Carlo (MCMC) sampling method requires to derive the posterior distributions of soil parameters—making it impractical for use.

For an optimization-based updating procedure, the updated prediction performance mainly depends on the searching ability of the adopted optimization methods. In previous studies, deterministic optimization techniques (e.g., the Levenberg–Marquardt method [11, 12], the Broyden–Fletcher–Goldfarb–Shanno (BFGS) method [13], and the Nelder–Mead simplex [14]) have usually been employed [14–17]. However, because of their intrinsic drawbacks, e.g., only capable of searching for a local minimum [18–22] and strongly dependent on user's skills [23–26], the deterministic optimization techniques are not suitable for solving optimization problems in excavation. To overcome the disadvantages of deterministic optimization algorithms (e.g., only capable of searching for a local minimum, strongly dependent on user's skills), stochastic optimization techniques have adopted to identify soil parameters in excavation: these include genetic algorithms (GAs by Holland [27]) [28], differential evolution algorithms (DEs, by Price and Storn [29]) [30], adaptive multi-objective method (AMALGAM) [31], and the combination of BPNN and NSGA-II (by Deb et al. [44]). Except for the cases conducted by Huang et al. [31] and Sun et al. [32], the rest of the studies were performed as a single-objective problem and only one type of field observation was used in the back analysis. However, the case adopted AMALGAM in Huang et al. [31] requires a high computational cost for performing a single multi-objective optimization run with 10,000 numerical model evaluations. To consider the effects of both vertical and horizontal observations on the updating parameters and simultaneously save the computational cost, an efficient stochastic multi-objective optimization is more suitable.

Compared to Bayesian updating methods, the finite element method (FEM) is commonly used as an analysis tool in optimization-based updating processed, as it can give an accurate prediction of lateral wall deflection induced by an excavation with a conventional soil model [15, 30, 33]. However, it is generally recognized that it is more difficult to predict ground settlement than lateral wall deflection, even with well-measured soil parameters from laboratory tests or field observations. Some studies [34–36] have innovatively considered the small-strain stiffness behavior of soils and obtained accurate predictions. In addition, due to its natural deposition, the soil always exhibits naturally inherent cross-anisotropy of elasticity [37, 38], which has a significant effect on lateral wall deflection and ground movement in an excavation [39, 40]. Due to the integrity that exists between the soil and the structure in an excavation, associating the structures' parameters (e.g., stiffness of the diagram wall, of supporting structures) with the soil parameters in the optimization could give more accurate predictions of the wall and ground responses. This identification of both structure and soil parameters could further consider the influence of construction quality and other environmental factors. However, few works simultaneously identify the soil parameters related to the small-strain stiffness, the anisotropy of elasticity, and the effects of the structures in one updating process for both the Bayesian updating procedure and the optimization-based updating procedure. Therefore, simultaneously incorporating stochastic multi-objective optimization (MOOP) with the identification of soil- and structure-related parameters in an excavation is a desirable methodology.

In this paper, a framework that combines stochastic multi-objective optimization and the observed field data to update soil and structural parameters is presented for forward prediction of wall and ground responses in later excavation stages. First, a framework for multi-objective optimization-based updating against excavation is proposed. Next, an enhanced multi-objective differential evolution algorithm is developed and employed in the updating process. Then, an elastoplastic soil model accounting for small-strain stiffness and anisotropic elasticity behaviors of clays is proposed for use. Finally, the proposed procedure is applied to a well-instrumented deep excavation. The observed wall deflection and ground surface settlement in the early stages are used as objectives to identify soil and structural parameters for the optimization and the wall and ground responses in the subsequent excavation stages are predicted with the updated parameters accordingly. The process begins with the first excavation stage and is repeated stage by stage as the excavation proceeds until the entire excavation is completed. The effectiveness and feasibility of the proposed method is demonstrated and its advantages are thoroughly discussed.

9.3 Framework for Multi-objective Optimization-Based Updating

9.3.1 *Procedure of Parameter Identification*

Parameter identification through optimization consists of: (1) the formulation of an error function measuring the difference between numerical and observational results, and (2) the selection of an optimization strategy that makes possible the search for the optimal solutions.

9.3.1.1 Error Function

As the first step in the formulation of an error function, an expression for the individual norm has to be established. In general, the individual norm is based on Euclidean measures between discrete points, composed of the experimental and the numerical results. To make the error independent of the type of test and the number of measurement points, an advanced error function proposed by Levasseur et al. [41] was adopted in this study. The average difference between the measured and the simulated results is expressed using the least square method,

$$\mathrm{Error}(\mathbf{x}) = \sqrt{\frac{1}{N}\sum_{i=1}^{N}\left(\frac{U_i^{\mathrm{obs}} - U_i^{\mathrm{num}}}{U_i^{\mathrm{obs}}}\right)^2} \times 100 \tag{9.1}$$

where $\mathbf{x}$ is an n-sized vector of unknown model input parameters, and N is the number of observation points. U_i^{obs} is the value corresponding to the ith point of observed data, and U_i^{num} is the value corresponding to the ith point of simulated data.

The next step is to formulate a final norm, a total error function, based on the individual norms computed using Eq. (4.5). When multiple types of measurements are to be simultaneously considered in the back analysis of excavation, the generalized objective function can be expressed as a multi-objective optimization form,

$$\min[\mathbf{Error}(\mathbf{x})] = \min\begin{bmatrix}\mathrm{Error}_1(\mathbf{x}),\ \mathrm{Error}_2(\mathbf{x}), \\ \ldots, \mathrm{Error}_i(\mathbf{x}), \ldots, \mathrm{Error}_m(\mathbf{x})\end{bmatrix} \tag{9.2}$$

where $\mathbf{x}$ is the vector of parameters; $\mathbf{Error}(\mathbf{x})$ is the vector of objective error functions; $\mathrm{Error}_i(\mathbf{x})$ with $i = 1, 2, \ldots, m$ is an ith objective error; m is the number of objectives involved in the optimization.

Generally, for a braced excavation, the lateral wall deflection and ground movement are two extremely important indicators that reflect the influence of soil–structure interaction on the excavation. To identify the soil and structure parameters, the objective error function should involve these two important indicators,

$$\min[\mathbf{Error}(\mathbf{x})] = \min[\mathrm{Error}_{\mathrm{WD}}(\mathbf{x}), \mathrm{Error}_{\mathrm{GM}}(\mathbf{x})] \tag{9.3}$$

where $\mathrm{Error}_{\mathrm{WD}}(\mathbf{x})$ is the objective error based on the lateral **W**all **D**eflection in each excavation stage; $\mathrm{Error}_{\mathrm{GM}}(\mathbf{x})$ is the objective error based on the **G**round **M**ovement in each excavation stage.

The final goal for a multi-objective algorithm is to find Pareto front—a set of optimal solutions in multidimensional objective space. For a real engineering problem, the most suitable solution can be determined based on the knowledge of the Pareto frontier with a posteriori selection criterion predefined by the user, e.g., the solution corresponding to the optimal prediction of wall deflection or the optimal prediction of ground settlement depending on its importance.

9.3.1.2 Parameter Identification

Figure 9.1 shows the identification procedure based on the successive use of two different parts: the FEM model for simulating excavation and the optimization process for finding the Pareto solutions. For the optimization program, any powerful multi-objective optimization algorithm can be employed to find the optimal Pareto solutions. Some available open-source multi-objective algorithms (such as NSGA-II) can be found in http://yarpiz.com/. In the following sections, an efficient and fast convergence differential evolution-based, multi-objective algorithm will be introduced to enhance the optimization performance in this study. For the numerical simulation process, the users can adopt different FEM analysis tools to conduct the excavation according to their requirements, such as PLAXIS [15, 16], or ABAQUS [30, 31], or other such tools. In this study, the ABAQUS is employed.

It is generally understood that using a well-distributed sampling to generate the initial population can increase the robustness of the algorithm and avoid premature convergence. This initial population is governed by the number of individuals, their domain (range), and the method controlling the distribution of the individuals within their domain. In accordance with the approach taken by Poles et al. [42], the SOBOL sampling method should also be an alternative way in the optimization algorithm to generate the initial population. SOBOL is a deterministic algorithm that imitates the behavior of a random sequence, through which a uniform sampling in the design space can be obtained.

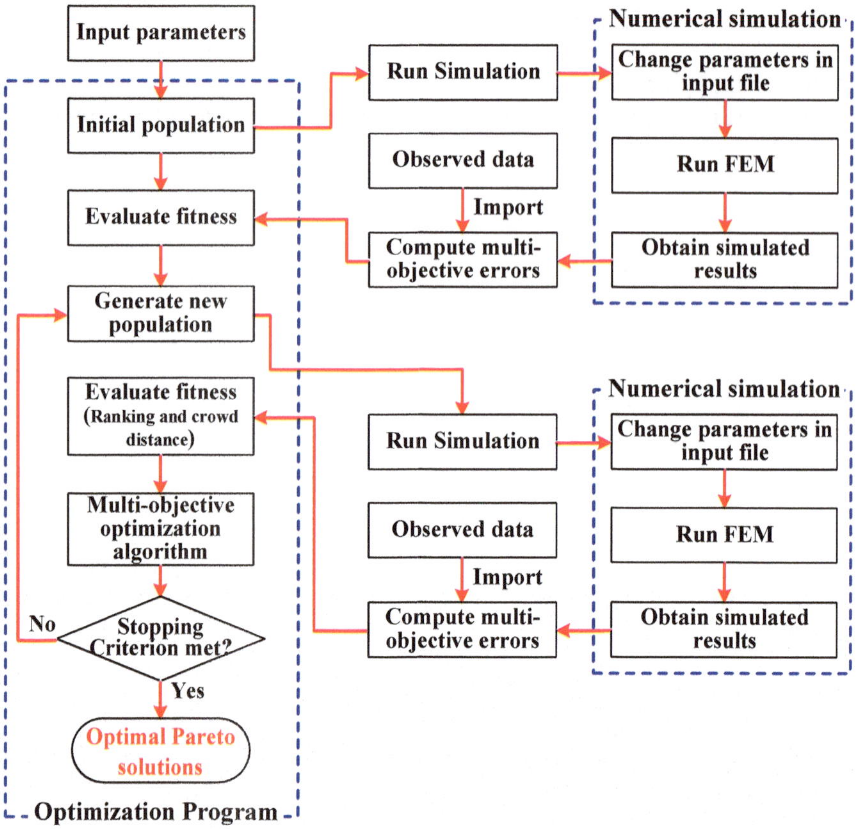

Fig. 9.1 Flow chart of multi-objective parameter identification

9.3.2 Procedure for Updating a Prediction

Figure 9.2 shows the proposed procedure of updating predictive deformation combined with parameter identification. For a deep excavation, after the initial excavation stage is conducted, the wall and ground responses are measured, based on which the parameters related to the soil and structure are first identified using the proposed identification procedure. Then with the identified parameters, the wall and ground responses in the subsequent stages of excavation can be predicted. Meanwhile, the influence of the predicted responses from the wall and ground on the adjacent buildings and utilities are evaluated. If the predicted deformation is too large and uncontrollable, some precautionary measures in the subsequent stage, e.g., adding temporary support, should be launched to protect the adjacent buildings and utilities. When the next stage of excavation is finished, the accuracy of the predictions is estimated by calculating the difference between predictions and observations. This

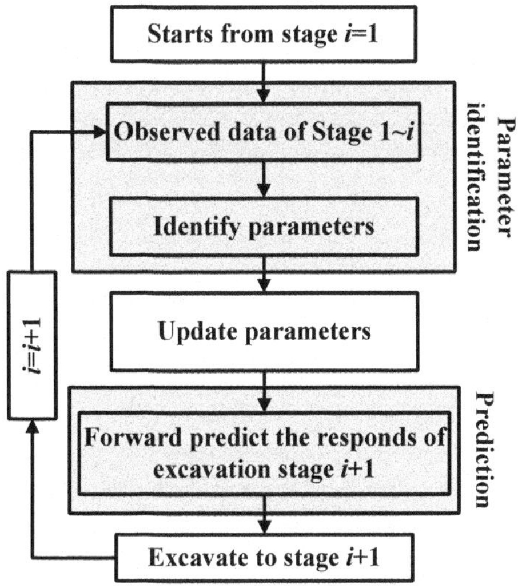

Fig. 9.2 Combination of parameter identification and updating prediction process

process is repeated stage by stage as the excavation proceeds along with the addition of further field observations to the optimization objective.

The proposed procedure is similar to the Bayesian updating procedure proposed by Juang et al. [5]. In Bayesian updating, the updated soil parameters are represented by their posterior distributions and sample statistics, which are obtained through the Markov chain Monte Carlo (MCMC) simulation. For the proposed multi-objective optimization-based updating procedure, the optimized parameters are selected from the Pareto front, which are obtained using the multi-objective optimization algorithm. Compared to Bayesian updating, the parameters in the Pareto front can form a distribution with fewer points due to the limited number of individuals in the population, while it can also consider the uncertainty involved in the braced excavation. The merits of the proposed procedure will be highlighted in a real excavation case combined with an enhanced multi-objective differential evolution algorithm and a soil model considering the small-strain stiffness feature.

9.4 Enhancement of Multi-objective Differential Evolution Algorithm

Generally, there are two goals in multi-objective optimization: (i) to discover better Pareto front, and (ii) to find diverse solutions. Satisfying these two goals is a challenging task for any algorithm adopted in a multi-objective optimization. In recent years, many algorithms for multi-objective optimizations have been proposed based on artificial intelligence (AI) and evolutionary algorithms (EAs), including the Strength Pareto Evolutionary Algorithm (SPEA2) by Zitzler et al. [43], the Non-dominated

Sorting Genetic Algorithm-II (NSGA-II) by Deb et al. [44], GeDEA-II by Da Ronco and Benini [45], and the recently proposed particle swarm optimization multi-objective algorithms with excellent performance [46, 47]. Among the EAs, the differential evolution (DE) has been extended to handle multi-objective optimization problems (MOPs) [48], such as Pareto differential evolution (PDE) [49], differential evolution for multi-objective optimization (DEMO) [50], Pareto-based multi-objective differential evolution [51], vector-evaluated differential evolution for multi-objective optimization (VEDE) [52], multi-objective differential evolution (MODE) [53], non-dominated sorting differential evolution (NSDE) [54], the third version of generalized differential evolution (GDE3) [55], multi-objective self-adaptive DE (MOSaDE) [56], adaptive differential evolution for multi-objective problems (ADEMO/D) [57], enhanced self-adaptive differential evolution with mixed crossover (ESDE-MC) [58], and differential evolution using clustering-based objective reduction (DECOR) [59]. However, these works mainly focused on finding more and accurate solutions with diversity while ignored the convergence speed. Thus, to improve the performance of DE-based multi-objective algorithms, an efficient multi-objective optimization algorithm under the framework of DE enhanced with a Simplex crossover is proposed in this section.

9.4.1 Differential Evolution Algorithm

The differential evolution (DE) algorithm, proposed by Price and Storn [29, 60], is a simple yet powerful population-based stochastic search technique, which is an efficient and effective global optimizer in the continuous search domain. Like other population-based optimization algorithms, DE involves two phases as well: initialization and evolution (mutation, crossover and selection). In the initialization phase, the DE population is generated randomly if nothing is known about the problem. In the evolution phase, DE uses a mutation operation as a search mechanism, a crossover to increase the diversity and a selection operation based on the differences in randomly sampled pairs of solutions in the population.

(1) *Mutation*

A new mutation strategy proposed by Zhang and Sanderson [61] was adopted, which places the perturbation at a location between a randomly chosen vector and the best-performing vector,

$$\mathbf{v}_i = x_i + F^i\left(x_{\text{best},i} - x_i\right) + F^i\left(x_{r1} - x_{r2}\right) \tag{9.4}$$

where the indices r_1 and r_2 are distinct integers uniformly chosen from the set $\{1, 2, \ldots, N_p\}$; N_p is the number of individuals in one generation. $(x_{r1} - x_{r2})$ is a difference vector to mutate the corresponding parent x_i; $x_{\text{best},i}$ is the best vector at current generation i, which is randomly chosen as one of the top $100p\%$ individuals in the current population, where $p \in (0, 1]$ is the proportion of best-selected individuals in whole population, and in this case p is set to 0.1; F^i is the mutation factor that is associated with x_i and is regenerated in each generation by a randomly uniform distribution within [0.5, 1.0].

(2) *Crossover*

After mutation, a binomial crossover operation forms the final trial/offspring vector

$$\mathbf{u}_{i,j} = \begin{cases} \mathbf{v}_{i,j}, & \text{if} \operatorname{rand}(0, 1) \leq \text{CR or } j = j_{\text{rand}} \\ \mathrm{x}_{i,j}, & \text{otherwise} \end{cases} \tag{9.5}$$

where rand(a, b) is a uniform random number on the interval $[a, b]$ and independently generated for each j and each i, $j_{\text{rand}} = \text{randint}(1, D)$ is an integer randomly chosen from 1 to D and newly generated for each i. D is the problem dimension; The crossover probability CR $\in [0, 1]$ roughly corresponds to the average fraction of vector components that are inherited from the mutation vector, and CR$= 0.3$ was employed in this study.

(3) *Selection*

To avoid the fast loss of the diversity, an elitism strategy was adopted to perform the selection. In this selection, 10% of the individuals with high fitness are selected from the parent and children and kept in the next generation. The remains are chosen by tournament selection from the mating pool composed of parent and children except for the 10% individuals.

9.4.2 *Simplex Crossover (SPX)*

The Simplex crossover (SPX) is a multi-parent recombination operator for generating a new individual, which is a search operator analogous to DE mutation and differs from the crossover of DE. In this study, the SPX for multi-objective algorithm proposed by Da Ronco and Benini [45] was adopted, which was modified from the original Simplex theory, to further enhance the local search capability without penalizing the exploration of the search space and thus improve the convergence performance. The offspring vector is formed by using SPX as follows:

$$\xi_i = (1 + \text{Refl}) \cdot M - \text{Refl} \cdot x_i^2 \tag{9.6}$$

where M is the centroid of x_i^1, which can be calculated in the following manner:

$$M = \left(\frac{1}{n}\right) \cdot x_i^1 \tag{9.7}$$

where x_i^1 and x_i^2 are two parents selected by the tournament to create the offspring. It is assumed that x_i^1 is the best fit individual of the two chosen parents to form the offspring. The Refl coefficient is set as equal to a random number [0, 1], n is the number of the remaining individuals, after the worst one has been excluded, and $n = 2$ is employed in this study according to the test results conducted by Da Ronco and Benini [45].

9.4.3 EMODE

For real geotechnical engineering, numerous calculations during optimization process would pose unacceptable computational cost. A possible solution is to improve the convergence speed for the adopted multi-objective algorithm and thus accomplish the purpose of cost saving in inverse analysis, which is the concern in this study. According to Da Ronco and Benini [45], the experimental results with test functions used in their studies showed that the SPX performs well in solving multi-objective problems, which can speed up the search direction to the true Pareto front. Therefore, the proposed multi-objective algorithm in this paper adopted the SPX operator to improve the convergence performance through a hybrid strategy combined with the DE mutation mentioned earlier. The DE mutation is parallel with SPX because of the similar search mechanism to generate new offspring. For simplicity's sake, the enhanced multi-objective differential evolution algorithm is called EMODE in the following sections.

Figure 9.3 shows the EMODE flowchart. First, an initial population of μ individuals was generated using the SOBOL sampling method and the generation number is set to zero. Second, a mating pool of 2λ individuals is formed, each individual having the same probability of being selected. Third, λ offspring are generated by DE mutation or SPX according to their possibilities "p_m" and "p_s". After the DE mutation and SPX, some bits of the offspring are also randomly selected to DE crossover with a probability CR. Next, the repeat individuals in the whole population of $\mu + \lambda$ individuals are removed and replaced with new randomly generated individuals. Last, the values of the $\mu + \lambda$ individuals in the objective function are evaluated and the non-dominated sorting procedure presented by Deb et al. [44] is performed to assign the ranks to the solutions according to the objectives of the multi-objective optimization (MOOP). The whole population of $\mu + \lambda$ individuals is processed to determine the value of the distance-based genetic diversity measure for each individual. The non-dominated sorting procedure incorporates a diversity preservation mechanism, which estimates the density of solutions in the objective space, and the crowded comparison operator, which guides the selection process toward a uniformly spread Pareto frontier. Then, the best μ solutions among the parents and offspring, according to the ranks assigned previously, are selected for survival and the remaining λ are eliminated. The iteration stops when the convergence criterion or the maximum generation number is reached.

Fig. 9.3 Flowchart of proposed EMODE

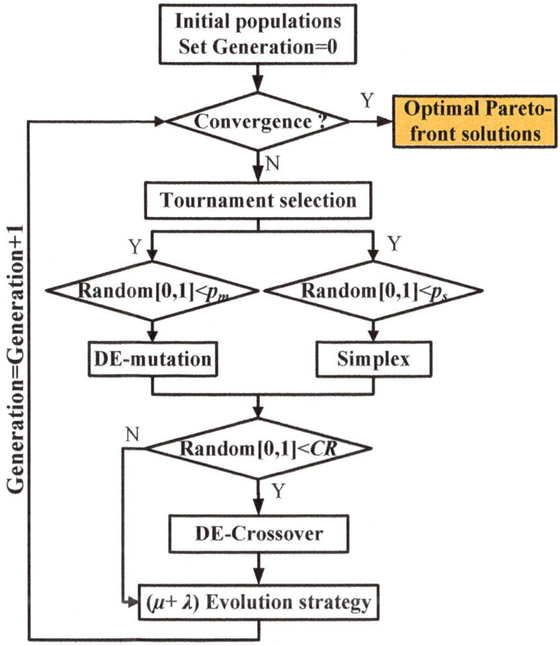

For any algorithm, the analysis of computational complexity is important, which determines number of computations needed for a given optimization problem. Considering the complexity of one iteration of the entire algorithm, the basic operations and their worst-case complexities are as follows: non-dominated sorting $O(M(2N)^2)$ (where M is the number of objectives and N is the population size), crowding-distance assignment $O(M(2N)\log(2N))$, etc. The overall complexity of the EMODE algorithm is $O(MN^2)$, which is governed by the non-dominated sorting part of the algorithm and similar to NSGA-II.

9.5 Mathematical Validation for EMODE

9.5.1 *Test Problems*

The performance of the proposed algorithm was examined on a set of 12 widely used unconstrained benchmarks covering a broad range of problems with different features (concave, convex, disconnected, deceptive, etc.). For problems having two objectives, the test suites ZDT (ZDT1, ZDT2, ZDT3, ZDT4, and ZDT6) [62] were selected. Moreover, to increase the complexity of testing, seven problems of DTLZ family [63] with three objectives (DTLZ 1–7) were *c* of these problems can be found in Deb et al. [63, 64]. The details of all selected benchmark tests are listed in Appendix I.

Appendix I: Benchmark tests for multi-objective optimization algorithm

Problems	Definition	Constraints
ZDT1	$f_1(x) = x_1,$ $f_2(x) = g(x)\left(1 - \sqrt{x_1/g(x)}\right),$ $g(x) = 1 + \frac{9}{n-1}\sum_{i=2}^{n} x_i$	$n = 30$ and $x_i \in [0, 1]$
ZDT2	$f_1(x) = x_1,$ $f_2(x) = g(x)\left(1 - (x_1/g(x))^2\right),$ $g(x) = 1 + \frac{9}{n-1}\sum_{i=2}^{n} x_i$	$n = 30$ and $x_i \in [0, 1]$
ZDT3	$f_1(x) = x_1,$ $f_2(x) = g(x)\left(1 - \sqrt{x_1/g(x)} - \frac{x_1}{g(x)}\sin(10\pi x_1)\right),$ $g(x) = 1 + \frac{9}{n-1}\sum_{i=2}^{n} x_i$	$n = 30$ and $x_i \in [0, 1]$
ZDT4	$f_1(x) = x_1,$ $f_2(x) = g(x)\left(1 - \sqrt{x_1/g(x)}\right),$ $g(x) = 1 + 10(n-1) + \sum_{i=2}^{n}\left(x_i^2 - 10\cos(4\pi x_i)\right)$	$n = 10$ and $x_1 \in [0, 1]$, $x_i \in [-5, 5]$ for $i = 2, \ldots, 9$
ZDT6	$f_1(x) = 1 - \exp(-4x_1)\sin^6(6\pi x_1),$ $f_2(x) = g(x)\left(1 - (f_1(x)/g(x))^2\right),$ $g(x) = 1 + \frac{9}{n-1}\sum_{i=2}^{n} x_i$	$n = 10$ and $x_i \in [0, 1]$
DTLZ1	$f_1(x) = 0.5x_1x_2(1 + g(x)),$ $f_2(x) = 0.5x_1(1 - x_2)(1 + g(x)),$ $f_3(x) = 0.5(1 - x_1)(1 + g(x)),$ $g(x) = 100\left[5 + \sum_{i=3}^{n}\left((x_i - 0.5)^2 - \cos(20\pi(x_i - 0.5))\right)\right]$	$n = 7$ and $x_i \in [0, 1]$
DTLZ2	$f_1(x) = (1 + g(x))\cos(0.5\pi x_1)\cos(0.5\pi x_2),$ $f_2(x) = (1 + g(x))\cos(0.5\pi x_1)\sin(0.5\pi x_2),$ $f_3(x) = (1 + g(x))\sin(0.5\pi x_1),$ $g(x) = \sum_{i=3}^{n}(x_i - 0.5)^2$	$n = 12$ and $x_i \in [0, 1]$
DTLZ3	$f_1(x) = (1 + g(x))\cos(0.5\pi x_1)\cos(0.5\pi x_2),$ $f_2(x) = (1 + g(x))\cos(0.5\pi x_1)\sin(0.5\pi x_2),$ $f_3(x) = (1 + g(x))\sin(0.5\pi x_1),$ $g(x) = 100\left[10 + \sum_{i=3}^{n}\left((x_i - 0.5)^2 - \cos(20\pi(x_i - 0.5))\right)\right]$	$n = 12$ and $x_i \in [0, 1]$

(continued)

(continued)

Problems	Definition	Constraints
DTLZ4	$f_1(x) = (1 + g(x))\cos\left(0.5\pi x_1^{100}\right)\cos\left(0.5\pi x_2^{100}\right),$ $f_2(x) = (1 + g(x))\cos\left(0.5\pi x_1^{100}\right)\sin\left(0.5\pi x_2^{100}\right),$ $f_3(x) = (1 + g(x))\sin\left(0.5\pi x_1^{100}\right),$ $g(x) = \sum_{i=3}^{n}(x_i - 0.5)^2$	$n = 12$ and $x_i \in [0, 1]$
DTLZ5	$f_1(x) = (1 + g(x))\cos(0.5\pi\theta_1)\cos(0.5\pi\theta_2),$ $f_2(x) = (1 + g(x))\cos(0.5\pi\theta_1)\sin(0.5\pi\theta_2),$ $f_3(x) = (1 + g(x))\sin(0.5\pi\theta_1),$ $g(x) = \sum_{i=3}^{n}(x_i - 0.5)^2, \theta_1 = \pi(1 + 2x_i g(x))/(4(1 + g(x)))$	$n = 12$ and $x_i \in [0, 1]$
DTLZ6	$f_1(x) = (1 + g(x))\cos(0.5\pi\theta_1)\cos(0.5\pi\theta_2),$ $f_2(x) = (1 + g(x))\cos(0.5\pi\theta_1)\sin(0.5\pi\theta_2),$ $f_3(x) = (1 + g(x))\sin(0.5\pi\theta_1),$ $g(x) = \sum_{i=3}^{n} x_i^{0.1}, \theta_1 = \pi(1 + 2x_i g(x))/(4(1 + g(x)))$	$n = 12$ and $x_i \in [0, 1]$
DTLZ7	$f_1(x) = x_1, f_2(x) = x_2,$ $f_3(x) = (1 + g(x)) \cdot h(f_1, f_2, g(x)),$ $g(x) = 1 + 9\sum_{i=3}^{n} x_i/20,$ $h(f_1, f_2, g(x)) = 3 - \sum_{i=1}^{2} f_i \cdot (1 + \sin(3\pi f_i))/(1 + g(x))$	$n = 12$ and $x_i \in [0, 1]$

To demonstrate the performance, the proposed algorithm was compared with four state-of-the-art multi-objective algorithms, i.e., the Non-dominated Sorting Genetic Algorithm-II (NSGA-II) [44], the improved Strength Pareto Evolutionary Algorithm (SPEA2) [43], the region-based Pareto-Envelope-based Selection Algorithm (PESA-II) [65], and the third version of Generalized Differential Evolution (GDE3) [55] for same benchmarks. The four state-of-the-art algorithms can be found in the "PlatEMO" [66] which is an open-source and free MATLAB-based platform (http://bimk.ahu.edu.cn/index.php?s=/Index/Software/index.html). For all algorithms, the number of initial population was set to 100. Since other selected algorithms use the random sampling method to generate the initial population in benchmark testing, the SOBOL in the proposed EMODE was replaced by the random sampling method for a fair comparison. The maximum generation was set to 250 for all problems for a fair comparison. Then, the maximum number of function evaluations is 25,000 for all benchmarks. The parameter settings of all selected algorithms are summarized in Table 9.1.

Table 9.1 Parameter settings for all selected algorithms

MOOP algorithms	Parameter settings
NSGA-II	$N = 100, p_c = 0.9, p_m = 1/n, \eta_c = 20, \eta_m = 20$
SPEA2	$N = 100, p_c = 0.9, p_m = 1/n, \eta_c = 20, \eta_m = 20$
PESA-II	$N = 100, p_c = 0.9, p_m = 1/n, \eta_c = 20, \eta_m = 20$
GDE3	$N = 100$, CR= 0.3, $F = 0.5$
EMODE	$N = 100$, CR= 0.3, $F = \text{rand}[0.5, 1.0], p_m = 0.8, p_s = 1.0$

9.5.2 *Performance Metrics*

The performance indicator was used to demonstrate the effectiveness of the proposed algorithm for MOOPs. For the selected test problems, their true Pareto fronts are known. Thus, three widely used performance metrics (the general distance (GD), the inverted generational distance (IGD), and the hypervolume (HV)) were adopted in this study.

1. Generational distance (GD)

The concept of generational distance was introduced by Van Veldhuizen and Lamont [67] to compute the distance between the non-dominated solutions set PF and the Pareto-optimal front PF*. This indicator is defined as:

$$\text{GD}\left(\text{PF}, \text{PF}^*\right) = \frac{\sqrt{\sum_{v \in \text{PF}} d(v, \text{PF}^*)^2}}{|\text{PF}|} \tag{9.8}$$

where |PF| is the number of the obtained solutions in PF and $d(v, \text{PF}^*)$ is the Euclidean distance between v and the nearest point in PF*. The metric of GD can be used to evaluate how far these obtained solutions are from those in the Pareto-optimal set on PF. A smaller value of GD indicates a better convergence to the true PF.

2. Inverted generational distance (IGD)

The inverted generational distance (IGD) proposed by Zitzler et al. [68] is defined as:

$$\text{GD}\left(\text{PF}, \text{PF}^*\right) = \frac{\sqrt{\sum_{v \in \text{PF}^*} d(v, \text{PF})^2}}{|\text{PF}^*|} \tag{9.9}$$

where $d(v, \text{PF})$ is the minimum Euclidean distance between solution v and the obtained solution in PF; $|\text{PF}^*|$ is the number of solutions in PF*. If $|\text{PF}^*|$ is big

Table 9.2 Mean (SD) of GD for all selected algorithms on 12 benchmark tests

Problems	NSGA-II	GDE3	PESA-II	SPEA2	EMODE
ZDT1	1.6954e−3 (5.75e−4)	5.3878e−2 (2.20e−2)	5.4141e−3 (6.72e−3)	2.8560e−3 (2.16e−3)	**2.3069e−04 (4.22e−05)**
ZDT2	2.2522e−3 (2.38e−3)	3.4090e−1 (1.67e−1)	3.0434e−2 (7.42e−2)	2.4280e−3 (1.82e−3)	**9.6135e−05 (4.02e−06)**
ZDT3	2.5626e−3 (2.56e−3)	5.4816e−2 (1.97e−2)	1.0538e−2 (1.35e−2)	2.6313e−3 (2.15e−3)	**1.6612e−04 (1.32e−05)**
ZDT4	**5.3360e−3 (6.35e−3)**	1.8006e+0 (4.47e−1)	1.3550e−2 (1.41e−2)	1.0240e−2 (1.42e−2)	6.0229e−03 (1.93e−02)
ZDT6	1.0416e−3 (7.13e−4)	3.1014e−2 (3.44e−2)	5.6360e−3 (1.49e−2)	1.6397e−3 (1.18e−3)	**6.9599e−05 (4.26e−06)**
DTLZ1	4.4054e−2 (5.44e−2)	8.5456e−3 (1.43e−2)	2.6449e−2 (2.73e−2)	3.5400e−2 (3.77e−2)	**2.5241e−04 (6.19e−06)**
DTLZ2	**5.6456e−4** (3.13e−5)	2.1215e−3 (3.88e−4)	6.0415e−4 (5.00e−5)	5.6523e−4 (5.53e−5)	7.1616e−04 (**2.28e−05**)
DTLZ3	6.0065e−1 (3.58e−1)	4.8569e+1 (1.44e+0)	1.3300e−2 (6.08e−3)	6.9583e−1 (4.68e−1)	**7.7106e−04 (5.33e−05)**
DTLZ4	5.9734e−4 (6.44e−5)	1.8645e−3 (1.84e−4)	5.8382e−4 (3.68e−5)	**4.8427e−4** (1.56e−4)	7.1800e−04 (**2.63e−05**)
DTLZ5	4.4973e−5 (3.69e−5)	1.1617e−4 (3.70e−5)	6.9501e−5 (6.56e−5)	5.4443e−5 (5.45e−5)	**8.8777e−06 (3.36e−07)**
DTLZ6	4.8460e−6 (2.18e−7)	4.7400e−6 (**1.56e−7**)	4.8392e−6 (3.14e−7)	**4.7192e−6** (2.83e−7)	8.8227e−06 (5.21e−07)
DTLZ7	2.3007e−3 (3.93e−4)	3.5565e−3 (6.62e−4)	1.5963e−3 (6.84e−4)	**1.5090e−3 (2.54e−4)**	3.0061e−03 (4.57e−04)
w/t/l	2/9/1	0/9/3	0/11/1	3/8/1	7/5/0

Table 9.3 Mean (SD) of IGD for all selected algorithms on 12 benchmark tests

Problems	NSGA-II	GDE3	PESA-II	SPEA2	EMODE
ZDT1	1.3710e−1 (1.23e−1)	2.6138e−1 (8.66e−2)	4.0414e−1 (1.52e−1)	6.7059e−2 (8.57e−2)	**4.3584e−03 (1.18e−04)**
ZDT2	3.7619e−1 (1.05e−1)	8.4592e−1 (2.68e−1)	5.2503e−1 (1.77e−1)	3.7760e−1 (9.31e−2)	**4.7817e−03 (8.20e−04)**
ZDT3	1.2784e−1 (1.44e−1)	2.4950e−1 (6.92e−2)	5.3103e−1 (2.38e−1)	6.5538e−2 (4.86e−2)	**5.1356e−03 (1.33e−04)**
ZDT4	6.8092e−2 (6.92e−2)	3.4695e+0 (7.35e−1)	2.9472e−1 (1.60e−1)	1.0736e−1 (1.21e−1)	**5.8605e−02 (1.71e−03)**
ZDT6	1.1322e−2 (6.29e−3)	**3.1010e−3 (2.61e−5)**	1.9780e−2 (1.11e−2)	1.6322e−2 (1.07e−2)	5.7463e−03 (3.12e−04)
DTLZ1	3.5484e−1 (4.09e−1)	6.2306e−2 (7.50e−2)	2.1905e−1 (2.07e−1)	3.0582e−1 (3.07e−1)	**2.5733e−02 (9.00e−04)**
DTLZ2	7.0502e−2 (3.62e−3)	6.8638e−2 (2.22e−3)	6.6790e−2 (3.71e−3)	**5.3036e−2 (3.87e−4)**	6.4215e−02 (2.08e−03)
DTLZ3	5.7712e+0 (3.22e+0)	1.6987e+1 (3.44e+0)	1.2088e−1 (5.38e−2)	6.4335e+0 (4.58e+0)	**6.3334e−02 (1.98e−03)**
DTLZ4	6.8818e−2 (3.00e−3)	6.7725e−2 (1.04e−3)	6.4610e−2 (2.69e−3)	1.7790e−1 (2.72e−1)	**6.3407e−02 (1.45e−03)**
DTLZ5	5.8977e−3 (4.04e−4)	4.6605e−3 (9.75e−5)	1.9133e−2 (1.05e−2)	**4.1440e−3 (4.76e−5)**	5.6687e−03 (2.65e−04)
DTLZ6	6.0261e−3 (2.12e−4)	4.4648e−3 (7.46e−5)	1.8343e−2 (3.39e−3)	**4.0977e−3 (1.89e−5)**	7.0364e−03 (2.88e−04)
DTLZ7	**7.8247e−2** (4.63e−3)	8.0126e−2 (4.46e−3)	2.1783e−1 (2.17e−1)	7.8605e−2 (7.43e−2)	8.3902e−02 **(3.19e−03)**
w/t/l	1/10/1	1/8/3	0/11/1	3/8/1	7/5/0

Table 9.4 Mean (SD) of HV for all selected algorithms on 12 benchmark tests

Problems	NSGA-II	GDE3	PESA-II	SPEA2	EMODE
ZDT1	7.5760e−1 (8.93e−2)	4.7894e−1 (1.13e−1)	5.2914e−1 (1.46e−1)	7.9905e−1 (6.78e−2)	**8.7122e−01 (1.32e−04)**
ZDT2	1.8177e−1 (6.15e−2)	1.3651e−2 (4.00e−2)	1.1775e−1 (6.41e−2)	1.7658e−1 (5.00e−2)	**5.3758e−01 (9.17e−04)**
ZDT3	8.4654e−1 (1.67e−1)	5.6387e−1 (1.04e−1)	3.8385e−1 (2.60e−1)	9.1946e−1 (6.53e−2)	**1.0241e+00 (7.80e−05)**
ZDT4	7.8103e−1 (9.66e−2)	0.0000e+0 (0.00e+0)	5.7376e−1 (1.88e−1)	7.2709e−1 (1.72e−1)	**8.0511e−01 (1.99e−01)**
ZDT6	4.1890e−1 (9.52e−3)	**4.3342e−1 (5.01e−5)**	4.0429e−1 (1.61e−2)	4.1152e−1 (1.60e−2)	4.3052e−01 (3.36e−04)
DTLZ1	6.2660e−2 (6.30e−2)	1.1946e−1 (3.36e−2)	7.1597e−2 (6.20e−2)	5.9320e−2 (5.89e−2)	**1.3829e−01 (4.96e−04)**
DTLZ2	7.1315e−1 (4.87e−3)	6.9733e−1 (5.65e−3)	6.9387e−1 (1.50e−2)	**7.4592e−1 (9.10e−4)**	7.2490e−01 (2.81e−03)
DTLZ3	0.0000e+0 (0.00e+0)	0.0000e+0 (0.00e+0)	5.1966e−1 (3.15e−02)	0.0000e+0 (0.00e+0)	**7.1966e−01 (5.15e−03)**
DTLZ4	7.1586e−1 (4.57e−3)	7.0330e−1 (3.80e−3)	7.1657e−1 (8.64e−3)	6.6676e−1 (1.81e−1)	**7.2282e−01 (5.27e−03)**
DTLZ5	1.3269e−1 (8.39e−5)	1.3291e−1 (6.88e−5)	1.2220e−1 (7.15e−3)	**1.3306e−1 (4.49e−5)**	1.3267e−01 (1.05e−04)
DTLZ6	1.3266e−1 (1.65e−4)	**1.3328e−1 (1.86e−5)**	1.2304e−1 (5.79e−3)	1.3315e−1 (2.71e−5)	1.3209e−01 (1.24e−04)
DTLZ7	1.5950e+0 (1.36e−2)	1.5699e+0 (1.32e−2)	1.5054e+0 (1.28e−1)	**1.6275e+0** (5.27e−2)	1.6014e+00 **(1.11e−02)**
w/t/l	0/11/1	2/7/3	0/11/1	3/8/1	7/5/0

enough, the value of IGD can measure both the convergence and the diversity of PF to a certain extent. A smaller value of IGD means a better convergence and diversity to the true PF.

3. Hypervolume (HV)

Hypervolume indicator (HV) [62, 69] is a measure that incorporates both the optimality of a solution set and its spread in the objective space, defined as:

$$\mathrm{HV} = \left[\bigcup_i \mathrm{vol}_i | v_i \in \mathrm{PF}_{\mathrm{known}} \right] \tag{9.10}$$

where v_i is a non-dominated vector in $\mathrm{PF}_{\mathrm{known}}$ and vol_i is the hypervolume between the reference point and vector v_i. In the case of two or three objectives, HV is the area or volume covered by a set of non-dominated solutions with comparison to a reference point. The HV can be used to evaluate the performance of different algorithms in terms of the convergence to the Pareto approximation set and the coverage.

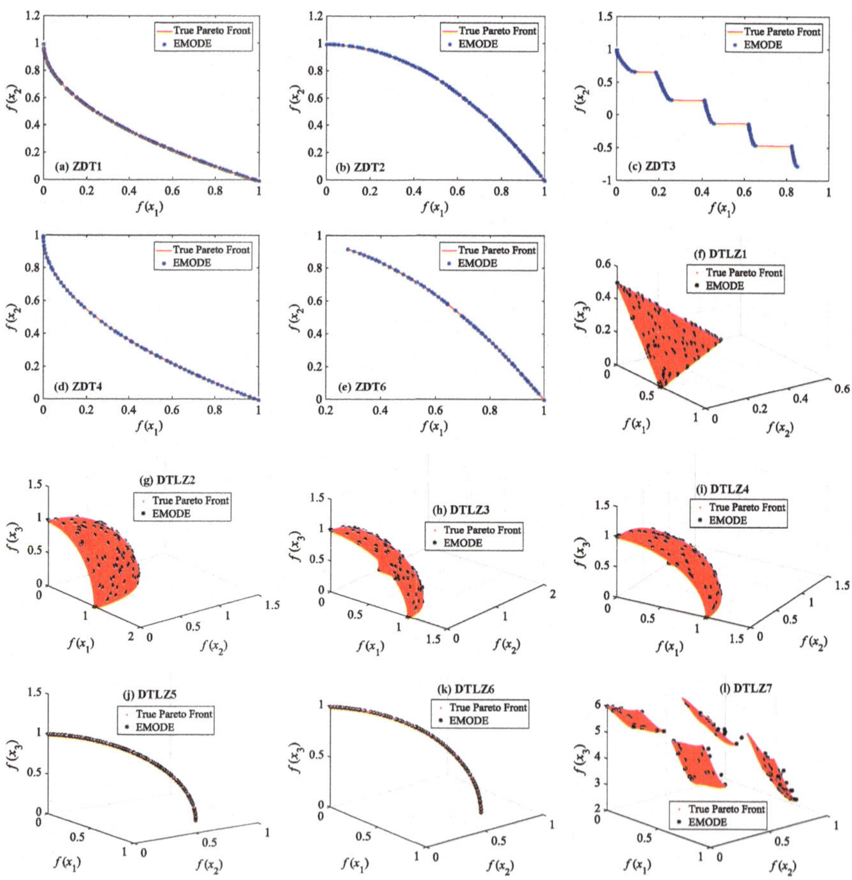

Fig. 9.4 Comparison of Pareto fronts by using EMODE and the true Pareto fronts on all test problems

9.5.3 Results

To avoid randomness, all algorithms independently run 20 times on each test problem. The statistical values including mean value (Mean) and standard deviation (SD) are presented in Tables 9.2, 9.3, and 9.4. All results are summarized as "*w*/*t*/*l*", which denotes that the selected algorithm wins on *w* functions, ties on *t* functions, and loses on *l* functions, compared with its corresponding competitors. The best values among these algorithms are marked with bold font.

Table 9.2 shows the values of GD indicator for all algorithms on all test problems. It can be found that the proposed EMODE obtains better results on ZDT1, ZDT2,

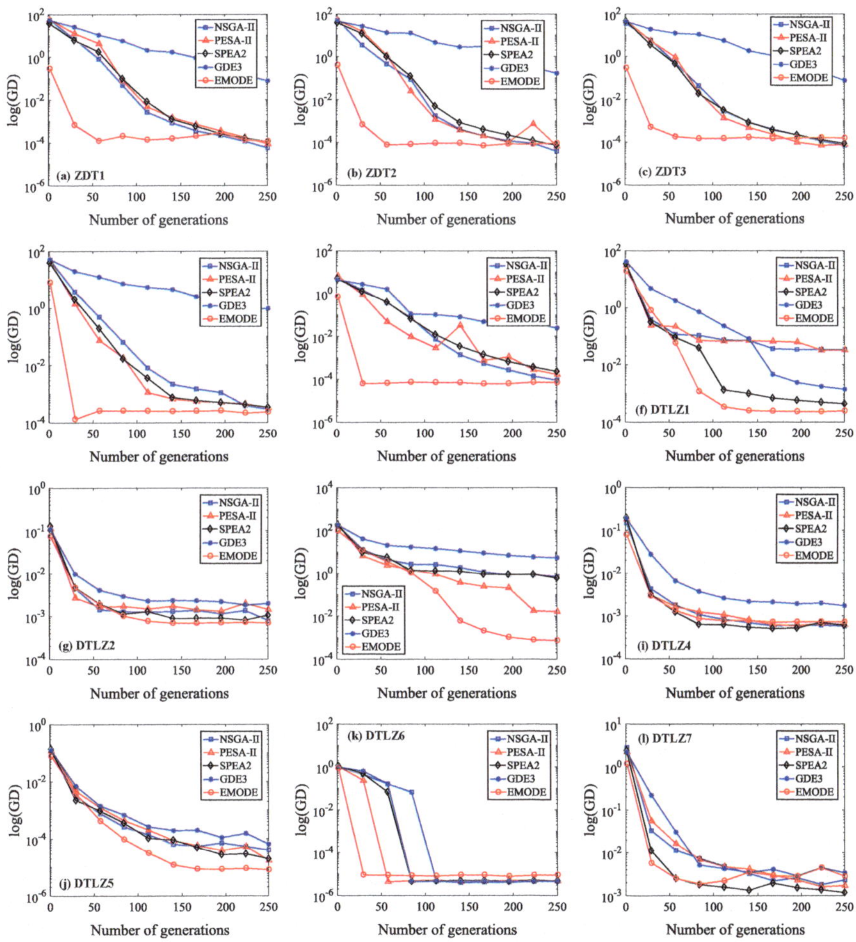

Fig. 9.5 Comparison of GD for EMODE, NSGA-II, PESA-II, SPEA2, and GD3 on 12 benchmark problems

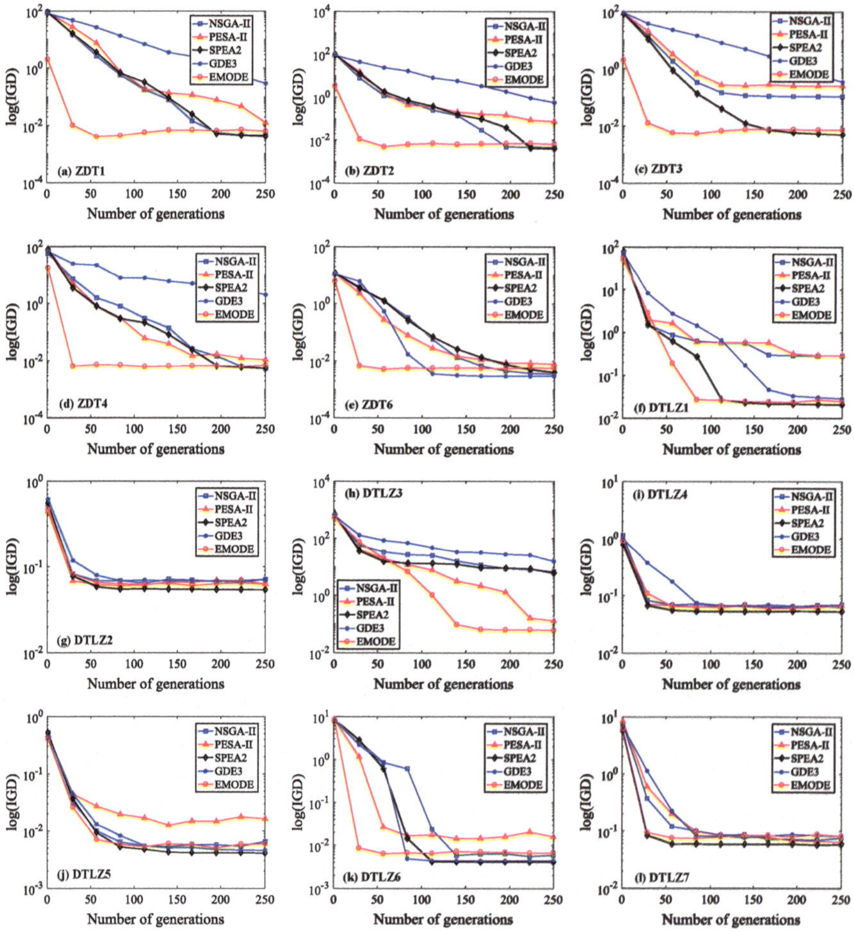

Fig. 9.6 Comparison of IGD for EMODE, NSGA-II, PESA-II, SPEA2, and GD3 on 12 benchmark problems

ZDT3, ZDT6, DTLZ1, DTLZ3, and DTLZ5 than other algorithms, and similar results for the remaining test problems, demonstrating a better performance of EMODE compared to SPEA2, PESA-II, NSGA-II, and GDE3. Table 9.3 shows the results of IGD indicator, and Table 9.4 shows the results of HV indicator for all algorithms on all test problems. Similar results compared to those obtained on GD indicator are found, which indicates that the EMODE provides better approximation Pareto fronts when considering both convergence and diversity on these 12 problems, especially for ZDT4 and DTLZ3. Furthermore, the standard deviation values by EMODE are smaller than those by other algorithms on 8 problems in terms of three indicators, demonstrating the excellent stability of EMODE. Figure 9.4 shows the comparison of Pareto fronts by EMODE and the true Pareto fronts for all test problems. The

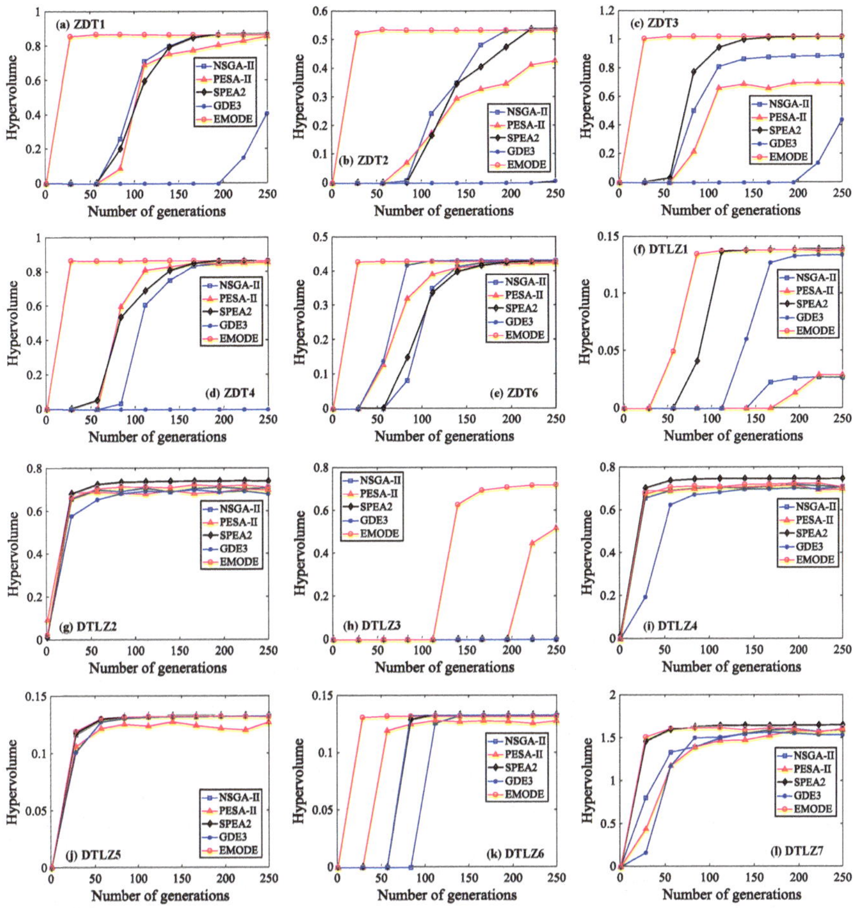

Fig. 9.7 Comparison of HV for EMODE, NSGA-II, PESA-II, SPEA2, and GD3 on 12 benchmark problems

obtained Pareto fronts are very close to the true Pareto fronts, demonstrating the effectiveness of proposed EMODE.

To show the performance of convergence speed of EMODE, the indictors of GD, IGD, and HV with increasing the generation numbers are shown in Figs. 9.5, 9.6 and 9.7. For all test problems, the proposed EMODE has a faster convergence speed compared to NSGA-II, SPEA2, PESA-II, and GDE3 judging from three indicators, which demonstrates the SPX has a significant effect on improving the performance of DE-based multi-objective algorithms.

Overall, all comparisons demonstrate that the proposed EMODE performs well in obtaining solutions that close to the true Pareto front with comparable convergence and diversity in most of test problems. Therefore, adopting the proposed EMODE to conduct parameter identification and update predictive deformation using field observations in an excavation would be a good choice.

9.6 Enhanced Soil Model and Its Finite Element Implementation

Generally, soft clays usually exhibit a significant degree of anisotropy, which develops during deposition. Neglecting anisotropy in soil behavior may lead to highly inaccurate predictions of soil response under loading [21, 70–76]. Therefore, an anisotropic elastoplastic model is necessary to simulate the behavior of clay during an excavation. As a result, the S-CLAY1 model developed by Wheeler et al. [70], was adopted and extended in the proposed updating framework. A brief introduction to S-CLAY1 is presented in Appendix II.

Apart from the anisotropic plasticity, the soil always exhibits a naturally inherent cross-anisotropy of elasticity, which has a significant effect on the lateral wall deflection and ground movement in an excavation [39]. The anisotropic elastic behavior is considered by adopting the following matrix of elastic stiffness:

$$\begin{bmatrix} \dot{\varepsilon}_x \\ \dot{\varepsilon}_y \\ \dot{\varepsilon}_z \\ \dot{\varepsilon}_{xy} \\ \dot{\varepsilon}_{yz} \\ \dot{\varepsilon}_{zx} \end{bmatrix} = \begin{bmatrix} 1/E_v & -\upsilon'_{vv}/E_v & -\upsilon'_{vv}/E_v & 0 & 0 & 0 \\ -\upsilon'_{vv}/E_v & 1/E_h & -\upsilon'_{vh}/E_h & 0 & 0 & 0 \\ -\upsilon'_{vv}/E_v & -\upsilon'_{vh}/E_h & 1/E_h & 0 & 0 & 0 \\ 0 & 0 & 0 & 1/2G_{vh} & 0 & 0 \\ 0 & 0 & 0 & 0 & \left(1+\upsilon'_{vh}\right)/E_h & 0 \\ 0 & 0 & 0 & 0 & 0 & 1/2G_{vh} \end{bmatrix} \begin{bmatrix} \dot{\sigma}'_x \\ \dot{\sigma}'_y \\ \dot{\sigma}'_z \\ \dot{\sigma}'_{xy} \\ \dot{\sigma}'_{yz} \\ \dot{\sigma}'_{zx} \end{bmatrix} \tag{9.11}$$

with $E_h = nE_v, \upsilon'_{vh} = \sqrt{n}\upsilon'_{vv}$, where E_v and E_h are the vertical and horizontal Young's modulus, respectively, υ'_{vv} and υ'_{vh} are the vertical and horizontal Poisson's ratio, respectively, and G_{vh} is the shear modulus. For the stress-controlled isotropic compression with an incremental stress of $\dot{\sigma}'_x = \dot{\sigma}'_y = \dot{\sigma}'_z = \dot{p}'$, one has

$$\dot{\varepsilon}_v = \dot{\varepsilon}_x + \dot{\varepsilon}_y + \dot{\varepsilon}_z = \left(1 - 4\upsilon'_{vv} + 2/n - 2\upsilon'_{vv}/\sqrt{n}\right)\frac{\dot{p}'}{E_v} \tag{9.12}$$

where "x" is the horizontal direction, "y" is the vertical direction, and "z" is the direction perpendicular to the "x–y" plane.

Based on the definition of bulk modulus, $K = \delta p'/\delta\varepsilon_v$, the vertical Young's modulus can be obtained as follows, with the shear modulus, G_{vh}:

$$E_v = \left(1 - 4\upsilon'_{vv} + 2/n - 2\upsilon'_{vv}/\sqrt{n}\right)\left(\frac{1 + e_0}{\kappa}\right)p' \tag{9.13}$$

$$G_{vh} = \frac{\sqrt{n}E_v}{2\left(1 + \sqrt{n}\upsilon'_{vv}\right)} \tag{9.14}$$

where e_0 is initial void ratio, κ is swelling index.

Then, one additional parameter, n, needs to be identified for the anisotropic elasticity in comparison with the isotropic elasticity.

As mentioned previously, the small-strain stiffness behavior significantly influences the prediction of excavation-induced settlement in an excavation [34–36]. According to Hardin and Drnevich [77], the shear modulus G was expressed as:

$$G = \frac{G_0}{(1 + \gamma/\gamma_{\text{ref}})^2} \tag{9.15}$$

where the shear strain γ can be replaced by the deviatoric strain ε_d for the extension. For simplicity's sake, adopting the shear modulus at 0.1% of strain $G_{0.1\%}$ as the input parameter, the initial shear modulus G_0 can be obtained:

$$G_0 = (1 + 0.001/\gamma_{\text{ref}})^2 G_{0.1\%} \tag{9.16}$$

For clayey soils, $G_{0.1\%}$ can be calculated using swelling index κ as:

$$G_{0.1\%} = \frac{(1 - 2\upsilon)}{2(1 + \upsilon)}\frac{(1 + e_0)}{\kappa}p' \tag{9.17}$$

Thus, the shear modulus G can be expressed as:

$$G = \frac{(1 + 0.001/\gamma_{\text{ref}})^2}{(1 + \varepsilon_d/\gamma_{\text{ref}})^2}\frac{(1 - 2\upsilon)}{2(1 + \upsilon)}\frac{(1 + e_0)}{\kappa}p' \tag{9.18}$$

where $\gamma_{\text{ref}} = 7\varepsilon_{70}/3$ with ε_{70} the deviatoric strain level corresponding to 70% of maximum stiffness.

Isotropic and K_0 compression tests with unloading–reloading loops on different clays indicated the increasing of κ due to the volumetric strain [78, 79]. To take into account the effect of volumetric strain for describing this stiffness degradation, a new relationship was proposed by modifying the swelling index κ:

$$\kappa_s = \frac{3(1 - 2\upsilon)(1 + e_0)}{2(1 + \upsilon)}\left(1 + P_{\text{rev}}\frac{3}{7}\left|\frac{\varepsilon^*_{\text{eq}}}{\varepsilon_{70}}\right|\right)^2\frac{p_{\text{at}}}{G_{\text{ref0}}} \tag{9.19}$$

with,

$$G_{\text{ref0}} = \frac{3(1 - 2\upsilon)(1 + e_0)}{2(1 + \upsilon)}\left(1 + P_{\text{rev}}\frac{3}{7}\left|\frac{0.001}{\varepsilon_{70}}\right|\right)^2\frac{p_{\text{at}}}{\kappa} \tag{9.20}$$

Table 9.5 Parameters of proposed model and their determination methods

Parameters	Definition	Determination
ϑ	Poisson's ratio	Values varying from 0.1 to 0.4
e_0	Initial void ratio	–
κ	Swelling index	Measured from the e–log p' curve (unloading–reloading cycle)
λ	Compression index	Measured from the e–log p' curve (loading stage superior to preconsolidation pressure)
M_c	Slope of critical state line in compression	Measured from p'–q curve at critical state
p'_{c0}	Initial size of yield surface	Calculated from the preconsolidation pressure from oedometer test
ε_{70}	Equivalent strain at 70% of maximum stiffness	Measured from the shear modulus degradation curve
n	The ratio of E_h/E_v	Based on directional tests at small strain

where κ_s is the modified swelling index under a low level of strain; e_0 is initial void ratio; ϑ is Poisson's ratio; P_{rev} is the controlling factor for stiffness, $P_{rev} = 1$ for loading conditions while $P_{rev} = 2$ for unloading conditions; $\varepsilon^*_{eq} = \sqrt{(\varepsilon_v)^2 + (\varepsilon_d)^2}$ is the average equivalent strain accounting for the effect of both shear strain and volumetric strain (e.g., see the isotropic loading–unloading–reloading cycle in Hattab and Hicher [79]); ε_{70} is the strain level corresponding to 70% of the maximum stiffness; p_{at} is atmospheric pressure, $p_{at} = 101.325$ kPa; G_{ref0} is the reference secant stiffness corresponding to a 0.001 strain level, and the degradation of shear modulus is cut at G_{ref0}. After obtaining the modified swelling index κ_s, the elastic modulus E_v in Eq. (9.13) can be calculated by replacing κ with κ_s; thus, the influence of small-strain on the elastic modulus is achieved still with κ as input parameter.

Overall, the proposed new model was achieved by incorporating the anisotropic elasticity and the small-strain stiffness into the original S-CLAY1 extended from the modified Cam-Clay (MCC) model. The proposed soil model has eight parameters in total, which can be divided into three groups: (1) the MCC parameters; (2) the anisotropic-elasticity-related parameters; and (3) the small-strain stiffness-related parameters. All parameters and their conventional determination methods are presented in Table 9.5.

In order to conduct the simulation of an excavation in the updating procedure, the proposed model was numerically implemented into the finite element code ABAQUS as a user-defined material. The stresses and hardening parameters were implicitly updated using the cutting plane algorithm proposed by Ortiz and Simo [80]. The implementation was verified by triaxial test and biaxial test with multiple elements before being applied to the excavation simulation. Figure 9.8 shows the geometries

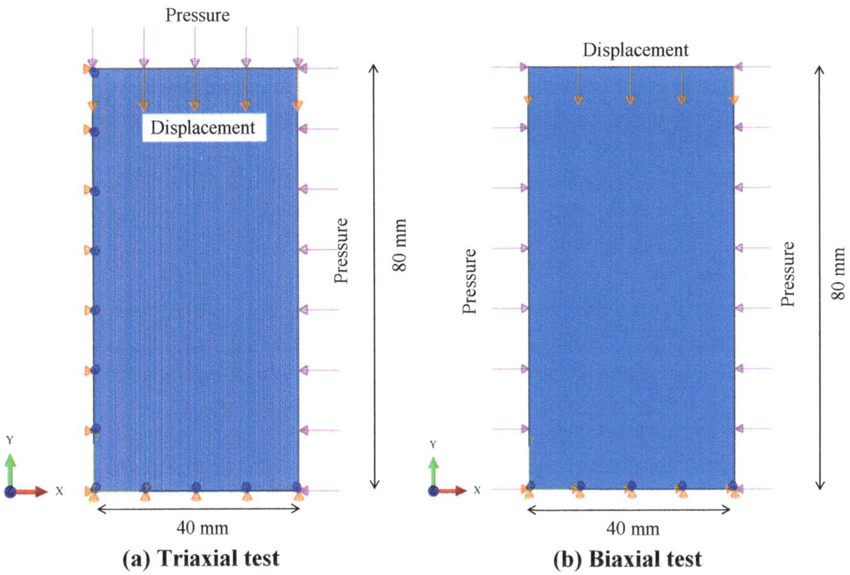

Fig. 9.8 Geometries and boundary conditions of two verification tests: **a** triaxial test; **b** biaxial test

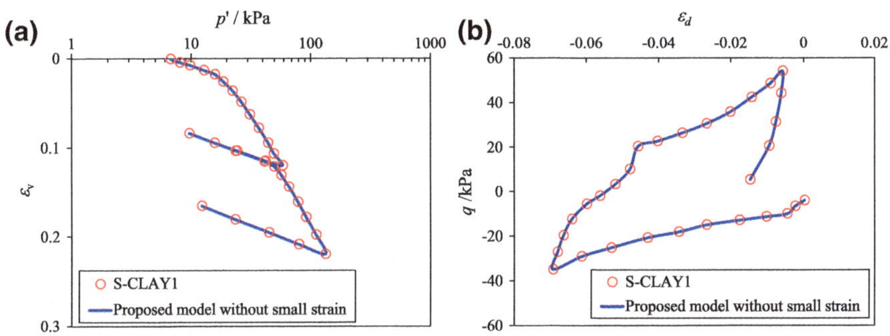

Fig. 9.9 Comparison of verification results for triaxial test: **a** $p' - \varepsilon_\mathrm{v}$; **b** $\varepsilon_\mathrm{d} - q$

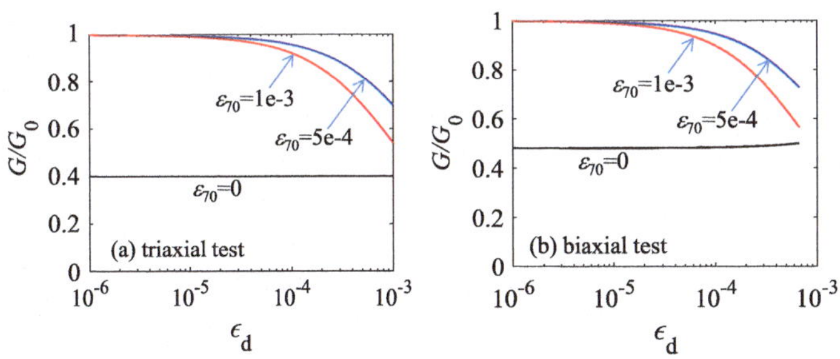

Fig. 9.10 Degradation evolution of normalized shear modulus with increasing of deviatoric strain

and boundary conditions of FEM models for triaxial and biaxial tests. To show the correctness of model implementation, the triaxial test under complex stress path conducted by Karstunen and Koskinen [81] was selected and simulated using the basic model S-CLAY1 (without considering small-strain stiffness) with the same parameters. As shown in Fig. 9.9, for the comparison, a good agreement between FE simulation and that of Karstunen and Koskinen [81] demonstrates that the model implementation is correct. To show the "enhanced" characteristics of the proposed model, the triaxial and biaxial tests with various values of ε_{70} were simulated. Figure 9.10 shows the evolution of normalized shear modulus G/G_0 with deviatoric strain for both tests. It can be seen that the proposed model can well reproduce the degradation of small-strain stiffness for soft clays while the S-CLAY1 does not have such ability. Therefore, the importance of small-strain stiffness for predicting the ground surface settlement in excavation can be highlighted by the proposed model.

9.7 EMODE-Based Updating Predictions in Excavation

The Taipei National Enterprise Center (TNEC), located in Taipei, is a well-documented excavation site for which field-monitoring and soil-testing data have been recorded accurately and completely [39, 82, 83]. Thus, the TNEC excavation was used to illustrate the multi-objective optimization framework for updating soil parameters and predicting deformations based on the observed wall and/or ground responses in the staged excavation.

9.7.1 Numerical Simulation of the TNEC Excavation

At TNEC, the excavation width is 41.2 m, and the length of the 0.9-m-thick diaphragm wall is 35 m. The TNEC excavation was performed using the top-down construction method in seven stages (with a final excavation depth of 19.7 m), in which the wall was supported by 150-mm-thick concrete floor slabs. The soil profile and the excavation depth in each of the seven stages are shown in Fig. 9.11. The ground consists of six alternating silty sand (SM) and silty clay (CL) layers overlying a thick gravel formation. The CL layers have low plasticity for most part and are slightly over-consolidated soft to medium clay. The gravel layer is located 46 m below the ground surface. More information about the TNEC excavation can be found in Ou et al. [82, 83].

Figure 9.12 shows the geometry and the finite element mesh of the TNEC excavation. In the simulations, because of the geometric symmetry, only half of the excavation was modeled under a plane strain condition. Note that the proposed constitutive model of soil is three-dimensional, while it would be switched to two-dimension with $\varepsilon_z = 0$ through the user-defined material (UMAT) in ABAQUS when simulating the excavation under plane strain condition. The overall model size was 100 m

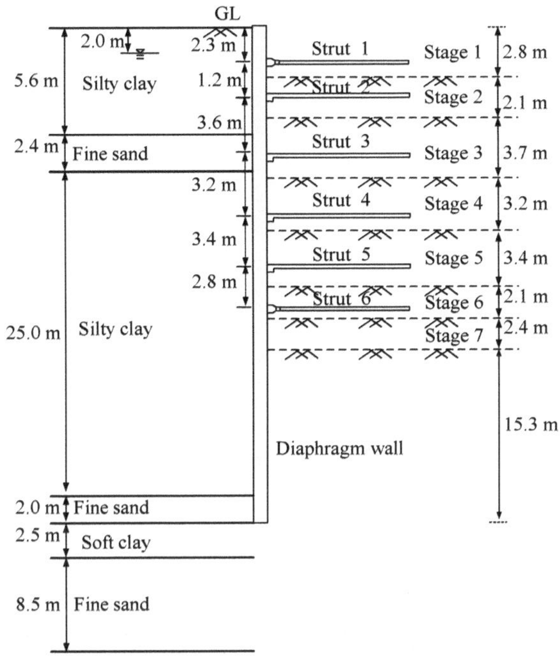

Fig. 9.11 Soil profile and the excavation depth in each of the seven stages in TNEC

Table 9.6 Parameters of MC used in finite element analysis

Depth (m)	γ (kN/m^3)	ϕ' (°)	E (MPa)	K_0	ϑ
5.6–8.0	18.9	30	6.8	0.49	0.3
33.0–35.0	19.6	33	26.5	0.49	0.3
37.5–46.0	19.6	35	30.0	0.47	0.3

Note: c and ψ are zero for three layers of sand

long and 45 m high, which is considered large enough to avoid boundary constraints. The horizontal displacement was fixed on both left and right sides. The horizontal and vertical displacements were fixed on the bottom. The excavation was conducted in seven stages according to Ou et al. [82, 83]. The retaining structure, including the retaining wall and the struts, were assumed to be linear elastic. The element type of the soil was a four-node bilinear rectangular element. A spring element and a two-node linear beam element were adopted for the strut and the retaining wall, respectively. According to Ou et al. [82, 83], Poisson's ratio of the diaphragm wall was assumed to be 0.2; The average nominal axial stiffness per unit width for the first strut level and the second strut level were 14,980 and 64,363 kN/m/m, respectively. The friction coefficient of the wall–soil interface, μ, was assumed to be a constant value of 0.273 due to the slight effect on wall deflection [30, 31]. In this study, a drained condition was assumed and the effective stress constitutive models were used

Table 9.7 Parameters of proposed model used in finite element analysis

Depth (m)	γ (kN/m^3)	λ	κ	e_0	M	POP (kPa)	K_0	ϑ	ε_{70}	n
0–5.6	18.3	0.10	0.012	1.067	1.09	20	1.0	[0.1, 0.4]	[10^{-6}, 10^{-3}]	[0.1, 3.0]
8.0–33.0	18.9	0.15	0.018	0.820	1.02	1.0	0.70	[0.1, 0.4]	[10^{-6}, 10^{-3}]	[0.1, 3.0]
35.0–37.5	18.2	0.10	0.012	0.824	1.05	1.0	0.52			

for both sandy and clayey soils, which was consistent with many previous studies [30, 51, 83].

In the finite element analysis, the sand layers were modeled by the Mohr–Coulomb (MC) model without considering the small-strain stiffness due to the thin thickness. The parameters of the MC are summarized in Table 9.6 according to previous studies [30, 304].

Three clay layers were modeled by the proposed soil model. To focus on identifying the important parameters which are difficult to determine using laboratory tests, the parameters corresponding to MCC measured from conventional oedometer and triaxial tests for three layers were not considered in the optimization procedure. These parameters are referenced from previous studies [34, 39, 84], as summarized in Table 9.7.

9.7.2 Parameter Identification and Forward Prediction

Due to the importance of small-strain stiffness and the anisotropic elasticity of natural clay, the small-strain stiffness-related parameter ε_{70}, and the anisotropic-elasticity-related parameter n, together with a Poisson's ratio ϑ of three clay layers were deemed the optimization variables, which were identified by using EMODE based on field observations stage by stage as excavation proceeded. Since the clay layer at a depth of 35.0–37.5 m, is far from the excavation zone, the parameters ε_{70}, n, and ϑ were considered to be identical to those in the above clay layer. The wide ranges of ε_{70}, n, and ϑ with their intervals for three clay layers are also summarized in Table 9.7.

During excavation, the large bending moment of the diaphragm wall and the occurrence of cracks in the concrete can result in a reduction of stiffness in the concrete structures [30, 31, 83]. Furthermore, poor quality of joint installations between temporal horizontal struts and the diaphragm wall also influences the support performance. Thus, these factors also have a significant influence on the wall and ground responses in the excavation. Note that the construction quality for the joint between temporal horizontal struts and the diaphragm wall can be reflected by the change in preloading force in temporal horizontal struts. Therefore, in order to consider the effect of the structure on the deformation of excavation, the stiffness of the diaphragm wall, E_{wall}, the axial stiffness of the concrete floor slabs, E_{slab}, and the preloading force of the temporal horizontal struts, F_{struts}, were also considered to be optimization variables and identified along with the soil parameters in the proposed EMODE based on field observations stage by stage. The ranges and intervals of three structurally related parameters in the optimization are presented in Table 9.8.

Overall, ten parameters in total, i.e., the small-strain-stiffness-related parameter, ε_{70}, the anisotropic-elastic-related parameter, n, and Poisson's ratio, ϑ, for two clay layers ($2 \times 3 = 6$), one parameter related to the stiffness of the diaphragm wall, E_{wall}, one parameter related to the axial stiffness of the concrete floor slabs, E_{slab}, and the parameter for the preloading force, F_{struts}, for the first and other temporal horizontal

Table 9.8 Ranges and intervals of three structurally related parameters in optimization

Parameters	E_{wall} (kPa)	E_{slab} (kPa)	F_{struts} (kN)
Ranges	$[10^2, 10^9]$	$[10^2, 10^9]$	$[10^0, 10^3]$
Intervals	1	1	1

struts, respectively ($1 \times 2 = 2$), were set for optimization and prediction updates stage by stage.

In this case, the EMODE-based optimization framework with two objectives was adopted. One objective represents the error of wall deflection between observation and prediction, noted as "Objective1" and the other represents that of ground settlement noted as "Objective2". The two Objectives are defined as:

$$\text{Objective1} = \sqrt{\frac{1}{N}\sum_{i=1}^{N}\left(\frac{(\text{wall deflection})_i^{\text{obs}} - (\text{wall deflection})_i^{\text{num}}}{(\text{wall deflection})_i^{\text{obs}}}\right)^2} \times 100$$

$$\text{Objective2} = \sqrt{\frac{1}{N}\sum_{i=1}^{N}\left(\frac{(\text{ground settlement})_i^{\text{obs}} - (\text{ground settlement})_i^{\text{num}}}{(\text{ground settlement})_i^{\text{obs}}}\right)^2} \times 100 \tag{9.21}$$

where N is the number of observation points. "obs" and "num" represents the observed data and simulated data, respectively.

For EMODE, the initial population size was taken to be ten times the number of decision variables for a uniform EMODE testing environment. Thus, a population size of 100 was used for each generation. The maximum generation was set to 50, which was reasonable for balancing calculation cost and the accuracy of the results. Other settings were the same as those used in the benchmark testing. For comparison, the same updating process was conducted by original multi-objective DE (i.e., the third version of generalized differential evolution (GDE3) in this study). Other settings for GDE3 are same as that used in the benchmark testing.

Figure 9.13 shows the updating procedure for the TNEC excavation. Since the ground settlement for Stage 1 is unavailable, the updating procedure began from the Stage 2. Prior to the current stage of excavation, both the wall deflections and ground settlements of the former stages were used as the objective data to identify the soil and structure parameters. With the identified parameters corresponding to Pareto solutions, the possible ranges for the wall and ground responses for the next unexcavated stage were predicted prior to the next stage of excavation. The possible ranges of deformation that were predicted could be considered as indicators for assessing the safety of the subsequent excavations and the serviceability of adjacent buildings [4], which is useful in practice. As noted previously, the above process was conducted as the excavation proceeded until the excavation was finished (up to Stage 7).

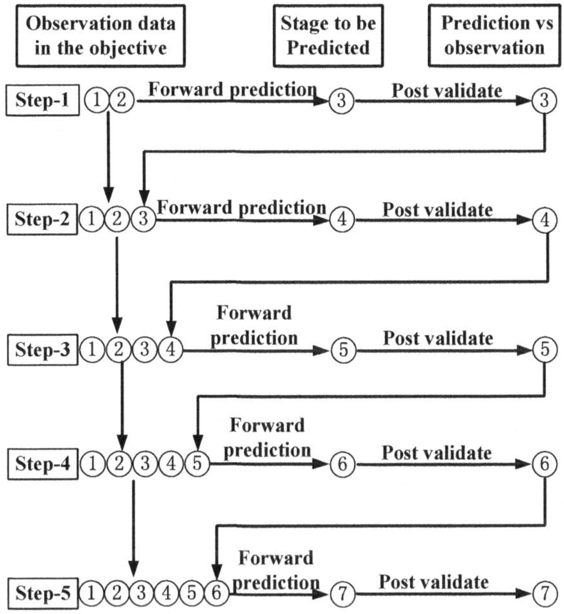

Fig. 9.13 Updating procedure for TNEC excavation

9.7.3 *Results and Discussion*

Following the procedure shown in Fig. 9.1, the parameters for the soil and structure in Stage 2 were first identified by observation, and then were followed by Stages 3 to 6. The corresponding Pareto-optimal solutions for each optimization method are shown in Fig. 9.14. Note that the obtained Pareto fronts in later stages are not as smooth as those obtained in earlier stages. This is due to more measurements involved and more soil layers excavated in objectives. Similar phenomena can be found in Huang et al. [31]. Compared to GDE3, the optimal solutions obtained were distributed uniformly on the Pareto front, which indicates that the proposed EMODE has outstanding ability in detecting all possible solutions for a complex multi-objective optimization problem. It is noted that the approximate Pareto front for each stage was not smooth compared to those obtained in the mathematical benchmark tests, which is reasonable for real engineering problems.

In the early stages, the identification of parameters and updates to excavation predictions are not meaningful because of the inevitable changes in the deformation pattern [5]. Therefore, the predicted performance in the early stages may be inadequate. Although there are discrepancies in the predicted maximum wall deflection and ground settlement in the early stages, the deformation shape of the diagram wall and the zone of influence in ground settlement induced by excavation were well predicted. The good agreement for the ground settlement also indicates the importance of considering the small-strain stiffness characteristics. Compared to previous studies using MCC for clay layers [31], the enhanced soil model generally improves the

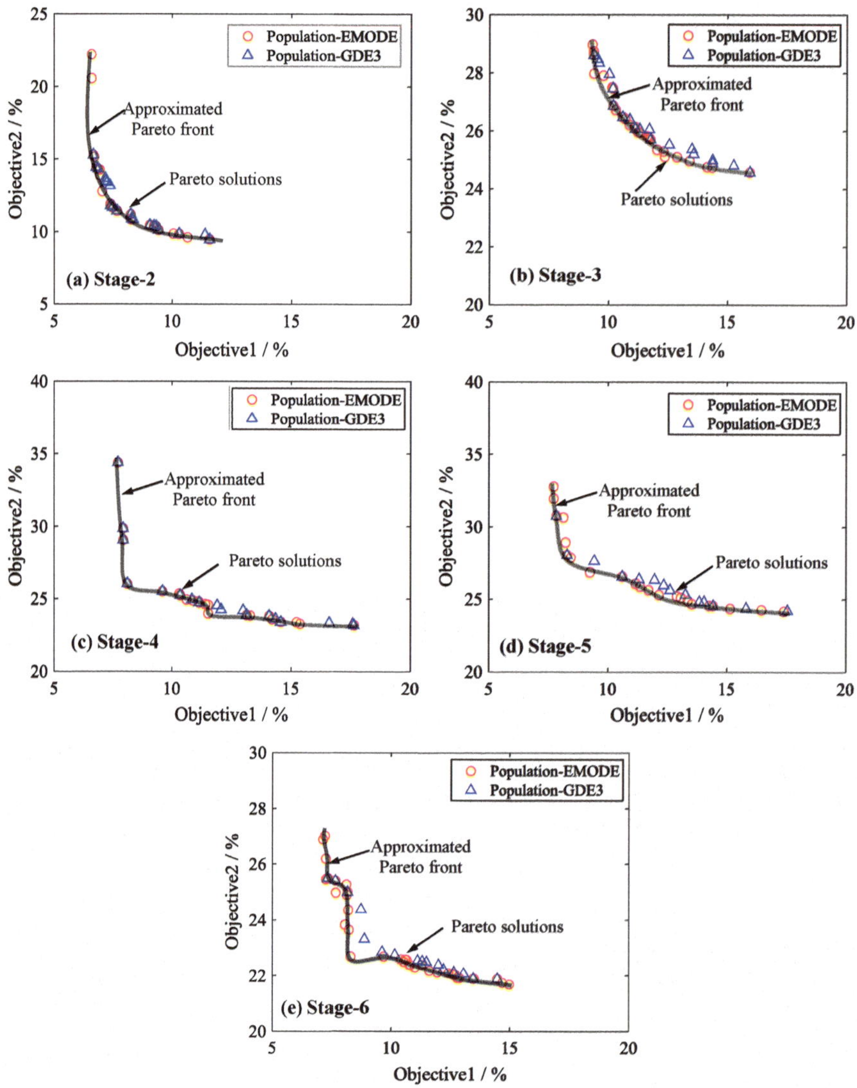

Fig. 9.14 Pareto-optimal solutions obtained based on the observations of stage from 2 to 6

deformation behavior of the excavation. Moreover, only the maximum values of wall deflection and ground settlement were captured in studies using the Bayesian updating method [5] without showing the full deformations, and this was significantly improved upon in the proposed MOOP-based updating method.

Using the updated Pareto-optimal parameters of each stage, the wall deflection and ground surface settlement of subsequent stage were predicted prior to the excavation. For a given stage, the mean and standard deviation values of predictions in terms

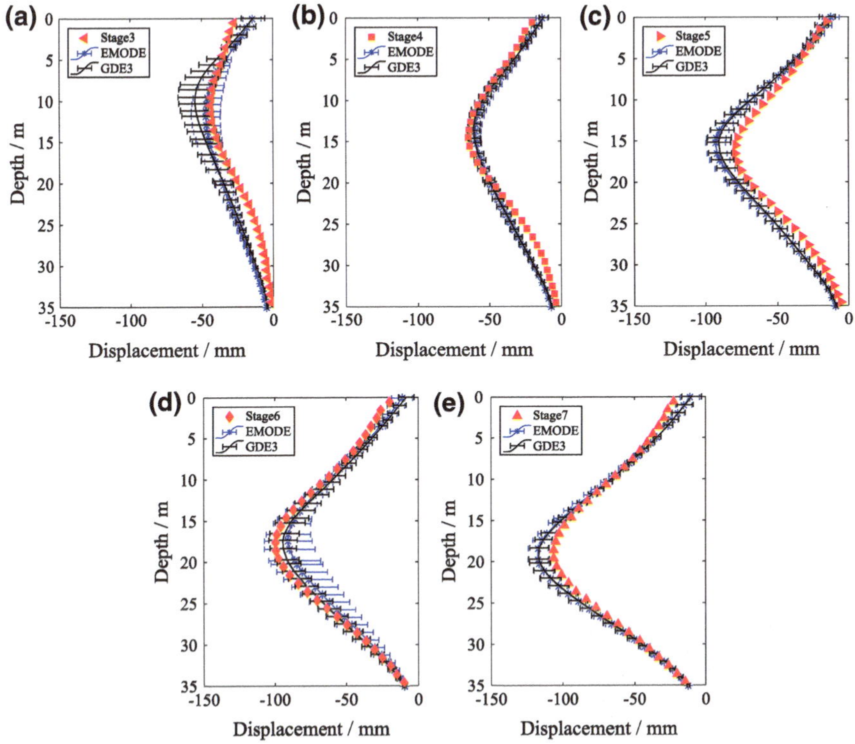

Fig. 9.15 Measured and predicted wall deflection using Pareto-optimal parameter values based on the previous stage of excavation

of wall deflection and ground settlement were computed and plotted. Note that the proposed updating procedure starts from stage 2. Thus the predictions begin from stage 3. Figure 9.15 shows the comparison of wall deflection between the observations and predictions with error bars, and Fig. 9.16 for the ground surface settlement. It can be seen that the performance of all predictions is acceptable and improved with the process of excavation for both multi-objective algorithms. The predictions of wall deflection by EMODE are approximately identical between both algorithms, and the predictions of ground settlement by EMODE are more accurate than those by GDE3. Therefore, the proposed EMODE is more effective with good performance. Note that for the predictions of ground surface settlement in early stages 2 and 3 are more different from observations than later stages, which is similar to that in the Bayesian updating [4, 5]. Thus, the effectiveness of optimization updating in the early stages is compromised, however, which is acceptable in practice since the deformations are small.

Figure 9.17 shows the comparisons of optimal parameters from stages 2 to 6 in the TNEC excavation for both used algorithms. The x-axis of the graph, 1 to 7, denotes the seven stages of the excavation. For EMODE, a reasonable narrow

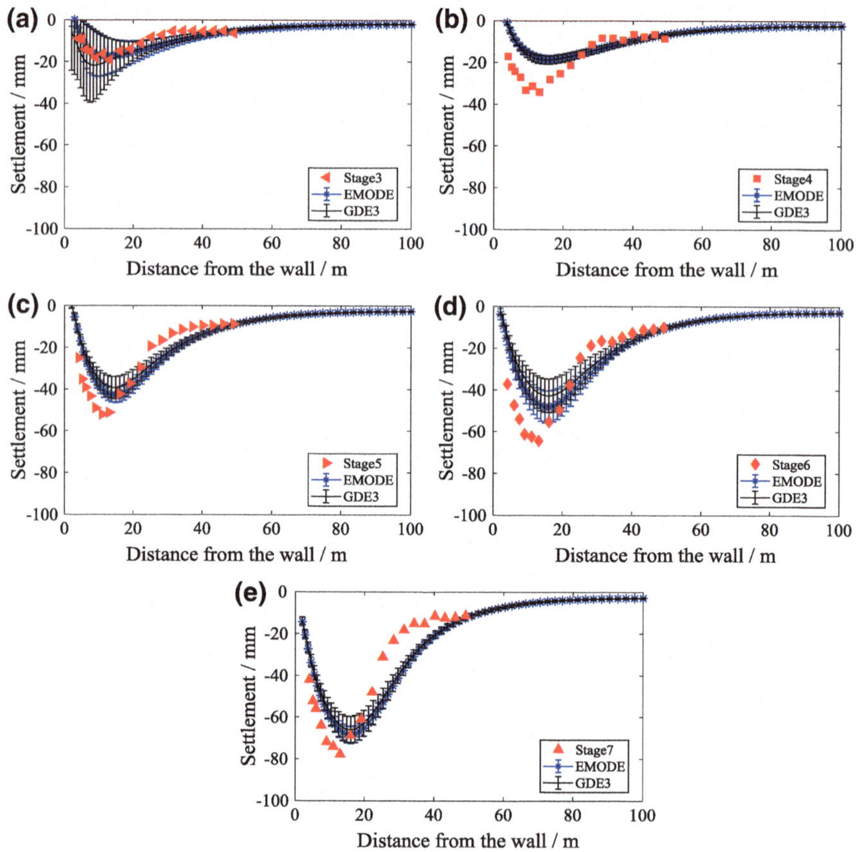

Fig. 9.16 Measured and predicted ground surface settlement using Pareto-optimal parameter values based on the previous stage of excavation

range from 0.25 to 0.35 of Poisson's ratio was obtained for three clay layers. The cross-anisotropic elasticity ratio n for the layer 0–5.6 m is 1.5–2.0 and for the layer 8.0–33 m is around 1.5, which are consistent with the test results obtained by [39]. A small range of ε_{70} from 5×10^{-4} to 10^{-3} was obtained for each excavation stage, which is slightly smaller than that from the empirical equation using the Atteberg limits $\varepsilon_{70} = 5 \times 10^{-5}\exp(0.0693 I_p)$ according to Vucetic and Dobry [85]. For the stiffness of the diagram wall, it's value degrades from 60 to 7 GPa with progression of the excavation stages, which accords with the reduction in the concrete stiffness resulting from the large bending moment of the diaphragm wall and the occurrence of cracks in the concrete [30, 31]. For the stiffness of slabs, an average value of 2×10^5 kPa was obtained, which is reasonable according to Zhao et al. [30]. The forces for preloading for two struts obtained from the optimization are also within a reasonable range [39]. All the optimal parameters demonstrate that the proposed MOOP-based parameter identification procedure combined with the advanced constitutive model

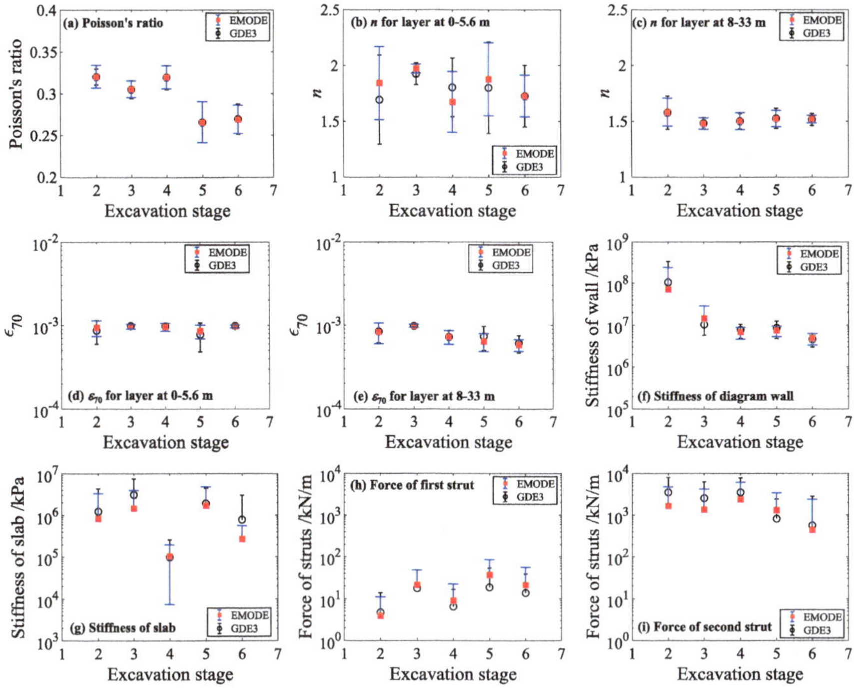

Fig. 9.17 Optimal parameters of stage from 2 to 6 for TNEC excavation

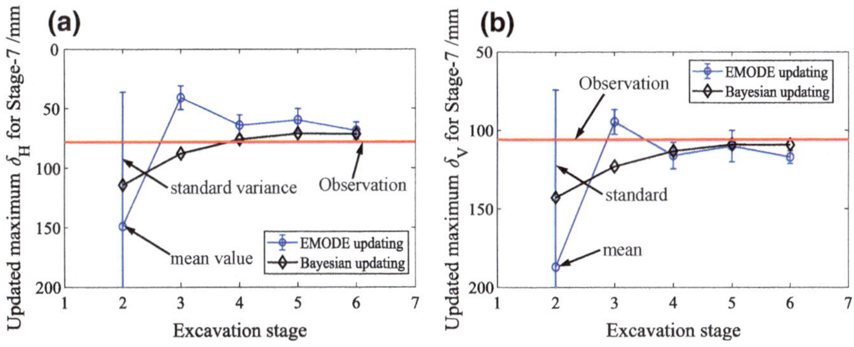

Fig. 9.18 Comparison between proposed optimization-based updating and observations for **a** the wall deflection and **b** the settlement predictions at a target depth of 19.7 m using the parameters obtained from one of the previous excavation stages (2, 3, 4, 5, or 6)

is effective. Compared to EMODE, the ranges of all parameters obtained by GDE3 are also reasonable. The accuracy of these identified parameters should be further evaluated using the performance of predictions in terms of wall deflection and ground settlement.

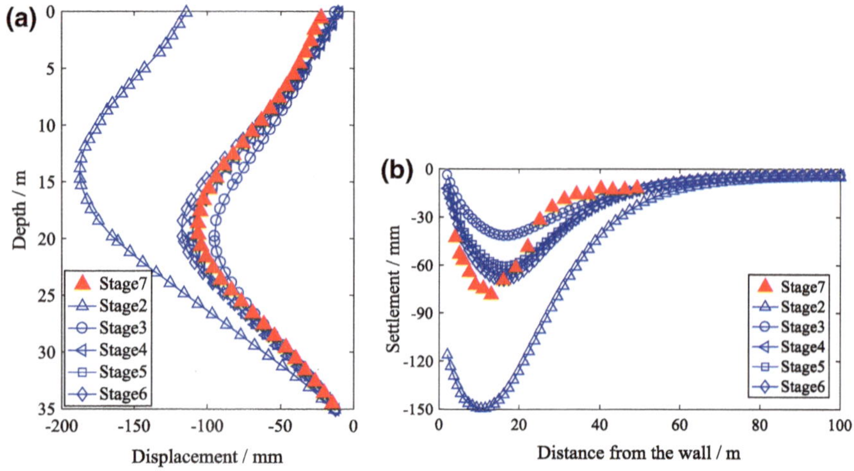

Fig. 9.19 Comparison between the predicted mean values and observations for **a** the wall deflection and **b** the settlement at a target depth of 19.7 m using the parameters obtained from one of the previous excavation stages (2, 3, 4, 5, or 6)

To evaluate the performance for effectiveness and efficiency, the wall deflection and the settlement predictions at a target depth of 19.7 m using the parameters obtained from one of the previous Stages (2, 3, 4, 5, or 6) of the excavation were first predicted. The updated mean values and standard variances in predictions in terms of maximum wall deflection and ground settlement were compared to those obtained by Bayesian updating by Juang et al. [5], shown in Fig. 9.18. Note that the updated mean values and standard variances in predictions are not monotonic because of soil uncertainties and monitoring accuracy of field data. The mean predicted values for wall deflection along with the excavation depth and for ground settlement with distance far away from the wall are shown in Fig. 9.19. By progressively conducting the excavation and involving a number of observations into the objective, the predictions for wall deflection and settlement in Stage 7 gradually became accurate and close to the observations, which resembles the Bayesian updating [5]. Moreover, compared to the Bayesian updating, not only the accurate, updated maximum predictions of wall deflection and settlement, but also the deformation shape of the wall and the zone of influence in the excavation-induced settlement were achieved with the proposed MOOP-based updating method.

The identified parameters in Stage 3 were then employed to predict the responses of Stage 4 to Stage 7. All predictions of wall deflection and ground settlement were compared to the results by Huang et al. [31] using modified Cam-Clay model: Fig. 9.20 shows the comparison of wall deflection and Fig. 9.21 shows the comparison of ground settlement. It can be seen that the SCLAY1-SS combined with EMODE gives more accurate predictions of wall deflection and approximately identical predictions of ground settlement. All results demonstrate that the SCLAY1-SS

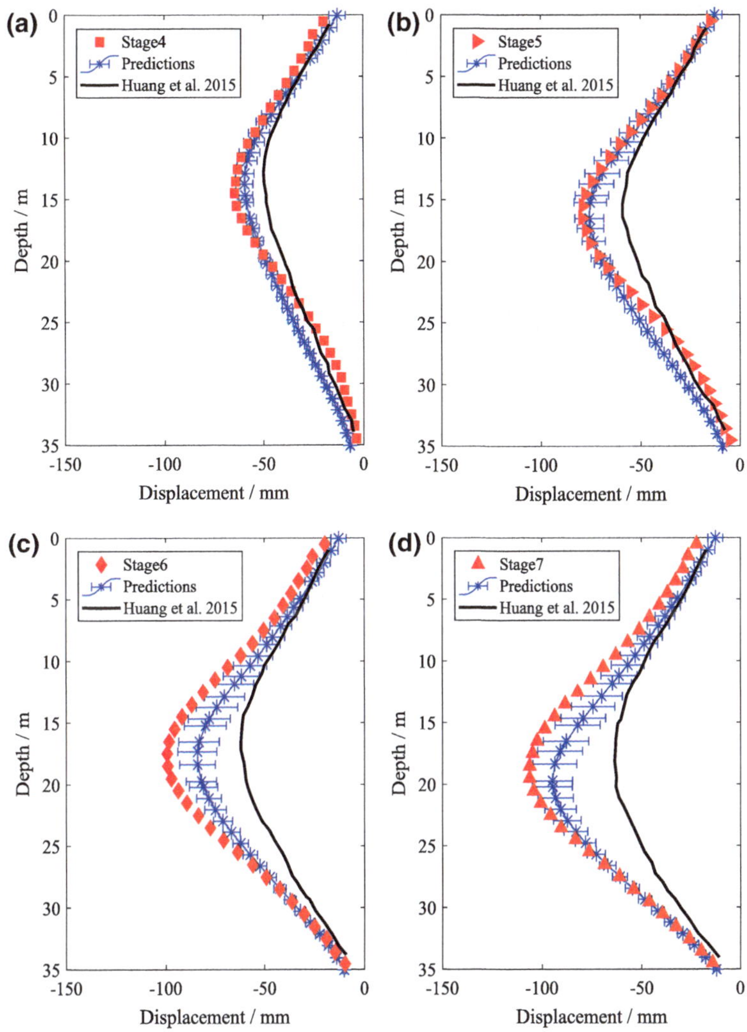

Fig. 9.20 Comparison between the predictions and observations for wall deflection at Stage 4 to Stage 7 using the parameters obtained from Stage 3 by EMODE and GED3

combined with EMODE is more efficient in MOOP-based parameter identification from excavation than the others.

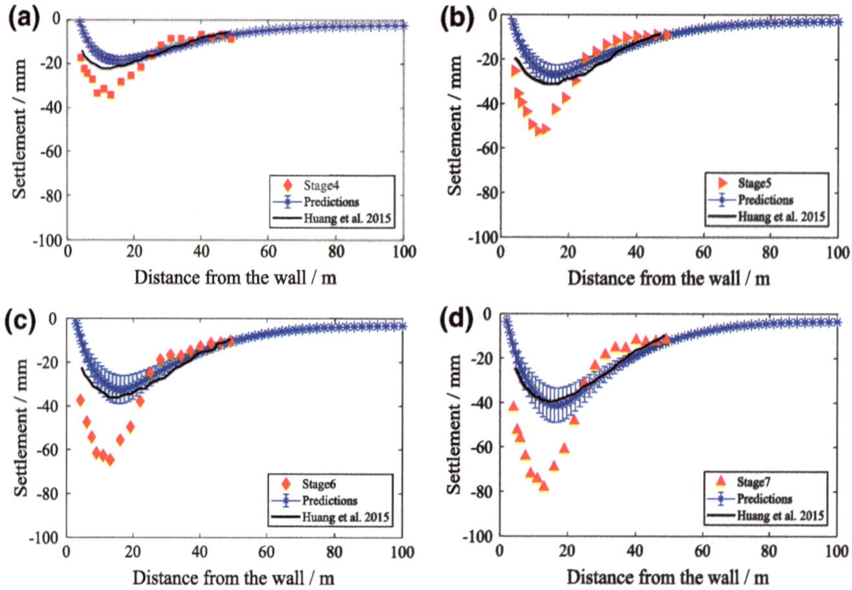

Fig. 9.21 Comparison between the predictions and observations for ground settlement at Stage 4 to Stage 7 using the parameters obtained from Stage 3 by EMODE and GED3

9.8 Limitations

First, the wall and ground responses in excavation may be influenced by many factors, such as construction quality (workmanship), temperature variation [5], and soil creep [82], in addition to the soil–wall interaction mechanism. However, apart from the construction quality, other factors were not considered in this case.

Second, the soils in the excavation are assumed to be drained. As the construction of an excavation usually takes several months, the soil behaviors are more likely partially drained [30]. A coupled soil–water analysis is more appropriate for such a condition.

These limitations are out of the scope of this study and will be considered in future studies.

9.9 Summary

An effective multi-objective optimization (MOOP)-based updating framework was established for identifying the soil- and structure-related parameters in a multistage-braced excavation. The proposed framework involves (1) an enhanced multi-objective differential evolution algorithm (EMODE), (2) a finite element code with an enhanced elastoplastic model named SCLAY1-SS, and (3) an identification procedure for parameters using field measurements.

To improve the performance of multi-objective optimization, EMODE was developed incorporating a newly developed outstanding Simplex to improve the searching ability and convergence speed. The proposed EMODE was validated by comparison with NSGA-II on five classical benchmark tests, which demonstrated that EMODE can detect more Pareto solutions and has a faster convergence speed.

To improve the performance in simulating soil behavior, an advanced soil model called SCLAY1-SS was proposed that considers the anisotropic yield surface, the anisotropic elasticity, and the small-strain stiffness. In the proposed framework, the finite element analysis associated with SCLAY1-SS was used as the analysis tool to simulate the excavation, which is different from traditional Bayesian updates using a semi-empirical model.

Then, based on the proposed framework using EMODE and SCLAY1-SS, an identification procedure for parameters using field measurements was proposed. The effectiveness and flexibility of the proposed framework were examined through a case study at the TNEC excavation in seven stages. Apart from the soil parameters, the stiffness of the diagram wall and the slab, and the preloading force of the temporal horizontal struts were also identified based on all known observations of wall deflection and ground surface settlement induced by excavation. Using the identified Pareto-optimal parameters, the subsequent wall deflection and ground surface settlement were predicted prior to the next stage of excavation. Compared to Bayesian updates, not only the maximum deformation, but also the deformation shape of the diagram wall and the zone of influence for settlement can be accurately predicted using the proposed MOOP-based updating framework.

All the comparisons demonstrate that the proposed framework is effective and efficient for updating soil and structure parameters in a multistage excavation. More details can be found in Jin et al. [86].

References

1. Shen S-L, Wu H-N, Cui Y-J, Yin Z-Y (2014) Long-term settlement behaviour of metro tunnels in the soft deposits of Shanghai. Tunn Undergr Space Technol 40(309–323)
2. Xu Y-S, Shen S-L, Ren D-J, Wu H-N (2016) Analysis of factors in land subsidence in Shanghai: a view based on a strategic environmental assessment. Sustainability 8(6):573
3. Shen S, Wang Z, Cheng W (2017) Estimation of lateral displacement induced by jet grouting in clayey soils. Géotechnique 1–10
4. Hsiao EC, Schuster M, Juang CH, Kung GT (2008) Reliability analysis and updating of excavation-induced ground settlement for building serviceability assessment. J Geotech Geoenviron Eng 134(10):1448–1458
5. Hsein Juang C, Luo Z, Atamturktur S, Huang H (2012) Bayesian updating of soil parameters for braced excavations using field observations. J Geotech Geoenviron Eng 139(3):395–406
6. Qi X-H, Zhou W-H (2017) An efficient probabilistic back-analysis method for braced excavations using wall deflection data at multiple points. Comput Geotech 85(186–198)

7. Wang L, Ravichandran N, Juang CH (2012) Bayesian updating of KJHH model for prediction of maximum ground settlement in braced excavations using centrifuge data. Comput Geotech 44(3):1–8
8. Špačková O, Straub D (2013) Dynamic Bayesian network for probabilistic modeling of tunnel excavation processes. Comput Aided Civ Infrastruct Eng 28(1):1–21
9. Zhang R, Mahadevan S (2000) Model uncertainty and Bayesian updating in reliability-based inspection. Struct Saf 22(2):145–160
10. Sun H, Betti R (2015) A hybrid optimization algorithm with Bayesian inference for probabilistic model updating. Comput Aided Civ Infrastruct Eng 30(8):602–619
11. Moré JJ (1978) The Levenberg-Marquardt algorithm: implementation and theory. Numerical analysis. Springer, pp 105–116
12. Marquardt DW (1963) An algorithm for least-squares estimation of nonlinear parameters. J Soc Ind Appl Math 11(2):431–441
13. Head JD, Zerner MC (1985) A Broyden—Fletcher—Goldfarb—Shanno optimization procedure for molecular geometries. Chem Phys Lett 122(3):264–270
14. Nelder JA, Mead R (1965) A simplex method for function minimization. Comput J 7(4):308–313
15. Calvello M, Finno RJ (2004) Selecting parameters to optimize in model calibration by inverse analysis. Comput Geotech 31(5):410–424
16. Finno RJ, Calvello M (2005) Supported excavations: observational method and inverse modeling. J Geotech Geoenviron Eng 131(7):826–836
17. Rechea C, Levasseur S, Finno R (2008) Inverse analysis techniques for parameter identification in simulation of excavation support systems. Comput Geotech 35(3):331–345
18. Jin Y-F, Yin Z-Y, Shen S-L, Hicher P-Y (2016) Selection of sand models and identification of parameters using an enhanced genetic algorithm. Int J Numer Anal Methods Geomech 40(8):1219–1240
19. Jin Y-F, Yin Z-Y, Shen S-L, Zhang D-M (2017) A new hybrid real-coded genetic algorithm and its application to parameters identification of soils. Inverse Prob Sci Eng 25(9):1343–1366
20. Yin Z-Y, Jin Y-F, Shen JS, Hicher P-Y (2018) Optimization techniques for identifying soil parameters in geotechnical engineering: comparative study and enhancement. Int J Numer Anal Methods Geomech 42(1):70–94
21. Yin Z-Y, Jin Y-F, Shen S-L, Huang H-W (2017) An efficient optimization method for identifying parameters of soft structured clay by an enhanced genetic algorithm and elastic–viscoplastic model. Acta Geotech 12(4):849–867
22. Deep K, Thakur M (2007) A new mutation operator for real coded genetic algorithms. Appl Math Comput 193(1):211–230
23. Papon A, Riou Y, Dano C, Hicher PY (2012) Single-and multi-objective genetic algorithm optimization for identifying soil parameters. Int J Numer Anal Methods Geomech 36(5):597–618
24. Yin ZY, Hicher PY (2008) Identifying parameters controlling soil delayed behaviour from laboratory and in situ pressuremeter testing. Int J Numer Anal Methods Geomech 32(12):1515–1535
25. Ye L, Jin Y-F, Shen S-L, Sun P-P, Zhou C (2016) An efficient parameter identification procedure for soft sensitive clays. J Zhejiang Univ Sci A 17(1):76–88
26. Jin Y-F, Yin Z-Y, Wu Z-X, Zhou W-H (2018) Identifying parameters of easily crushable sand and application to offshore pile driving. Ocean Eng 154(416–429)
27. Holland J (1989) Genetic algorithms in search, optimization and machine learning. Addison-Wesley, Reading, MA
28. Hashash YMA, Levasseur S, Osouli A, Finno R, Malecot Y (2010) Comparison of two inverse analysis techniques for learning deep excavation response. Comput Geotech 37(3):323–333
29. Price K, Storn R (1997) Differential evolution: a simple evolution strategy for fast optimization. Dr Dobb's J 22(4):18–24
30. Zhao B, Zhang L, Jeng D, Wang J, Chen J (2015) Inverse analysis of deep excavation using differential evolution algorithm. Int J Numer Anal Methods Geomech 39(2):115–134
31. Huang Z, Zhang L, Cheng S, Zhang J, Xia X (2014) Back-analysis and parameter identification for deep excavation based on Pareto multiobjective optimization. J Aerosp Eng 28(6):A4014007

32. Sun Y, Jiang Q, Yin T, Zhou C (2018) A back-analysis method using an intelligent multi-objective optimization for predicting slope deformation induced by excavation. Eng Geol 239(214–228)
33. Chen SG, Zhao J (2002) Modeling of tunnel excavation using a hybrid DEM/BEM method. Comput Aided Civ Infrastruct Eng 17(5):381–386
34. Kung TC, Ou CY, Juang CH (2009) Modeling small-strain behavior of Taipei clays for finite element analysis of braced excavations. Comput Geotech 36(1–2):304–319
35. Kung TC, Hsiao CL, Junag CH (2007) Evaluation of a simplified small-strain soil model for analysis of excavation-induced movements. Can Geotech J 44(6):726–736
36. Schädlich B, Schweiger HF (2012) Influence of anisotropic small strain stiffness on the deformation behavior of geotechnical structures. Int J Geomech 13(6):861–868
37. Jiang M, Li T, Chareyre B (2016) Fabric rates applied to kinematic models: evaluating elliptical granular materials under simple shear tests via discrete element method. Granular Matter 18(3):1–15
38. Jiang M, Sima J, Li L, Zhou C, Cui L (2017) Investigation of influence of particle characteristics on the non-coaxiality of anisotropic granular materials using DEM. Int J Numer Anal Methods Geomech 41(2):198–222
39. Teng FC, Ou CY, Hsieh PG (2014) Measurements and numerical simulations of inherent stiffness anisotropy in soft Taipei clay. J Geotech Geoenviron Eng 140(1):237–250
40. Jiang M, Shen Z, Utili S (2016) DEM modeling of cantilever retaining excavations: implications for lunar constructions. Eng Comput 33(2):366–394
41. Levasseur S, Malécot Y, Boulon M, Flavigny E (2008) Soil parameter identification using a genetic algorithm. Int J Numer Anal Methods Geomech 32(2):189–213
42. Poles S, Fu Y, Rigoni E (2009) The effect of initial population sampling on the convergence of multi-objective genetic algorithms. Multiobjective Programming and Goal Programming. Springer, Berlin, Heidelberg, pp 123–133
43. Zitzler E, Laumanns M, Thiele L (2001) SPEA2: improving the strength Pareto evolutionary algorithm
44. Deb K, Pratap A, Agarwal S, Meyarivan T (2002) A fast and elitist multiobjective genetic algorithm: NSGA-II. Evol Comput IEEE Trans 6(2):182–197
45. Da Ronco CC, Benini E (2013) A simplex crossover based evolutionary algorithm including the genetic diversity as objective. Appl Soft Comput 13(4):2104–2123
46. Panduro MA, Brizuela CA, Garza J, Hinojosa S, Reyna A (2013) A comparison of NSGA-II, DEMO, and EM-MOPSO for the multi-objective design of concentric rings antenna arrays. J Electromag Waves Appl 27(9):1100–1113
47. Nourbakhsh A, Safikhani H, Derakhshan S (2011) The comparison of multi-objective particle swarm optimization and NSGA II algorithm: applications in centrifugal pumps. Eng Optim 43(10):1095–1113
48. Alatas B, Akin E, Karci A (2008) MODENAR: Multi-objective differential evolution algorithm for mining numeric association rules. Appl Soft Comput 8(1):646–656
49. Abbass HA, Sarker R, Newton C (2001) PDE: a Pareto-frontier differential evolution approach for multi-objective optimization problems. In: 2001 congress on 2001 proceedings of the evolutionary computation. IEEE, pp 971–978
50. Babu B, Jehan MML (2003) Differential evolution for multi-objective optimization. In: The 2003 congress on evolutionary computation, 2003 CEC'03. IEEE, pp 2696–2703
51. Xue F, Sanderson AC, Graves RJ (2003) Pareto-based multi-objective differential evolution. In: The 2003 congress on evolutionary computation, 2003 CEC'03. IEEE, pp 862–869
52. Parsopoulos KE, Tasoulis DK, Pavlidis NG, Plagianakos VP, Vrahatis MN (2004) Vector evaluated differential evolution for multiobjective optimization. In: CEC2004 congress on evolutionary computation, 2004. IEEE, pp 204–211
53. Babu B, Anbarasu B (2005) Multi-objective differential evolution (MODE): an evolutionary algorithm for multi-objective optimization problems (MOOPs). In: Proceedings of international symposium and 58th annual session of IIChE2005

54. Angira R, Babu B (2005) Non-dominated sorting differential evolution (NSDE): an extension of differential evolution for multi-objective optimization. In: IICAI2005, pp 1428–1443
55. Kukkonen S, Lampinen J (2005) GDE3: the third evolution step of generalized differential evolution. In: The 2005 IEEE congress on evolutionary computation, 2005. IEEE, pp 443–450
56. Huang VL, Zhao SZ, Mallipeddi R, Suganthan PN (2009) Multi-objective optimization using self-adaptive differential evolution algorithm. In: IEEE Congress on evolutionary computation, 2009 CEC'09. IEEE, pp 190–194
57. Venske SM, Gonçalves RA, Delgado MR (2014) ADEMO/D: multiobjective optimization by an adaptive differential evolution algorithm. Neurocomputing 127(65–77)
58. Pulluri H, Naresh R, Sharma V (2017) An enhanced self-adaptive differential evolution based solution methodology for multiobjective optimal power flow. Appl Soft Comput 54(229–245)
59. Pal M, Saha S, Bandyopadhyay S (2018) DECOR: differential evolution using clustering based objective reduction for many-objective optimization. Inf Sci 423(200–218)
60. Storn R, Price K (1997) Differential evolution—a simple and efficient heuristic for global optimization over continuous spaces. J Global Optim 11(4):341–359
61. Zhang J, Sanderson AC (2009) JADE: adaptive differential evolution with optional external archive. Evol Comput IEEE Trans 13(5):945–958
62. Zitzler E, Deb K, Thiele L (2000) Comparison of multiobjective evolutionary algorithms: empirical results. Evolut Comput 8(2):173–195
63. Deb K, Thiele L, Laumanns M, Zitzler E (2005) Scalable test problems for evolutionary multiobjective optimization. Evolutionary multiobjective optimization. Springer, pp 105–145
64. Deb K (2001) Multi-objective optimization using evolutionary algorithms. Wiley
65. Corne DW, Jerram NR, Knowles JD, Oates MJ (2001) PESA-II: region-based selection in evolutionary multiobjective optimization. In: Proceedings of the genetic and evolutionary computation conference, GECCO'2001, Citeseer
66. Tian Y, Cheng R, Zhang X, Jin Y (2017) PlatEMO: A MATLAB platform for evolutionary multi-objective optimization [educational forum]. IEEE Comput Intell Mag 12(4):73–87
67. Van Veldhuizen DA, Lamont GB (1998) Multiobjective evolutionary algorithm research: a history and analysis. Citeseer
68. Zitzler E, Thiele L, Laumanns M, Fonseca CM, Da Fonseca VG (2003) Performance assessment of multiobjective optimizers: an analysis and review. IEEE Trans Evol Comput 7(2):117–132
69. Zitzler E, Deb K, Thiele L (1999) Comparison of multiobjective evolutionary algorithms on test functions of different difficulty. In: Wu AS (ed) Proceedings of the 1999 genetic and evolutionary computation conference workshop program, pp 121–122
70. Wheeler SJ, Näätänen A, Karstunen M, Lojander M (2003) An anisotropic elastoplastic model for soft clays. Can Geotech J 40(2):403–418
71. Yin ZY, Chang CS, Karstunen M, Hicher PY (2010) An anisotropic elastic-viscoplastic model for soft clays. Int J Solids Struct 47(5):665–677
72. Yin Z-Y, Yin J-H, Huang H-W (2015) Rate-dependent and long-term yield stress and strength of soft Wenzhou marine clay: experiments and modeling. Mar Georesour Geotechnol 33(1):79–91
73. Yin Z-Y, Zhu Q-Y, Yin J-H, Ni Q (2014) Stress relaxation coefficient and formulation for soft soils. Géotechnique Lett 4:45–51
74. Yin ZY, Karstunen M, Chang CS, Koskinen M, Lojander M (2011) Modeling time-dependent behavior of soft sensitive clay. J Geotech Geoenviron Eng 137(11):1103–1113
75. Karstunen M, Yin ZY (2010) Modelling time-dependent behaviour of Murro test embankment. Geotechnique 60(10):735–749
76. Jin Y-F, Yin Z-Y (2018) ErosLab: a modelling tool for soil tests. Adv Eng Softw 121(84–97)
77. Hardin B, Dmevich V (1972) Shear modulus and damping in soils: design equations and curves, ASCE, vol 98, no. SM6-7
78. Allman M, Atkinson J (1992) Mechanical properties of reconstituted Bothkennar soil. Géotechnique 42(2):289–301
79. Hattab M, Hicher P-Y (2004) Dilating behaviour of overconsolidated clay. Soils Found 44(4):27–40

80. Ortiz M, Simo J (1986) An analysis of a new class of integration algorithms for elastoplastic constitutive relations. Int J Numer Methods Eng 23(3):353–366
81. Karstunen M, Koskinen M (2008) Plastic anisotropy of soft reconstituted clays. Can Geotech J 45(3):314–328
82. Ou C-Y, Liao J-T, Lin H-D (1998) Performance of diaphragm wall constructed using top-down method. J Geotech Geoenviron Eng 124(9):798–808
83. Ou C-Y, Shiau B-Y, Wang I-W (2000) Three-dimensional deformation behavior of the Taipei National Enterprise Center (TNEC) excavation case history. Can Geotech J 37(2):438–448
84. Lim A, Ou CY, Hsieh PG (2010) Evaluation of clay constitutive models for analysis of deep excavation under undrained conditions. J Geoeng 5
85. Vucetic M, Dobry R (1991) Effect of soil plasticity on cyclic response. J Geotech Eng 117(1):89–107
86. Jin Y-F, Yin Z-Y, Zhou W-H, Huang H-W (2019) Multi-objective optimization-based updating of predictions during excavation. Eng Appl Artif Intell 78(102–123)

Chapter 10
Development of Geotechnical Optimization Platform EROSOPT

10.1 Introduction

ErosOpt is a practical and simple software for parameter identification, which gathers various common test types and a range of constitutive models. The tool also provides support for both research and teaching regarding the practice of optimization methods in the fields of geomechanics and geotechnics. Simple and clear interfaces enable great ease of use for engineers, while the friendly graphical interface help users to view and analyze results. Various constitutive models can be used with an open interface for a user-defined model. The performances of different optimization algorithms can be compared while their results can be discussed. ErosOpt is one of GeoInvention's (http://www.geoinvention.com/en/) latest software.

10.2 Development of ErosOpt

Figure 10.1 shows a schematic overview of the mixed-language programming for ErosOpt. The tool was programmed using an admixture of Microsoft Visual C#, MATLAB, and FORTRAN. The graphical user interface of ErosOpt was programmed in C#, the postprocessing (for plotting figures, exporting the results and reading the help documentation) was realized using MATLAB, and the constitutive models were programmed in FORTRAN. All MATLAB files were built as dynamic library files (*.dll) under the .NET Framework 4.0. The version used was MATLAB 2016b.

Z. Yin and Y. Jin, *Practice of Optimisation Theory in Geotechnical Engineering*,
https://doi.org/10.1007/978-981-13-3408-5_10

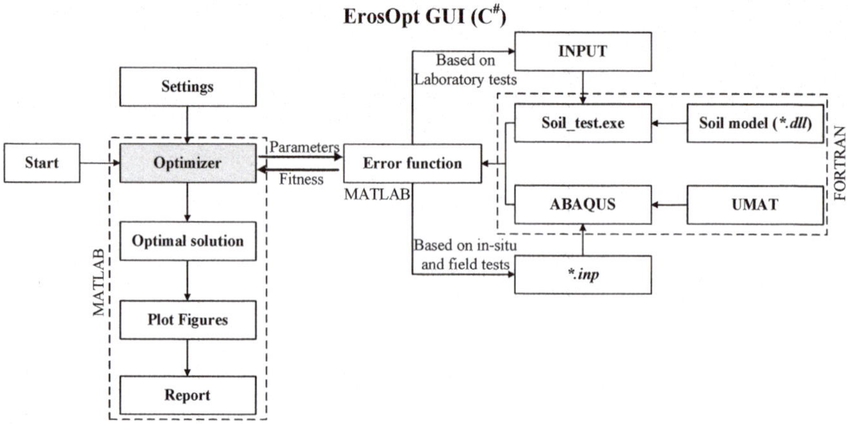

Fig. 10.1 Schematic overview of the mixed-language programming for ErosOpt

10.3 General Structure of ErosOpt

The general structure of ErosOpt is shown in Fig. 10.2, and the four main features are summarized in this section.

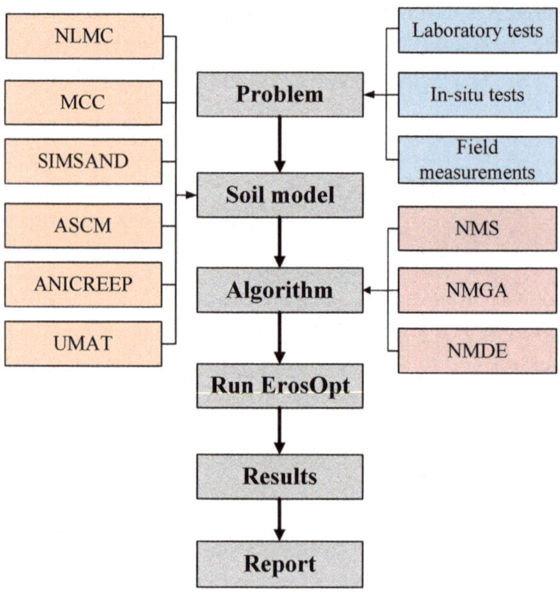

Fig. 10.2 General structure of ErosOpt

10.3.1 Dealing with Various Parameter Identification Problems

Three common types of parameter identification problem are covered by ErosOpt: (1) identification of parameters based on the results of laboratory tests; (2) identification of parameters based on the results of in situ tests; and (3) identification of parameters based on field measurements. For laboratory tests, the results of the oedometer test, triaxial test, or simple shear test are supported. In this study, the simulations of such laboratory tests were performed using the stress- and/or strain- control programmed in FORTRAN. For in situ tests, the result of a pressuremeter test (PMT) is merely allowed in the current version of the tool when using a widely adopted commercial finite element code, ABAQUS. Other types of in situ tests will be supported in the next version. For field measurements, any common measurements can be used for the parameter identification, such as the wall deflection and ground settlement for an excavation or the "*p–s*" curve for a footing. All test results can be imported into ErosOpt from the Excel file (*.xlxs) in a fixed format, as shown in Fig. 10.3. The user should prepare the test data according to the stated format before saving them as an Excel file.

All programs described above were invoked via MATLAB. Figure 10.4 shows an example of the use of MATLAB to invoke the FORTRAN program. The technique for invoking commercial software via MATLAB can be found in the documentation. For example, the command to invoke ABAQUS in MATLAB is "!abaqus job=<u>the name of inp</u> user=<u>the UMAT</u> int".

	A	B	C	D	E	F	G	H
1	IDtest=1	Loading time (h)	Axial strain (%)	Vertical stress (kPa)	-	Void ratio	-	Oedometer test
2	IDtest=2	Loading time (h)	Axial strain (%)	q (kPa)	p' (kPa)	Void ratio	Δu	Triaxial test
3	IDtest=3	Loading time (h)	Gamma (%)	Shear stress (kPa)	Vertical stress (kPa)	Void ratio	Δu	Simple shear test
4	2	1	0	0	200	0.73	0	
5			0.045	11.8042	191.9347333	0.73	10	
6			0.063	18.8418	189.2806	0.73	17	
7			0.084	26.746	182.9153333	0.73	27	
8			0.108	34.9975	173.6658333	0.73	38	
9			0.135	43.0627	164.3542333	0.73	49	
10			0.165	50.4509	154.8169667	0.73	62	
11			0.198	56.7708	147.9236	0.73	72	
12			0.234	60.13098446	139.0436615	0.73	81	
13			0.273	71.94666217	130.9822207	0.73	91	
14			0.315	70.90953109	120.6365104	0.73	102	
15			0.36	68.6855	112.8951667	0.73	111	
16			0.408	68.782	102.9273333	0.73	120	
17			0.459	66.50508446	98.16836149	0.73	125	
18			0.513	67.0076	89.33586667	0.73	133	
19			0.57	77.0552088	88.6850696	0.73	136	
20			0.63	65.57931554	78.85977185	0.73	143	

	A	B	C	D	E	F	G	H	I
1	IDtest=4	Loading time (s)	Radial strain (%)	Radial stress (kPa)	Vertical stress (kPa)	Void ratio	Pore pressure (kPa)	IDdrained=1,draind; IDdrained=2,undraind;	PMT test
2	4	1	0	33	47.14285714	0.756	0	1	
3			0.7612	48.437244	0	0.756	0		
4			1.4918	72.655866	0	0.756	0		
5			2.4847	108.9838	0	0.756	0		
6			3.1645	145.47537	0	0.756	0		
7			3.7907	172.6395	0	0.756	0		
8			4	176.12	0	0.756	0		
9			4.25	190.65	0	0.756	0		
10			4.6285	205.07658	0	0.756	0		
11			4.9923	212.07658	0	0.756	0		
12			5.6721	236.29521	0	0.756	0		
13			6.7694	266.56848	0	0.756	0		
14			7.8131	290.78711	0	0.756	0		
15			8.7552	309.11471	0	0.756	0		
16			10.374	336.27884	0	0.756	0		
17			11.94	354.44281	0	0.756	0		
18			13.613	369.66127	0	0.756	0		
19			15.491	387.82523	0	0.756	0		
20			17.113	403.04369	0	0.756	0		
21			19.15	412.04386	0	0.756	0		

Fig. 10.3 Format of test data used for identifying parameters in ErosOpt

```
function [epsa,q,p,e_model]=SimSand_error(parameter,optimalpara,posi,testtype,soilmodel)
%% Parameters
parameter(1)=e0;
parameter(posi)=optimalpara;
parameter(length(parameter):50)=0;
%% load
sigma_ini=confining;
sigma_end=confining;
strain_V=temp_ref/100;
%% Fortran programm
dos('Soil_test.exe');
%% import results
[epsa,q,p,e_model]=Output_simulation(testtype, soilmodel);
```

Fig. 10.4 Example to show the use of MATLAB to invoke the FORTRAN program in ErosOpt

10.3.2 Provision of a Variety of Constitutive Models of Soils

In ErosOpt, a total of five soil constitutive models (NLMC, MCC, SIMSAND, ASCM, and ANICREEP) are provided, which cover most commonly adopted mechanical models of soils. Other advanced soil models will be available in the next version of ErosOpt. The presented soil models will now be briefly described.

The nonlinear Mohr–Coulomb model (NLMC) was developed against the framework of Mohr–Coulomb, by implementing nonlinear elasticity, nonlinear plastic hardening, and a simplified 3D strength criterion [1]. This model is similar to the shearing part of the hardening soil model (HS). Meanwhile, the modified Cam-Clay (MCC) model was developed by researchers at the University of Cambridge based on the mechanical behavior of remolded clay [2] and is widely used in geotechnical analysis. The critical-state-based SIMple SAND model (SIMSAND) was devised on the basis of the NLMC by implementing the critical state concept and the cap mechanism [1, 3]. Elsewhere, the anisotropic structured clay model (ASCM) was developed using the MCC as its foundation and takes into account the behavior of intact clays because of its natural structure [4]. This model can be used to predict the mechanical behavior of soft structured clay, stiff clay, and artificial reinforced clay. The ANIsotropic CREEP model (ANICREEP) for natural soft clays was also based on the MCC, the overstress theory, and the different time-dependent behaviors of natural soft clays [5–7]. The ANICREEP can be applied to various natural soft clays, stiff clays, and artificial soils.

To improve the extensibility of the proposed tool, the user-defined material (UMAT) is supported, which allows the user to implement other soil models in ErosOpt. Figure 10.5 shows an interface module of UMAT written in FORTRAN language. A .dll file can be compiled by adopting the Intel FORTRAN 32 bit as the compiler tool. After this is done, the .dll file should be renamed "Umat.dll" and

```
!------------------------------------------------------------------------
    Subroutine Umat (IDtask,cm,deps,sig,hsv,CC,dTime)

      !DEC$ ATTRIBUTES DLLEXPORT, DECORATE, ALIAS:"Umat" :: Umat

      Implicit Double Precision (A-H, O-Z)
      Double Precision cm(50),hsv(50),deps(6),deps0(6),epsp(6)
      Double Precision Sig(6),CC(6,6),Adding the defitions...
      Integer    etype, IDtask
      Logical      ::    converged       ! convergence for iteration

!----------------------------------------
      ! parameters definition
        xNu      = cm(1)  ! Poisson's ratio
        dkappa   = cm(2)  ! Swelling index
        dlambda = cm(3)   ! Compression index
        ......Please add parameters the user needs
!----------------------------------------    ------------------------
      if (IDtask.eq. 1) then
        hsv = 0.
        hsv(1)=e0        ! size of yield surface
        hsv(2)=0.        ! PORE pressure
        hsv(3)=pm0       ! void ratio
        ......Please add state variables the user needs
     end if
!----------------------------------------    ------------------------
     if (IDtask.eq. 2) then

        call MATRIXDE (Sig,cm,hsv,CC)   !Elastic stiffness matrix

     end if
!--------------------------------------------
      if (IDtask.eq. 3) then

        ......Please update stress and state variables

       end if

    Return

    end Subroutine
```

Fig. 10.5 Interface of the user-defined material

placed in the same directory as the main program of ErosOpt. Then, the UMAT can be found in the tool. Note that the name of the subroutine must be "Umat" (changing this will lead to errors). IDtask is the task number. IDtask=1 is the initialization of the state variables; IDtask=2 calculates the elastic matrix; and IDtask=3 updates the stress and state variables. The cm is a vector with the material parameters; deps is the strain increment; sig is the stress; hsv is the state variables; and CC is the elastic matrix tensor. Other parameters and state variables can be defined by the user. "!DEC$ ATTRIBUTES DLLEXPORT, DECORATE, ALIAS: "Umat" :: Umat" is the statement of the subroutine name.

10.3.3 Provision of Various Efficient Optimization Algorithms

The ErosOpt provides three efficient optimizers: (1) the Nelder–Mead simplex (NMS); (2) the real-coded genetic algorithm (RCGA), which is enhanced by implementing the NMS, and named "NMGA" for simplicity; and (3) the differential evolution (DE) algorithm, which is enhanced by implementing the NMS, and named "NMDE" for simplicity. All these algorithms can used to effectively solve the parameter identification problems. The performances of the adopted algorithms have been validated by conducting several identifications [1, 8–12].

10.3.4 Provision of Visualization with Graphical Displays

The graphical user interface (GUI) of the developed software is composed of six interface objects: Main Form, Problem Form, Soil Model Form, Algorithm Form, Results Form, and Report Form. The Main Form interacts with the user and connects to the other forms, the functionalities of which can be discerned from their names. For example, the Problem Form defines the type of parameter identification problem and imports the corresponding test data as the objective; the Algorithm Form allows the user to select the algorithm and assign settings accordingly; the Result Form shows the best solution and the comparison between the objectives and the optimal simulations; and the Report Form provides the user with a concise report that summarizes the user's selections and the most favorable obtained results.

10.4 Installation and Operating Environment

The main program of ErosOpt is an executable file, which can run directly in most Windows systems with minor request for operating environment. The results can be obtained directly by using the ErosOpt platform. Note that the system needs a version of Microsoft .NET Framework 4.0 or newer. If the version of Microsoft .NET Framework is older than 4.0, the user can download the advanced version from https://www.microsoft.com/en-hk/download/details.aspx?id=17851. To guarantee the normal operation of ErosOpt without installing the FORTRAN program, three FORTRAN environment files "libifcoremd.dll", "libmmd.dll", and "msvcr100.dll" are provided in the installation package. **Note that the current version needs the instal-**

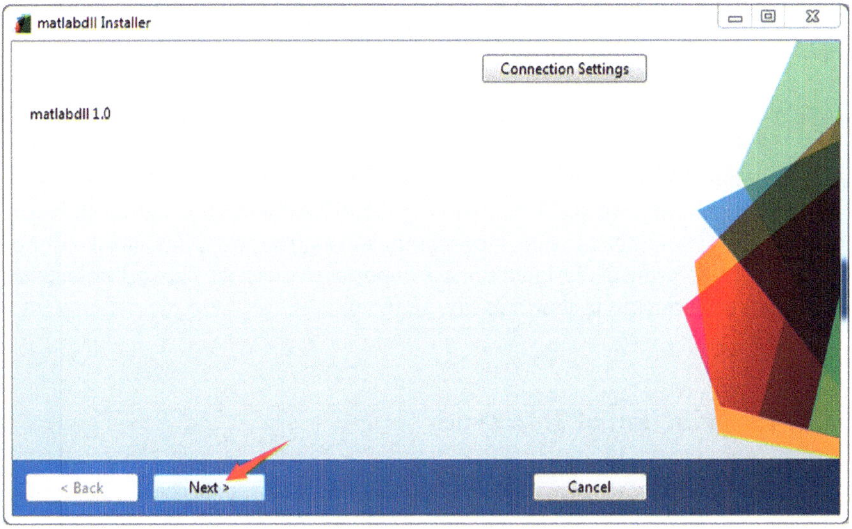

Fig. 10.6 Start interface of the MATLAB environment installation

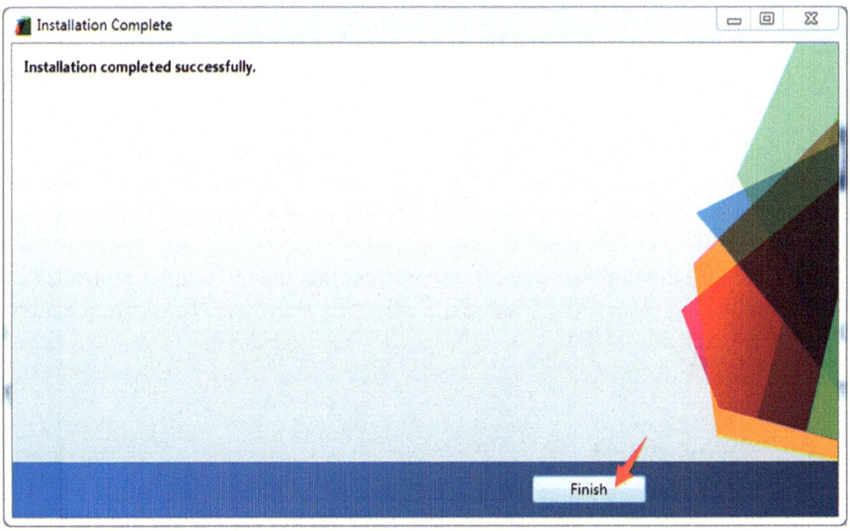

Fig. 10.7 Completed Interface of the MATLAB environment installation

lation of "Intel Fortran" (it is recommended to install Visual_studio_2010 and Intel.Visual.Fortran.Composer.XE.2011).

The MATLAB environment is needed for plotting the results. The version of MATLAB environment adopted is "MCR_R2016b_win64_installer.exe", which is free to download from the official Web site and free for use. For convenience, this program named "Matlab_env.exe" is already provided in the installation package.

The start interface of the installation of MATLAB environment is shown in Fig. 10.6. By clicking "Next" step, the program will automatically download required files and install. During the installation, it is important to keep the network connected. The completed interface is shown in Fig. 10.7.

10.5 Introduction of Test Types

There are three test types available in the platform, laboratory tests (oedometer test, triaxial test, and simple shear test), in situ tests (pressuremeter test), and field measurements (in developing), as shown in Fig. 10.8.

10.5.1 Oedometer Test

In the platform, the oedometer test is simulated as one-dimensional compression test, where the lateral deformation is constrained to be zero and only the vertical deformation is allowed, ($\varepsilon_2 = \varepsilon_3 = 0$ and $\varepsilon_1 > 0$), as shown in Fig. 10.9. The lateral stress necessarily keeps changing during the loading process because of the restriction of lateral deformation. Therefore, it is convenient that the test can be controlled by pure strain loading ($\varepsilon_2 = \varepsilon_3 = 0$ and $d\varepsilon_1 > 0$), or by strain and stress mixed loading ($\varepsilon_2 = \varepsilon_3 = 0$ and $d\sigma_1 > 0$).

10.5.2 Triaxial Test

Only the consolidated drained and undrained triaxial tests are available in this version. For conventional consolidated drained triaxial compression test, the soil sample is first consolidated to a given confining pressure, and then the axial load is increased up to the failure of the sample ($d\sigma_a' = d\sigma_1' > 0$ or $d\varepsilon_a = d\varepsilon_1 > 0$) while keeping the confining pressure constant $\left(d\sigma_r' = d\sigma_2' = d\sigma_3' = 0\right)$. The slope of this loading

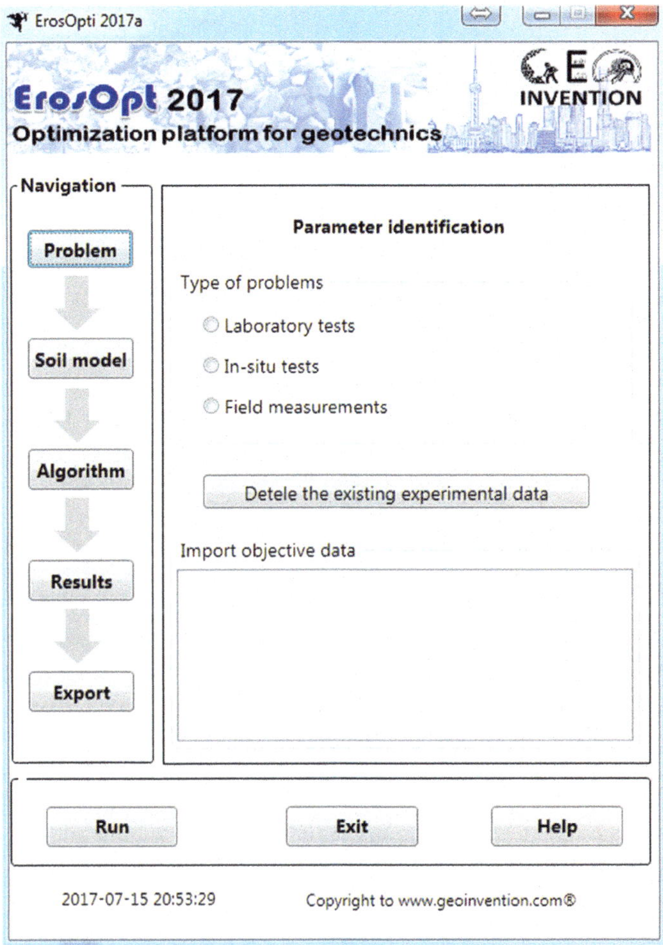

Fig. 10.8 Three test types available in ErosOpt

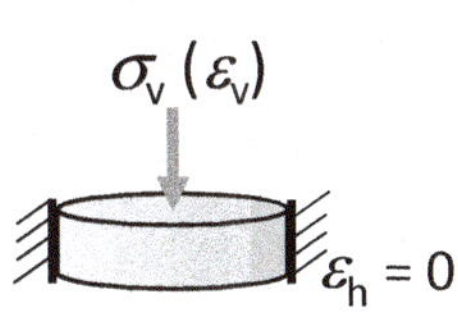

Fig. 10.9 Schematic diagram of an oedometer test

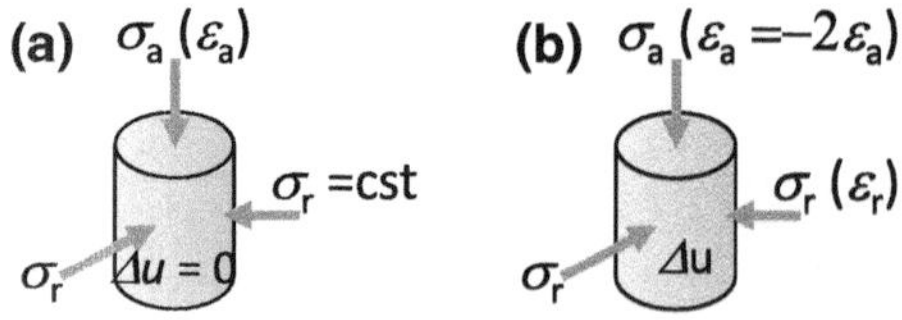

Fig. 10.10 Schematic diagram of triaxial test for **a** drained test and **b** undrained test

path in the p'-q plane is $\mathrm{d}q/\mathrm{d}p' = 3$, which is noted as the conventional triaxial compression path (CTC). Another approach to conduct this test is reducing the axial load till the sample reaches failure ($\mathrm{d}\sigma_a' = \mathrm{d}\sigma_3' < 0$ 或 $\mathrm{d}\varepsilon_a = \mathrm{d}\varepsilon_3 < 0$) while keeping the confining pressure constant $\left(\mathrm{d}\sigma_r' = \mathrm{d}\sigma_1' = \mathrm{d}\sigma_2' = 0\right)$. The slope of this loading path is $\mathrm{d}q/\mathrm{d}p' = -3$, which is the conventional triaxial extension path. The above stress schematic diagrams are shown in Fig. 10.10a.

In conventional consolidated undrained triaxial compression test (Fig. 10.10b), the increment of total confining stress is kept constant ($\mathrm{d}\sigma_\mathrm{r} = 0$). Thus, the slope of the loading path on the p-q plane is still 3 $\left(\mathrm{d}q/\mathrm{d}p = 3\right)$. In $p(p')$-q plane, the horizontal distance between the total stress path and the effective stress path is excess water pore pressure. The excess water pore pressure is always positive for the normal consolidated soil during the loading process; thus, the effective stress path is to the left of the total stress path in p'-q plane. While for over-consolidated soil, the excess water pore pressure is negative during the post loading process. Therefore, the effective stress path is to the right of the total stress path. Under the conventional confining pressure, both the soil particle and the water are considered to be incompressible, which makes it possible to fulfill the undrained condition by keeping the volumetric strain constant ($\mathrm{d}\varepsilon_v = 0 \Rightarrow \mathrm{d}\varepsilon_a = -2\mathrm{d}\varepsilon_r$). In this way, the compression or extension depends on the increasing or decreasing of the axial strain.

10.5.3 Simple Shear Test

When the soil subjected to shear stress reaches the critical state, the soil will slide along a surface which leads to the failure. In order to study such phenomena, a simple shear test (shown in Fig. 10.11a) has been developed and used (equivalent to direct shear and ring shear in Fig. 10.11b, c). In this simple shear test, the shear strain (γ) is defined as the ratio of the horizontal displacement to the sample height. Under the loading of vertical shear strain, the shear stress, vertical stress, and vertical displacement can be obtained from a simple shear test. There are two ways to conduct

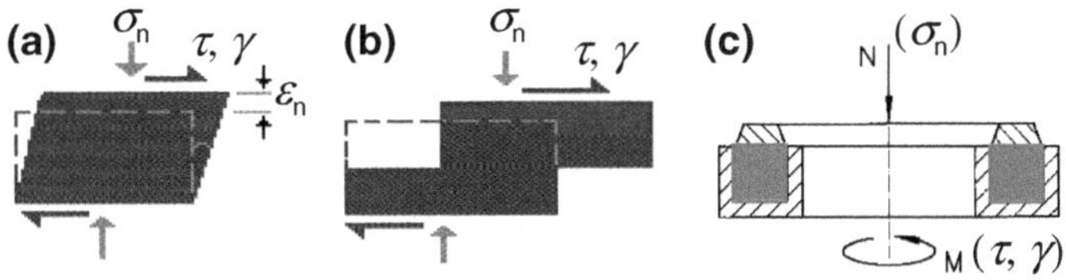

Fig. 10.11 Three typical shear tests: **a** simple shear test, **b** direct shear test, and **c** ring shear test

this simple shear test: (1) keeping a constant vertical load, which is the drained simple shear test, and (2) keeping the volume of the sample constant, which can be regarded as the undrained simple shear test.

10.5.4 Pressuremeter Test

A pressuremeter is a meter constructed to measure the "at-rest horizontal earth pressure". The probe of the pressuremeter is inserted into the borehole and is supported at test depth. The probe is an inflatable flexible membrane which applies even pressure to the walls of the borehole as it expands. As the pressure increases and the membrane expands, the walls of the borehole begin to deform. The pressure inside the probe is held constant for a specific period of time, and the increase in volume required to maintain the pressure is recorded. There are two types of tests that can be performed with the pressuremeter. The stress-controlled test increases pressure in equal increments while the strain-controlled test increases the volume in equal increments. The results of pressuremeter test allow engineers to design foundations that will be stable in these conditions.

In ErosOpt, the pressuremeter test simulated by ABAQUS is a displacement-controlled test and the small deformation is assumed. A 2D finite element model with an axisymmetric condition is created, as shown in Fig. 10.12. A total of 204 4-node reduced integration elements (CAX4RP) are used to simulate the soil. For reproducing the in situ conditions, the initial state of stress was defined by the K_0 condition. The initial vertical and horizontal stresses are, respectively, obtained from the import data. The same displacement as in a typical field test was applied, and at each step, the same displacement increment was applied. The final applied displacement is identical with the experimental data.

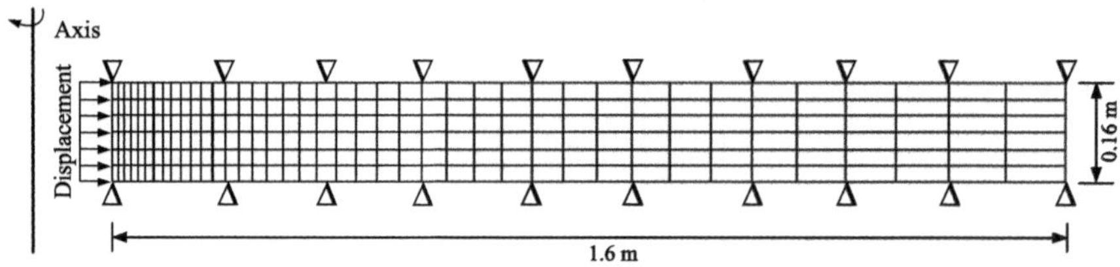

Fig. 10.12 ABAQUS model of pressuremeter test

10.6 Constitutive Models

10.6.1 Introduction to Constitutive Models

In this software, we provide five different constitutive models (Fig. 10.13): nonlinear Mohr–Coulomb (NLMC) model, modified Cam-Clay (MCC) model, critical-state-based SIMple SAND model (SIMSAND), anisotropic structured clay model (ASCM), and natural soft clay Anisotropic Creep model (ANICREEP). Moreover, the platform also provides an open access for user-defined models (UMAT), which may be useful for users to develop and test their own models (note that the micromechanics-based model—MicroSoil—will be open in next version).

Before introducing all adopted constitutive models, we will first introduce briefly the elastic stress–strain relationship and the three-dimensional strength criterion, which are common for the different models.

10.6.2 Elastic Constitutive Relation

(1) **Isotropic elasticity**

Due to the nonlinearity of the stress–strain behavior of soils, the elastic constitutive relation is normally expressed in incremental form using generalized Hooke's law:

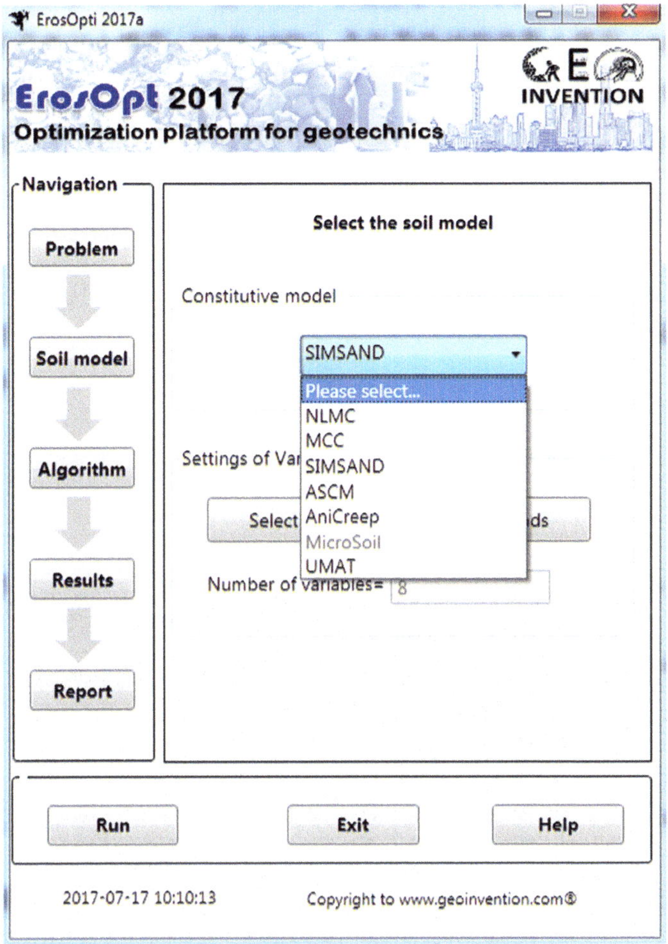

Fig. 10.13 Five constitutive models available in current version of the software

$$d\varepsilon_{ij} = \frac{1+\upsilon}{E}d\sigma'_{ij} - \frac{\upsilon}{E}d\sigma'_{kk}\delta_{ij} \tag{10.1}$$

or

$$d\sigma'_{ij} = \frac{E}{1+\upsilon} d\varepsilon_{ij} + \frac{\upsilon E}{(1+\upsilon)(1-2\upsilon)} d\varepsilon_{kk}\delta_{ij} \tag{10.2}$$

where two parameters are needed: Young's modulus E and Poisson's ratio υ.

In order to calculate the stress–strain relationship, we need to define a stiffness matrix for the material D. In most finite element codes, the engineering shear strain $(\gamma_{xy} = \varepsilon_{xy} + \varepsilon_{yx} = \partial u_x/\partial y + \partial u_y/\partial x)$ is used. Then, the elastic stiffness matrix with the stress–strain relationship in incremental form can be expressed as follows:

$$\begin{bmatrix} d\sigma'_{xx} \\ d\sigma'_{yy} \\ d\sigma'_{zz} \\ d\sigma'_{xy} \\ d\sigma'_{yz} \\ d\sigma'_{zx} \end{bmatrix} = \frac{E}{(1-2\upsilon)(1+\upsilon)} \begin{bmatrix} 1-\upsilon & \upsilon & \upsilon & 0 & 0 & 0 \\ \upsilon & 1-\upsilon & \upsilon & 0 & 0 & 0 \\ \upsilon & \upsilon & 1-\upsilon & 0 & 0 & 0 \\ 0 & 0 & 0 & 0.5-\upsilon & 0 & 0 \\ 0 & 0 & 0 & 0 & 0.5-\upsilon & 0 \\ 0 & 0 & 0 & 0 & 0 & 0.5-\upsilon \end{bmatrix} \begin{bmatrix} d\varepsilon_{xx} \\ d\varepsilon_{yy} \\ d\varepsilon_{zz} \\ d\gamma_{xy} \\ d\gamma_{yz} \\ d\gamma_{zx} \end{bmatrix} \tag{10.3}$$

which can also be written in the inverse way with an elastic flexibility matrix:

$$\begin{bmatrix} d\varepsilon_{xx} \\ d\varepsilon_{yy} \\ d\varepsilon_{zz} \\ d\gamma_{xy} \\ d\gamma_{yz} \\ d\gamma_{zx} \end{bmatrix} = \frac{1}{E} \begin{bmatrix} 1 & -\upsilon & -\upsilon & 0 & 0 & 0 \\ -\upsilon & 1 & -\upsilon & 0 & 0 & 0 \\ -\upsilon & -\upsilon & 1 & 0 & 0 & 0 \\ 0 & 0 & 0 & 2(1+\upsilon) & 0 & 0 \\ 0 & 0 & 0 & 0 & 2(1+\upsilon) & 0 \\ 0 & 0 & 0 & 0 & 0 & 2(1+\upsilon) \end{bmatrix} \begin{bmatrix} d\sigma'_{xx} \\ d\sigma'_{yy} \\ d\sigma'_{zz} \\ d\sigma'_{xy} \\ d\sigma'_{yz} \\ d\sigma'_{zx} \end{bmatrix} \tag{10.4}$$

According to experimental observations, for clays we can directly adopt the swelling index of the isotropic compression test ($\kappa = -\Delta e/\Delta \ln p'$) as the input parameter to calculate Young's modulus. Note that the swelling index from the oedometer test is slightly different, but acceptable as the value of this parameter.

$$K = \frac{1+e_0}{\kappa} p', \; E = 3K(1-2\upsilon) \tag{10.5}$$

For sand, the shear modulus is usually adopted as the input parameter to calculate Young's modulus. In the case that the isotropic compression curve is available, the bulk modulus can be directly measured to be an input parameter:

Table 10.1 Summary of elastic constants

	Shear G	Young's E	Constraint M	Bulk K	Lame λ	Poisson's υ
G, E	G	E	$\frac{G(4G-3)}{3G-E}$	$\frac{GE}{9G-3E}$	$\frac{G(E-2G)}{3G-E}$	$\frac{E-2G}{2G}$
G, M	G	$\frac{G(3M-4G)}{M-G}$	M	$M-\frac{4}{3}G$	$M-2G$	$\frac{M-2G}{2(M-G)}$
G, K	G	$\frac{9GK}{3K+G}$	$K+\frac{4}{3}G$	K	$K-\frac{2G}{3}$	$\frac{3K-2G}{2(3K+G)}$
G, λ	G	$\frac{G(3\lambda+2G)}{\lambda+G}$	$\lambda+2G$	$\lambda+\frac{2G}{3}$	λ	$\frac{\lambda}{2(\lambda+G)}$
G, υ	G	$2G(1+\upsilon)$	$\frac{2G(1-\upsilon)}{1-2\upsilon}$	$\frac{2G(1+\upsilon)}{3(1-2\upsilon)}$	$\frac{2G\upsilon}{1-2\upsilon}$	υ
E, K	$\frac{3KE}{9K-E}$	E	$\frac{K(9K+3E)}{9K-E}$	K	$\frac{K(9K-3E)}{9K-E}$	$\frac{3K-E}{6K}$
E, υ	$\frac{E}{2(1+\upsilon)}$	E	$\frac{E(1-\upsilon)}{(1+\upsilon)(1-2\upsilon)}$	$\frac{E}{3(1-2\upsilon)}$	$\frac{\upsilon E}{(1+\upsilon)(1-2\upsilon)}$	υ
K, λ	$\frac{3(K-\lambda)}{2}$	$\frac{9K(K-\lambda)}{3K-\lambda}$	$3K-2\lambda$	K	λ	$\frac{\lambda}{3K-\lambda}$
K, M	$\frac{3(M-K)}{4}$	$\frac{9K(M-K)}{3K+M}$	M	K	$\frac{3K-M}{2}$	$\frac{3K/M-1}{3K/M+1}$
K, υ	$\frac{3K(1-2\upsilon)}{2(1+\upsilon)}$	$2K(1-2\upsilon)$	$\frac{3K(1-\upsilon)}{1+\upsilon}$	K	$\frac{3K\upsilon}{1+\upsilon}$	υ

$$G = G_0 \cdot p_{at} \frac{(2.97-e)^2}{(1+e)} \left(\frac{p'}{p_{at}}\right)^n, \ E = 2G(1+\upsilon) \tag{10.6}$$

where e is the void ratio, p_{at} is the atmospheric pressure ($p_{\mathrm{at}} = 101.325$ kPa), and G_0 is the reference shear modulus, n is the parameters controlling the nonlinearity of the modulus with the applied mean effective stress. In the case of the lack of measurement of shear modulus, it is suggested to use the bulk modulus as input parameter from the isotropic compression test (which is easy to perform in laboratory). Then a typical value of Poisson's ratio $\nu = 0.25$ can be adopted to complete the input setting for elasticity.

Different elastic constants ($E, G, K, \upsilon, \lambda, M$) are related to each other. If we know two of them, we can calculate the others, as summarized in Table 10.1.

(2) **Cross-anisotropic elasticity**

During the natural sedimentation, soil exhibits a significant cross-anisotropy in elastic stiffness, friction angle and even critical state line. In this software, we consider the cross-anisotropic elasticity of Graham and Housley [13] for users to choose, expressed by:

$$\begin{bmatrix} \delta\varepsilon_{11} \\ \delta\varepsilon_{22} \\ \delta\varepsilon_{33} \\ \delta\varepsilon_{12} \\ \delta\varepsilon_{23} \\ \delta\varepsilon_{13} \end{bmatrix} = \begin{bmatrix} 1/E_v & -\upsilon'_{vv}/E_v & -\upsilon'_{vv}/E_v & 0 & 0 & 0 \\ -\upsilon'_{vv}/E_v & 1/E_h & -\upsilon'_{vh}/E_h & 0 & 0 & 0 \\ -\upsilon'_{vv}/E_v & -\upsilon'_{vh}/E_h & 1/E_h & 0 & 0 & 0 \\ 0 & 0 & 0 & 1/2G_{vh} & 0 & 0 \\ 0 & 0 & 0 & 0 & (1+\upsilon'_{vh})/E_h & 0 \\ 0 & 0 & 0 & 0 & 0 & 1/2G_{vh} \end{bmatrix} \begin{bmatrix} \delta\sigma'_{11} \\ \delta\sigma'_{22} \\ \delta\sigma'_{33} \\ \delta\sigma'_{12} \\ \delta\sigma'_{23} \\ \delta\sigma'_{13} \end{bmatrix} \tag{10.7}$$

where $E_h = nE_v$, $\upsilon'_{vh} = \sqrt{n}\upsilon'_{vv}$ with E_v and E_h representing vertical and horizontal Young's modulus; υ'_{vv} and υ'_{vh} are vertical and horizontal Poisson's ratio; G_{vh} is shear modulus. For a convenient utilization, the modification of the elastic modulus increment was obtained based on the stress-controlled isotropic compression test as follows:

$$\delta\varepsilon_v = \delta\varepsilon_{11} + \delta\varepsilon_{22} + \delta\varepsilon_{33} = \left(1 - 4\upsilon'_{vv} + \frac{2}{n} - \frac{2\upsilon'_{vv}}{\sqrt{n}}\right)\frac{\delta p'}{E_v} \tag{10.8}$$

According to $K = \delta p'/\delta\varepsilon_v$, the vertical Young's modulus E_v can be calculated by

$$E_v = \left(1 - 4\upsilon'_{vv} + \frac{2}{n} - \frac{2\upsilon'_{vv}}{\sqrt{n}}\right)\frac{1+e_0}{\kappa}p' \tag{10.9}$$

Then the shear modulus becomes

$$G_{vh} = \frac{\sqrt{n}E_v}{2\left(1+\sqrt{n}\upsilon'_{vv}\right)} \tag{10.10}$$

Thus, for cross-anisotropic elasticity, we need three input parameters E_v, υ'_{vv}, and n. Comparing to the isotropic elasticity, one extra parameter n is added for the cross-anisotropic elasticity. The K or κ can be obtained from the curve of isotropic compression test.

10.6.3 3D Strength Criterion

Two methods for modifying the strength in the stress space are introduced herein. These two methods are widely adopted in the models under the macro-mechanics framework.

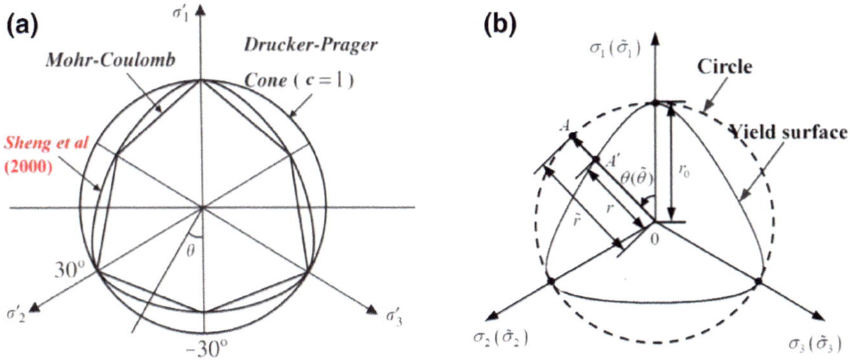

Fig. 10.14 3D strength criterion: **a** $g(\theta)$ modification and **b** transformation of stress space method

(1) **Modification of the Lode angle**

This method mainly works for some soil models which take the slope of the critical state line in p'-q plane as the main parameter. This method modifies the yield strength of different Lode angles by using $M = M_c g(\theta)$, namely $g(\theta)$ method. For example, the modification proposed by Sheng et al. [14] is as follows:

$$M = M_c \left[\frac{2c^4}{1 + c^4 + \left(1 - c^4\right) \sin 3\theta} \right]^{\frac{1}{4}} \tag{10.11}$$

where $c = M_e/M_c$ is the ratio of the critical state line in compression and extension conditions. For a friction angle independent of the Lode angel, $c = (3 - \sin\varphi) / (3 + \sin\varphi)$.

(2) **Transformation of stress space method**

Yao et al. [15–17] proposed a new method by transforming the strength failure plane in the principle stress space to the circular conical surface using a transformed stress tensor (see Fig. 10.14).

The equivalent relationship between the transformed stress tensor $\tilde{\sigma}_{ij}$ and the Cauchy stress tensor σ_{ij} is:

$$\tilde{\sigma}_{ij} = p'\delta_{ij} + \frac{\tilde{q}}{q}\left(\sigma_{ij} - p'\delta_{ij}\right) \tag{10.12}$$

The expression of strength has been obtained for two different typical strength criteria:

$$\tilde{q} = \frac{2I_1}{3\sqrt{(I_1I_2 - I_3)/(I_1I_2 - 9I_3)} - 1} \text{ (SMP criterion)} \tag{10.13}$$

$$\tilde{q} = 3p'\left\{1 - \frac{1}{2}\sqrt{\frac{I_3}{p'^3}}\left[\cos\left(\frac{1}{3}\cos^{-1}\left(-\sqrt{\frac{I_3}{p'^3}}\right)\right)\right]^{-1}\right\} \text{ (Lade criterion)} \tag{10.14}$$

To conveniently adopt the above two equations in the programming, we have modified them by defining $M = \tilde{q}/p'$,

$$M = \frac{6}{3\sqrt{(I_1I_2 - I_3)/(I_1I_2 - 9I_3)} - 1} \text{ (SMP criterion)} \tag{10.15}$$

$$M = 3\left\{1 - \frac{1}{2}\sqrt{\frac{I_3}{p'^3}}\left[\cos\left(\frac{1}{3}\cos^{-1}\left(-\sqrt{\frac{I_3}{p'^3}}\right)\right)\right]^{-1}\right\} \text{ (Lade criterion)} \tag{10.16}$$

In the current version of ErosOpt platform, the Eq. (10.15) is adopted in the constitutive model to modify the strength. Therefore, no extra parameter is needed as input as shown in Table 10.1.

10.6.4 Nonlinear Mohr–Coulomb Model—NLMC

Nonlinear Mohr–Coulomb model was developed under the framework of Mohr—Coulomb, implementing nonlinear elasticity, nonlinear plastic hardening, and a simplified three-dimensional strength criterion [1]. The model is similar to the shearing part of the hardening soil model (HS). The principle of the model is illustrated in Fig. 10.15. The basic constitutive equations are summarized in Table 10.2. Model parameters with their definitions are summarized in Table 10.3 (Fig. 10.16).

Note that it is generally considered the coefficient of earth pressure at rest $K_0 = 1 - \sin\varphi$. According to this assumption, Poisson's ratio can be obtained as a function of the friction angle since there is only elastic deformation in 1D compression:

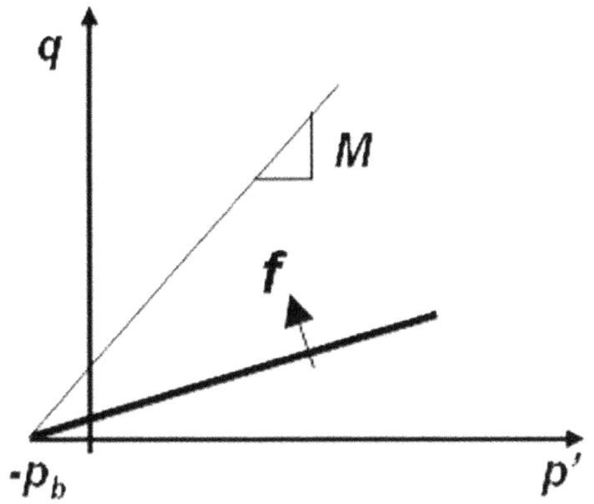

Fig. 10.15 Principle of nonlinear Mohr–Coulomb model

Table 10.2 Basic constitutive equations of NLMC

Components	Constitutive equations
Elasticity	$\delta\varepsilon_{ij}^{e} = \frac{1+\upsilon}{E}\delta\sigma'_{ij} - \frac{\upsilon}{E}\delta\sigma'_{kk}\delta_{ij}$ $E = E_0 p_{at}\frac{(2.97-e)^2}{(1+e)}\left(\frac{p'+p_b}{p_{at}}\right)^n$
Yield surface	$f = \frac{q}{p'+p_b} - H = 0$
Potential surface	$\frac{\partial g}{\partial p'} = M_{pt} - \frac{q}{p'+p_b}$ and $\frac{\partial g}{\partial s_{ij}} = \left\{1\ 1\ 1\ 1\ 1\ 1\right\}$ $M_{pt} = \frac{6\sin\phi_{pt}}{3-\sin\phi_{pt}}$ with $\phi_{pt} = \varphi - \psi$
Hardening rule	$H = \frac{M_p\varepsilon_d^p}{k_p+\varepsilon_d^p}$ with $M_p = \frac{6\sin\varphi}{3-\sin\varphi}$ $\delta p_b = -p_b\xi_b\delta\varepsilon_d^p$

$$\left.\begin{array}{r} \frac{\delta\varepsilon_v}{\delta\varepsilon_d} = \frac{\delta p/K}{\delta q/3G} = \frac{G}{K}\frac{(1+2K_0)}{(1-K_0)} \\ \text{Oedometer:}\frac{\delta\varepsilon_v}{\delta\varepsilon_d} = \frac{3}{2} \\ K_0 = 1-\sin\varphi \end{array}\right\} \Rightarrow \frac{G}{K} = \frac{3}{2}\frac{(1-K_0)}{(1+2K_0)} = \frac{3}{2}\frac{\sin\varphi}{(3-2\sin\varphi)} \tag{10.17}$$

Table 10.3 Model parameters and definitions of NLMC

Parameters	Definitions
e_0	Initial void ratio
υ	Poisson's ratio
E_0	Referential Young's modulus (dimensionless)
n	Elastic constant controlling nonlinear stiffness
φ	Friction angle
k_p	Plastic modulus constant
ψ	Dilatancy angle
e_{max}	Maximum void ratio (limit of dilation)
p_{b0}	Initial bonding adhesive stress
ξ_b	Constant controlling the degradation rate of bonding

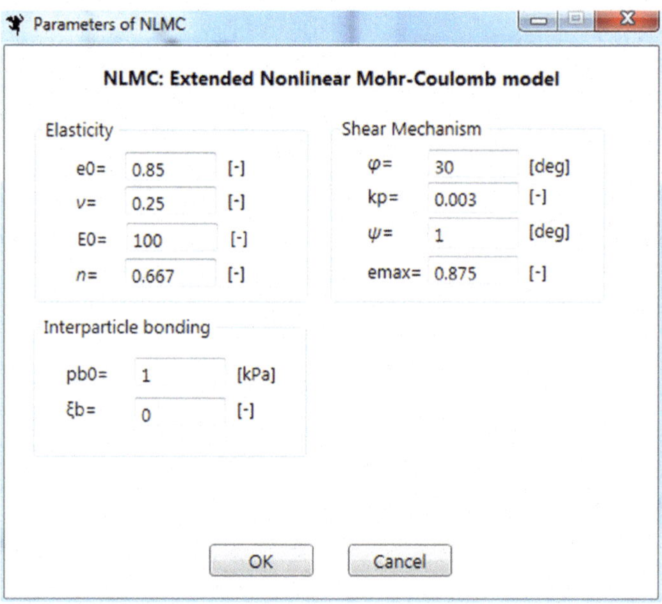

Fig. 10.16 Model parameters of NLMC in ErosOpt

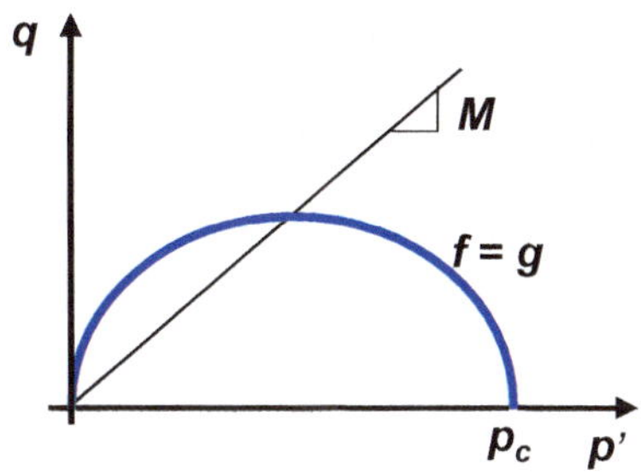

Fig. 10.17 Principle of modified Cam-Clay model

Table 10.4 Basic constitutive equations of MCC

Components	Constitutive equations
Elasticity	$\delta\varepsilon_{ij}^{e} = \frac{1}{2G}\delta s_{ij} + \frac{\kappa}{3(1+e_0)}\delta p'\delta_{ij}$
Yield surface	$f = \frac{q^2}{M^2} + p'^2 - p' p_c$
Potential surface	$g = f$
Hardening rule	$\delta p_c = p_c\left(\frac{1+e_0}{\lambda-\kappa}\right)\delta\varepsilon_v^p$

$$\upsilon = \frac{3 - 2\frac{G}{K}}{2\left(3 + \frac{G}{K}\right)} = \frac{1 - \sin\varphi}{2 - \sin\varphi} \tag{10.18}$$

This formulation provides a reference value of Poisson's ratio. In fact, for most soils $\upsilon = 0.2 - 0.3$ is the suggested value to be used.

10.6.5 Modified Cam-Clay Model—MCC

Modified Cam-Clay model was developed by researchers of the University of Cambridge according to the mechanical behavior of remolded clay [2] and is widely adopted in geotechnical analysis. The principle of the model is illustrated in Fig. 10.17. The basic constitutive equations are summarized in Table 10.4. Model parameters with their definitions are summarized in Table 10.5. Note that, to keep the original modified Cam-Clay model, the adopted strength criterion is von Mises criterion (Fig. 10.18).

Table 10.5 Model parameters and definitions of MCC

Parameters	Definitions
e_0	Initial void ratio
υ	Poisson's ratio
κ	Swelling index
λ	Compression index
M_c	Slope of the critical state line in the p'-q plane
p_{c0}	Initial size of the yield surface

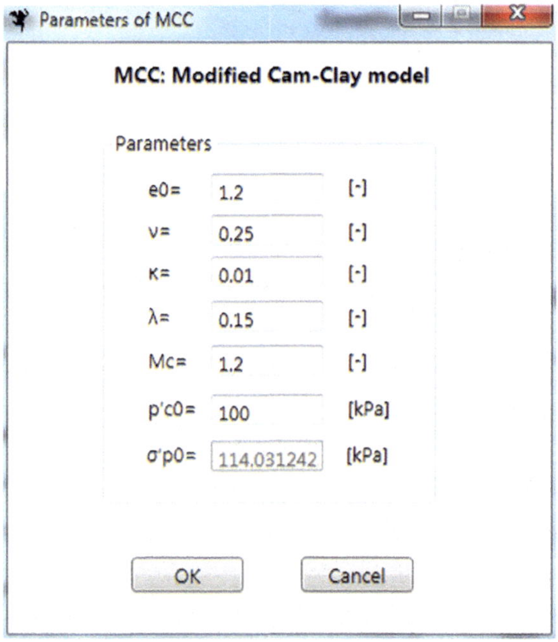

Fig. 10.18 Model parameters of MCC in ErosOpt

Note that the modified Cam-Clay model assumes a value for K_0 according to its stress–dilatancy (bigger than that of Jack):

$$\left.\begin{array}{r}\dfrac{\delta\varepsilon_v^p}{\delta\varepsilon_d^p}=\dfrac{M^2-\eta^2}{2\eta}\\ \text{Oedometer:}\dfrac{\delta\varepsilon_v}{\delta\varepsilon_d}=\dfrac{3}{2}\end{array}\right\}\Rightarrow \eta_{K0}=\frac{\sqrt{9+4M_c^2}-3}{2}\Rightarrow K_0=\frac{3-\eta_{K0}}{3+2\eta_{K0}}=\frac{9-\sqrt{9+4M_c^2}}{2\sqrt{9+4M_c^2}} \tag{10.19}$$

Thus, when we use the modified Cam-Clay model, the relationship between the preconsolidation pressure from the oedometer test and the initial size of the yield surface can be established as follows:

$$\left.\begin{aligned} f_{K0} = 0 \Rightarrow p_{c0} = \frac{q^2}{M^2 p'} + p' \\ q = (1 - K_0)\sigma_{p0} \\ p' = (1 + 2K_0)\sigma_{p0}/3 \end{aligned}\right\} \Rightarrow p_{c0} = \left[\frac{3(1 - K_0)^2}{(1 + 2K_0)M^2} + \frac{(1 + 2K_0)}{3}\right]\sigma_{p0} \quad (10.20)$$

Alternatively, σ_{p0} can also be an input parameter instead of p_{c0}.

10.6.6 Critical-State-Based Simple Sand Model—SIMSAND

The critical-state-based simple sand model was developed based on the nonlinear Mohr–Coulomb model through implementing the critical state concept, the cap mechanism [1, 3]. The principle of the model is illustrated in Fig. 10.19. The basic constitutive equations are summarized in Table 10.6. Model parameters with their definitions are summarized in Table 10.7.

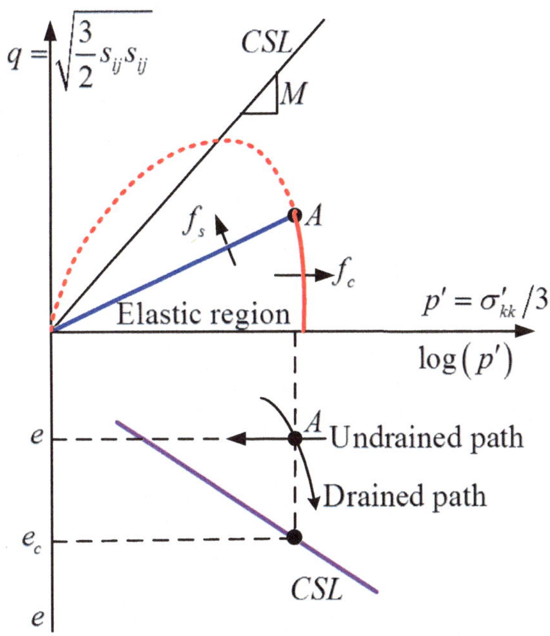

Fig. 10.19 Principle of SIMSAND

Table 10.6 Basic constitutive equations of SIMSAND

Components	Constitutive equations
Elasticity	$\delta\varepsilon_{ij}^{e} = \frac{1+\upsilon}{3K(1-2\upsilon)}\delta\sigma'_{ij} - \frac{\upsilon}{3K(1-2\upsilon)}\delta\sigma'_{kk}\delta_{ij}$ $K = K_0 p_{at}\frac{(2.97-e)^2}{(1+e)}\left(\frac{p'}{p_{at}}\right)^n$
Yield surface in shear	$f_s = \frac{q}{p'} - H$
Potential surface in shear	$\frac{\partial g_s}{\partial p'} = A_d\left(M_{pt} - \frac{q}{p'}\right); \frac{\partial g_s}{\partial s_{ij}} = \left\{1\ 1\ 1\ 1\ 1\ 1\right\}$
Hardening rule for shear	$H = \frac{M_p \varepsilon_d^p}{k_p + \varepsilon_d^p}$
Critical state line and interlocking effect	$e_c = e_{c0}\exp\left[-\lambda\left(\frac{p'}{p_{at}}\right)^{\xi}\right]$ $\tan\phi_p = \left(\frac{e_c}{e}\right)^{n_p}\tan\varphi; \tan\phi_{pt} = \left(\frac{e_c}{e}\right)^{-n_d}\tan\varphi$

Table 10.7 Parameters of SIMSAND

Parameters	Definitions
e_0	Initial void ratio
υ	Poisson's ratio
K_0	Referential bulk modulus (dimensionless)
n	Elastic constant controlling nonlinear stiffness
φ	Friction angle
e_{c0}	Initial critical state void ratio
λ	Constant controlling the nonlinearity of CSL
ξ	Constant controlling the nonlinearity of CSL
A_d	Constant of magnitude of the stress–dilatancy (0.5 ~ 1.5)
k_p	Plastic modulus-related constant (0.01 ~ 0.0001)
n_p	Peak friction angle-related constant ($\approx$1)
n_d	Phase transformation angle-related constant ($\approx$1)

Based on the SIMSAND model, the grain breakage effect has been further considered. The grain breakage can result in the increase of the compressive plastic strain, the change of grain size distribution, and the transformation of the critical state line [18, 19]. The related equations and corresponding parameters are defined in Tables 10.8 and 10.9 (Fig. 10.20).

Table 10.8 Additional constitutive equations considering the grain breakage effect

Components	Constitutive equations
Yield surface in compression	$f_c = \frac{1}{2}\left(\frac{q}{M_p p'}\right)^3 p' + p' - p_{c0}$
Potential surface in compression	$g_c = f_c$
Hardening rule in compression	$\delta p_c = p_c \frac{1+e}{(\lambda' - \kappa')e}\delta\varepsilon_v^p$ $\kappa' = (1+e_0)p' \Big/ \left[K_0 p_{at}\frac{(2.97-e)^2}{(1+e)}\left(\frac{p'}{p_{at}}\right)^n\right]$
Grain breakage-related formula	$B_r^* = \frac{w_p}{b+w_p}$ with $w_p = \int \left(p'\langle\delta\varepsilon_v^p\rangle + q\delta\varepsilon_d^p\right)$ $F(d) = \left(1 - B_r^*\right)F_0(d) + B_r^* F_u(d)$
Kinematics of CSL	$e_c = e'_{c0}\exp\left[-\lambda\left(\frac{p'}{p_{at}}\right)^{\xi}\right]$ $e'_{c0} = e_{cuf} + \left(e_{c0} - e_{cuf}\right)\exp\left(-\rho B_r^*\right)$

Table 10.9 Additional parameters related to grain breakage effect

Parameters	Definitions
λ'	Compression index under the plane of log$e$–logp'
p_{c0}	Initial size of the yield surface of grain breakage
b	Constant controlling the amount of grain breakage
ρ	Constant controlling the movement of CSL
e_{c0}	Virgin initial critical state void ratio before breakage
e_{cuf}	Fractal initial critical state void ratio due to breakage

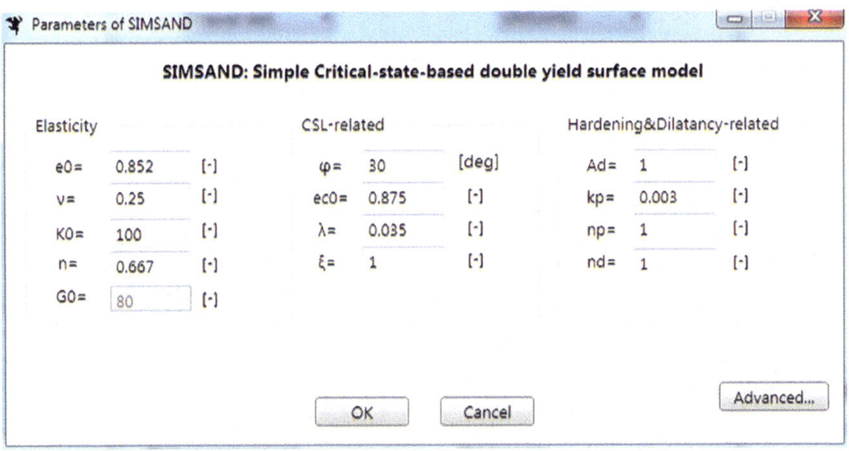

Fig. 10.20 Parameters of SIMSAND in ErosOpt

10.6.7 Anisotropic Structured Clay Model—ASCM

Anisotropic structured clay model was developed under the framework of the modified Cam-Clay model and considering the behavior of intact clays due to its structure [4]. The model can be used to predict the mechanical behavior of soft structured clay, stiff clay, and artificial reinforced clay. The principle of the model is illustrated in Fig. 10.21 Principle of ASCM. The basic constitutive equations are summarized in Table 10.10. Model parameters with their definitions are summarized in Table 10.11 (Fig. 10.22).

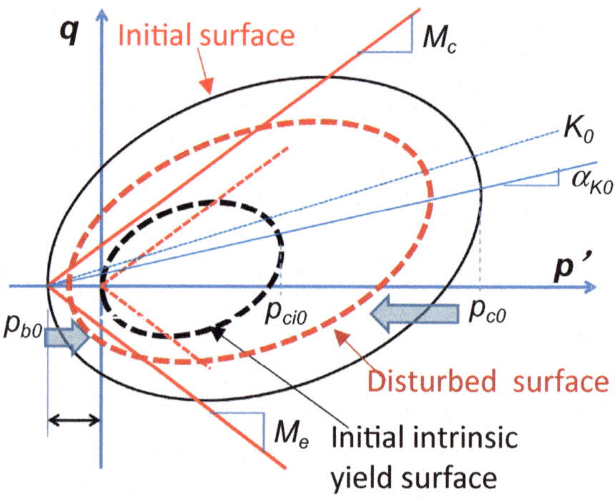

Fig. 10.21 Principle of ASCM

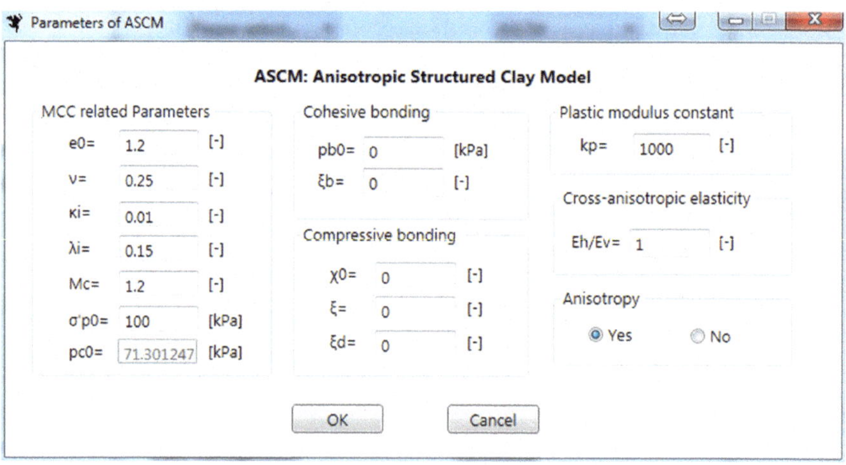

Fig. 10.22 Parameters of ASCM in ErosOpt

Table 10.10 Basic equations of ASCM

Components	Constitutive equations
Elasticity	$\delta\varepsilon_{ij}^{e} = \frac{1+\upsilon}{E}\delta\sigma_{ij}' - \frac{\upsilon}{E}\delta\sigma_{kk}'\delta_{ij}$ $E = 3K(1-2\upsilon)$ $K = \frac{1+e_0}{\kappa_i}\left[p' + p_{b0}(1-R_b) + (p_{c0} - p_{ci0})(1-R_c)\right]$
Yield surface	$f = \frac{3}{2}\frac{[s_{ij}-(p'+p_b)\alpha_{ij}]:[s_{ij}-(p'+p_b)\alpha_{ij}]}{\left(M^2-\frac{3}{2}\alpha_{ij}:\alpha_{ij}\right)} + (p'+p_b)(p'-p_c)$
Potential surface	$g = f$
Hardening rule	$\delta p_{ci} = p_{ci}\frac{1+e_0}{\lambda_i-\kappa_i}\delta\varepsilon_v^p$ $p_c = \frac{p_{ci}}{p_{ci0}}p_{c0}(1-R_c) + p_{ci}R_c$ with $R_c = 1-\exp\left(-\xi_c\sqrt{\varepsilon_{ij}^p\varepsilon_{ij}^p}\right)$ $p_b = p_{b0}(1-R_b)$ with $R_b = 1-\exp\left(-\xi_b\varepsilon_d^p\right)$ $\delta\alpha_{ij} = \omega\left[\left(\frac{as_{ij}}{p'+p_b} - \alpha_{ij}\right)\langle\delta\varepsilon_v^p\rangle - \alpha_{ij0}(1-R_\alpha)\omega_d\delta\varepsilon_d^p\right]$ with $R_\alpha =$ $1-\exp\left(-\xi_\alpha\varepsilon_d^p\right)$
Bounding surface rule	$\bar{\sigma}_{ij} = \beta\sigma_{ij}$ $K_p = \overline{K}_p + k_p\frac{1+e_0}{\lambda-\kappa}(\beta p')^3\left(1-\frac{1}{\beta}\right)$

Table 10.11 Parameters of ASCM

Parameters	Definitions
e_0	Initial void ratio
υ	Poisson's ratio
κ_i	Intrinsic swelling index (of remolded soil)
λ_i	Intrinsic compression index (of remolded soil)
M_c	Slope of the critical state line in the p'-q plane
p_{c0}	Initial size of the yield surface
α_{k0}	Initial inclination of the yield surface
a	Target inclination of the yield surface related to the volumetric strain
b	Target inclination of the yield surface related to the deviatoric plastic strain
ω	Absolute rotation rate of the yield surface
ω_d	Rotation rate of the yield surface related to the deviatoric plastic strain
k_p	Plastic modulus-related parameter in the bounding surface
χ_0	Initial bonding ratio ($p_{ci0} = p_{c0}/(1+\chi_0)$)
ξ	Degradation rate of the bonding ratio related to the plastic volumetric strain
ξ_d	Degradation rate of the bonding ratio related to the plastic deviatoric strain
p_{b0}	Initial interparticle bonding
ξ_b	Degradation rate of the interparticle bonding

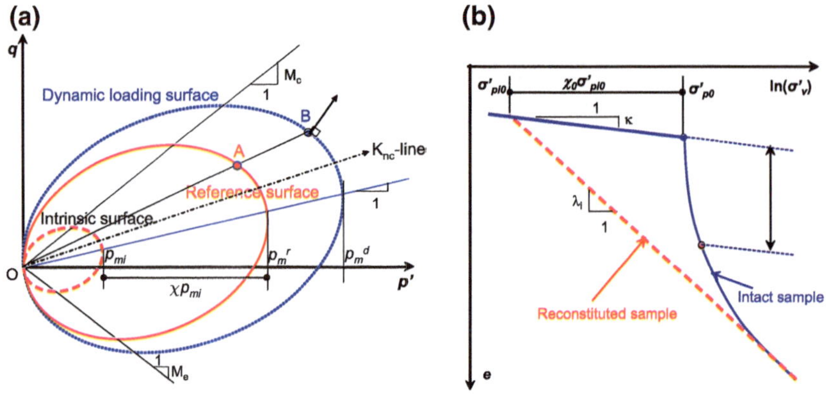

Fig. 10.23 Yield surface of ANICREEP: **a** $p' - q$, **b** 1D condition

Note that the anisotropy-related parameters are directly calculated by using the M_c:

$$a = 0.75, b = 0 \tag{10.21}$$

$$\alpha_{K0} = \eta_{K0} - \frac{M_c^2 - \eta_{K0}^2}{3} \text{ with } \eta_{K0} = \frac{3M_c}{6 - M_c} \tag{10.22}$$

$$\omega_d = \frac{-3(a\eta_{K0} - \alpha_{K0})}{2(b\eta_{K0} - \alpha_{K0})} = \frac{3\left[M_c^2 - \eta_{K0}^2 - 3(1-a)\eta_{K0}\right]}{2\left(\eta_{K0}^2 - M_c^2 + 3(1-b)\eta_{K0}\right)} \tag{10.23}$$

$$\omega = \frac{1 + e_0}{(\lambda_i - \kappa_i)} \ln \frac{10M_c^2 - 2\alpha_{K0}\omega_d}{M_c^2 - 2\alpha_{K0}\omega_d} \tag{10.24}$$

10.6.8 Anisotropic Creep Model for Natural Soft Clays—ANICREEP

The anisotropic creep model for natural soft clays is developed under the framework of the modified Cam-Clay model, the overstress theory, and the different time-dependent behaviors of natural soft clays [20, 21]. The ANICREEP can be applied to different soft clays, stiff clays, and artificial soils. Figures 10.23 and 10.24 show the principles it's the interface of parameters. Table 10.12 shows the basic equations. The model parameters are shown in Table 10.13.

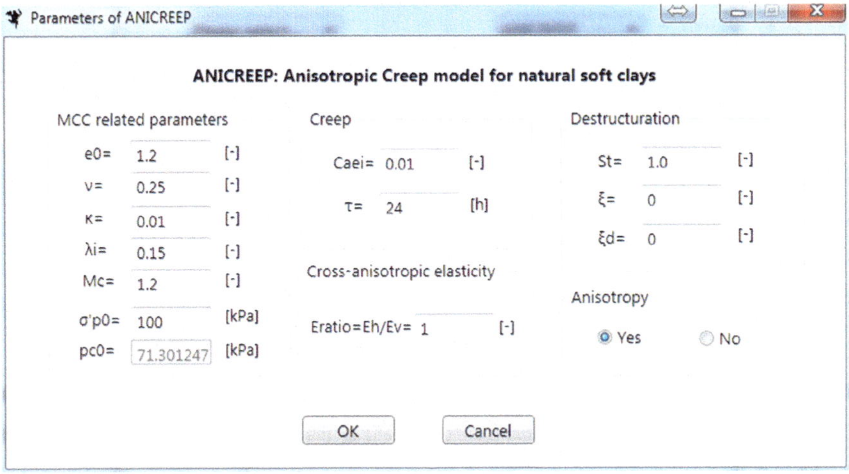

Fig. 10.24 Parameters of ANICREEP in ErosOpt

Table 10.12 Basic equations of ANICREEP

Components	Constitutive equations
Elasticity	$\dot{\varepsilon}^{e}_{ij} = \frac{1+\upsilon}{E}\dot{\sigma}'_{ij} - \frac{\upsilon}{E}\dot{\sigma}'_{kk}\delta_{ij},\ \delta\varepsilon^{e}_{ij} = \dot{\varepsilon}^{e}_{ij}\delta t$ $E = 3p'(1-2\upsilon)\frac{1+e_0}{\kappa}$
Reference yield surface	$f_r = \frac{3}{2}\frac{\left[s^r_{ij}-p'^r\alpha_{ij}\right]:\left[s^r_{ij}-p'^r\alpha_{ij}\right]}{\left(M^2-\frac{3}{2}\alpha_{ij}:\alpha_{ij}\right)p'^r}+p'^r - p^r_c$ (reference stresses)
Potential surface	$f_d = \frac{3}{2}\frac{\left[s_{ij}-p'\alpha_{ij}\right]:\left[s_{ij}-p'\alpha_{ij}\right]}{\left(M^2-\frac{3}{2}\alpha_{ij}:\alpha_{ij}\right)p'}+p' - p^d_c$ (current stresses)
Viscous plastic strain rate	$\dot{\varepsilon}^{vp}_{ij} = \mu\left(\frac{p^d_m}{p^r_m}\right)^{\beta}\frac{\partial f_d}{\partial \sigma'_{ij}},\ \delta\varepsilon^{vp}_{ij} = \dot{\varepsilon}^{vp}_{ij}\delta t$
Hardening rule	$\delta p_{ci} = p_{ci}\frac{1+e_0}{\lambda_i-\kappa_i}\delta\varepsilon^p_v$ $p_c = p_{ci}(1+\chi)$ with $\delta\chi = -\xi\chi\left(\left\|\delta\varepsilon^p_v\right\| + \xi_d\delta\varepsilon^p_d\right)$ $\delta\alpha_{ij} = \omega\left[\left(a\frac{s_{ij}}{p'} - \alpha_{ij}\right)\langle\delta\varepsilon^p_v\rangle - \omega_d\left(b\frac{s_{ij}}{p'} - \alpha_{ij}\right)\delta\varepsilon^p_d\right]$

Note that the involved anisotropy-related parameters are directly calculated using the slope of the critical state line M_c. The detailed information can be found in Yin et al. [5, 7].

Two parameters ξ and ξ_d controlling the degradation rate of the bond can be calculated by combining the isotropic compression test and the oedometer test:

Table 10.13 Parameters of ANICREEP

Parameters	Definitions
e_0	Initial void ratio
υ	Poisson's ratio
κ	Swelling index
λ_i	Intrinsic compression index (of remolded soil)
M_c	Slope of the critical state line in the p'-q plane
σ_{p0}	Initial reference preconsolidation pressure
$C_{\alpha ei}$	Intrinsic secondary compression index (remolded clay)
τ	Reference time (oedometer test $\tau = 24$h)
α_{k0}	Initial inclination of the yield surface
ω	Absolute rotation rate of the yield surface
ω_d	Rotation rate of the yield surface related to the deviatoric plastic strain
χ_0	Initial bonding ratio ($\chi_0 \approx S_t - 1$)
ξ	Absolute rate of the bond degradation
ξ_d	Relative rate of the bond degradation

$$\xi = \frac{-(1+e_0)}{\Delta e^{vp*}} \ln\left[\frac{\sigma'_v}{\chi_0^* \exp\left(\frac{\Delta e^{vp*}}{\lambda_i - \kappa}\right)\sigma'^*_{pi0}} - \frac{1}{\chi_0^*}\right] \tag{10.25}$$

$$\xi + \xi \cdot \xi_d \frac{2(\eta - \alpha)}{(M^2 - \eta^2)} = \frac{-(1+e_0)}{e^{vp}} \ln\left[\frac{\sigma'_f}{\chi_0 \exp\left(\frac{e^{vp}}{\lambda_i - \kappa}\right)\sigma'_{vi0}} - \frac{1}{\chi_0}\right] \tag{10.26}$$

10.6.9 User-Defined Material

In order to enrich the database of constitutive models, and to make it easier for users to write their own model, the ErosOpt platform provides an interface module of user-defined material. The interface module is written in FORTRAN language. A .dll file is compiled by adopting the Intel FORTRAN 32 bit in Visual Studio as the compiler tool. After finishing the compilation, the .dll file should be renamed to "Umat.dll" and should be put into the same directory with the main program of ErosOpt. Then,

```
!-----------------------------------------------------------------------
    Subroutine Umat (IDtask,cm,deps,sig,hsv,CC,dTime)

      !DEC$ ATTRIBUTES DLLEXPORT, DECORATE, ALIAS:"Umat" :: Umat

      Implicit Double Precision (A-H, O-Z)
      Double Precision cm(50),hsv(50),deps(6),deps0(6),epsp(6)
      Double Precision Sig(6),CC(6,6),Adding the defitions...
      Integer   etype, IDtask
      Logical     ::     converged       ! convergence for iteration

!---------------------------------------
      ! parameters definition
        xNu      = cm(1)   ! Poisson's ratio
        dkappa   = cm(2)   ! Swelling index
        dlambda = cm(3)    ! Compression index
        ......Please add parameters the user needs
!---------------------------------------    ----------------------
      if (IDtask.eq. 1) then
        hsv = 0.
        hsv(1)=e0          ! size of yield surface
        hsv(2)=0.          ! PORE pressure
        hsv(3)=pm0         ! void ratio
        ......Please add state variables the user needs
     end if
!---------------------------------------    ----------------------
     if (IDtask.eq. 2) then

        call MATRIXDE (Sig,cm,hsv,CC)  !Elastic stiffness matrix

     end if
!-------------------------------------------
      if (IDtask.eq. 3) then

        ......Please update stress and state variables

       end if

    Return

    end Subroutine
```

Fig. 10.25 Interface of the user-defined material

the user-defined material can be found in the platform and the user can use the UMAT to simulate different types of tests. The interface of UMAT is as follows (Fig. 10.25):

where the name of the subroutine must be "Umat" (changing this name will produce errors). IDtask is task number. IDtask=1 is the initialization of the state variables; IDtask=2 calculates the elastic matrix; IDtask=3 updates the stress and state variables. cm is a vector with the material parameters; deps is the strain increment; sig is stress; hsv are the state variables; CC is the elastic matrix tensor.

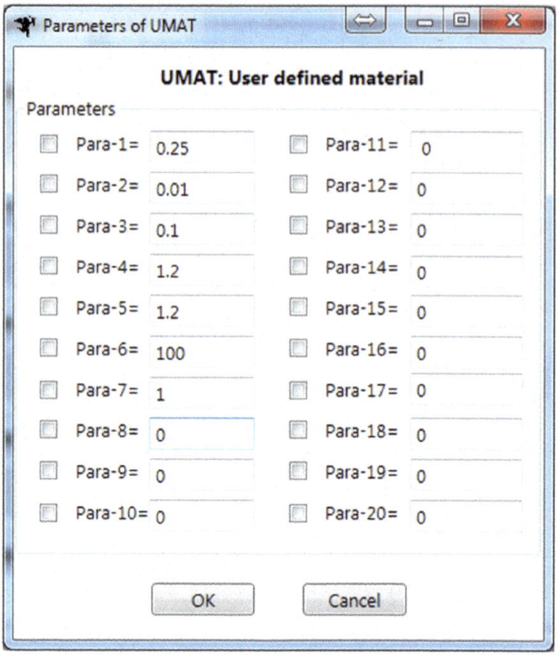

Fig. 10.26 Parameters of the user-defined material

"!DEC$ ATTRIBUTES DLLEXPORT, DECORATE, ALIAS:"Umat"::Umat" is the statement of the subroutine name. Other parameters and state variables are defined by the user. The total number of parameters in the platform is 20, as shown in Fig. 10.26 (note that the PMT test in current version is not supported by the user-defined model).

10.7 Operating Instructions

10.7.1 Problem Selection

Click the button "Problem"; the interface is shown in Fig. 10.27. Three optimization problems can be seen: (1) based on the laboratory tests; (2) based on in situ tests; and (3) based on filed measurement. The user can choose the specific problem according to their requirements. Note that the optimization based on the field measurements is being developed.

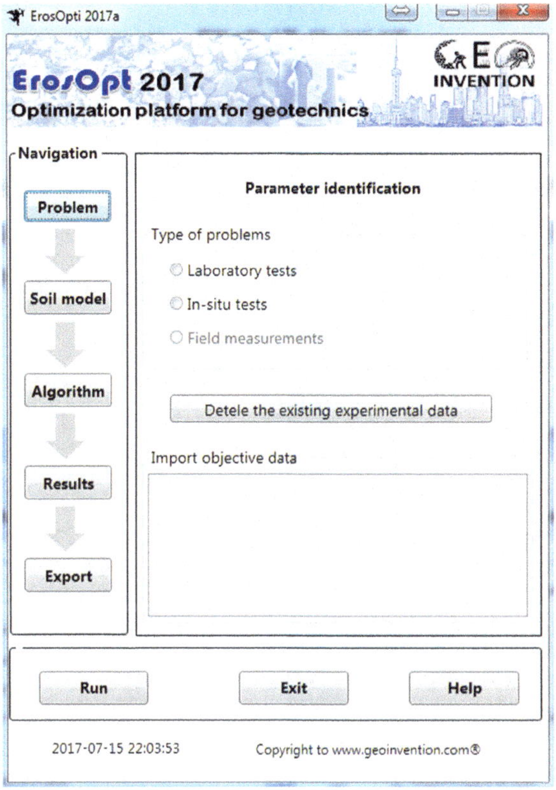

Fig. 10.27 Problem selection window

After selecting a problem, a window like Fig. 10.28 is shown in the screen. The objective data in the optimization are needed to be imported. The window for importing the objective data is shown in Fig. 10.29. The file for storing the objective data is Excel file with an extension .xlxs. For convenient, the rules for saving the data in excel are introduced, as shown in Fig. 10.30. For identifying the type of import test, a unique ID is given for each test used in the platform. In current version, "IDtest=1" represents the oedometer test; "IDtest=2" represents the triaxial test; "IDtest=3" represents the simple shear test; and "IDtest=4" represents the PMT test. The initial state for each simulation is set by using the import data (first line in the data file), including the initial stress data, the void ratio, and drainage condition. Therefore, the user should strictly save their data according the presented rules. For the triaxial test and simple shear test, the user can select the precompression stage before the shearing or not using the "PreComp".

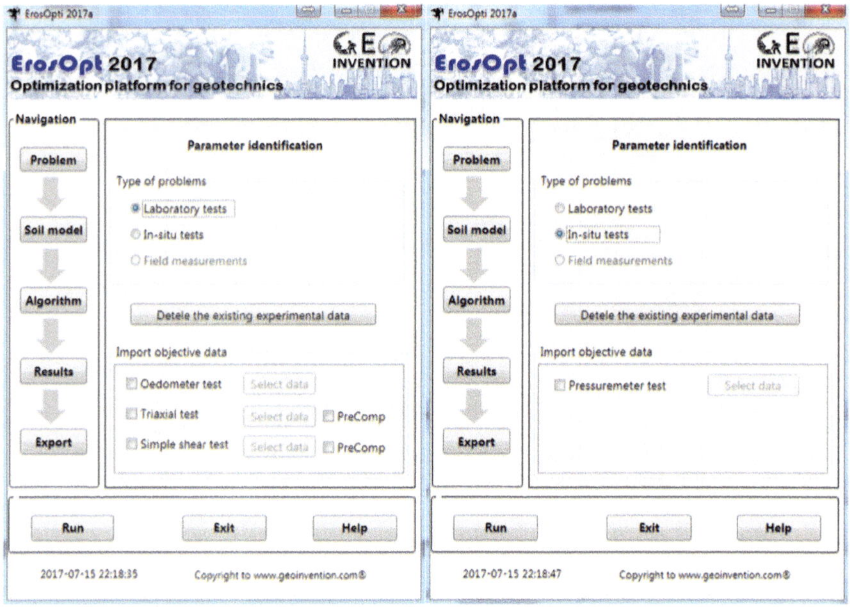

Fig. 10.28 Selection of test type

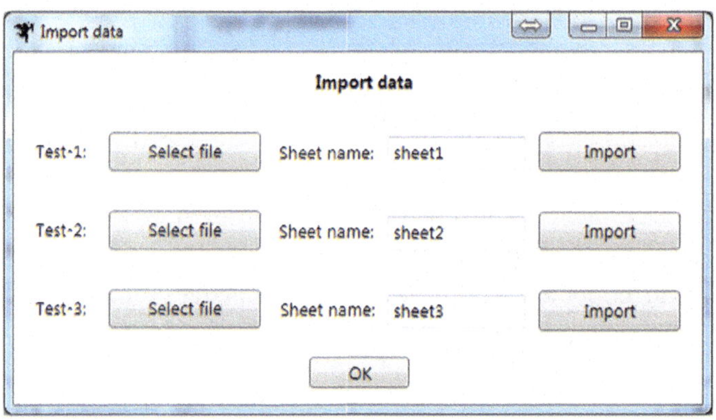

Fig. 10.29 Import the objective data

10.7.2 Selection of Constitutive Model

After the selection of test type and import the data, click the button "Soil model" to select the used constitutive model, as shown in Fig. 10.31. There are five common models and user-defined material provided for user to select. After selecting the soil model, in the "Settings of Variables", the user can find the "Select variables and set their bounds". Click it, a window like Fig. 10.32 is shown. The user can select the

(a)

	A	B	C	D	E	F	G	H
1	IDtest=1	Loading time (h)	Axial strain (%)	Vertical stress (kPa)	-	Void ratio	-	Oedometer test
2	IDtest=2	Loading time (h)	Axial strain (%)	q (kPa)	p' (kPa)	Void ratio	Δu	Triaxial test
3	IDtest=3	Loading time (h)	Gamma (%)	Shear stress (kPa)	Vertical stress (kPa)	Void ratio	Δu	Simple shear test
4	2	1	0	0	100	0.849384	0	
5			0.163132	23.196	104.5667	0.84935	0	
6			0.387853	45.33915	105.6237	0.849212	0	
7			0.438719	48.14074	107.7287	0.848617	0	
8			0.706825	50.10939	110.0673	0.847128	0	
9			0.748154	56.00972	115.1036	0.84558	0	
10			0.886976	56.13769	117.0347	0.843959	0	
11			1.050171	56.80892	118.0222	0.842336	0	
12			1.246217	59.07931	118.6526	0.840451	0	
13			1.401994	68.66907	118.6936	0.838305	0	
14			1.48783	70.22552	118.9148	0.835437	0	
15			1.74216	71.52784	119.669	0.833177	0	
16			1.88734	72.74699	120.2227	0.83109	0	
17			2.120476	76.06701	120.6448	0.829125	0	
18			2.264596	78.45864	121.4695	0.827167	0	
19			2.382224	78.79898	121.6009	0.825165	0	
20			2.583568	81.04807	121.963	0.823107	0	
21			2.780674	81.9989	121.4897	0.821221	0	
22			2.925854	84.93962	122.0225	0.819283	0	

(b)

	A	B	C	D	E	F	G	H	I
1	IDtest=4	Loading time (s)	Radial strain (%)	Radial stress (kPa)	Vertical stress (kPa)	Void ratio	Pore pressure (kPa)	IDdrained=1,draind; IDdrained=2,undraind;	PMT test
2	4	1	0	80	114.2857143	0.756	0	1	
3			0.135	81.819669	0	0.756	0		
4			1.0743	108.9838	0	0.756	0		
5			1.9121	136.31157	0	0.756	0		
6			2.5383	175.74865	0	0.756	0		
7			3.1645	212.07658	0	0.756	0		
8			4.1574	248.40452	0	0.756	0		
9			4.9923	275.73229	0	0.756	0		
10			5.6186	312.06022	0	0.756	0		
11			5.9853	318.11487	0	0.756	0		
12			6.6651	345.44264	0	0.756	0		
13			7.6044	384.71609	0	0.756	0		
14			8.6509	421.20766	0	0.756	0		
15			9.5366	448.37179	0	0.756	0		
16			10.634	475.69956	0	0.756	0		
17			11.836	493.86352	0	0.756	0		
18			12.775	512.02749	0	0.756	0		
19			13.872	545.40992	0	0.756	0		
20			15.023	566.68303	0	0.756	0		
21			16.225	578.79234	0	0.756	0		
22			17.531	596.95631	0	0.756	0		
23			18.524	618.06578	0	0.756	0		
24									

Fig. 10.30 Format of import data: **a** laboratory test and **b** in situ test

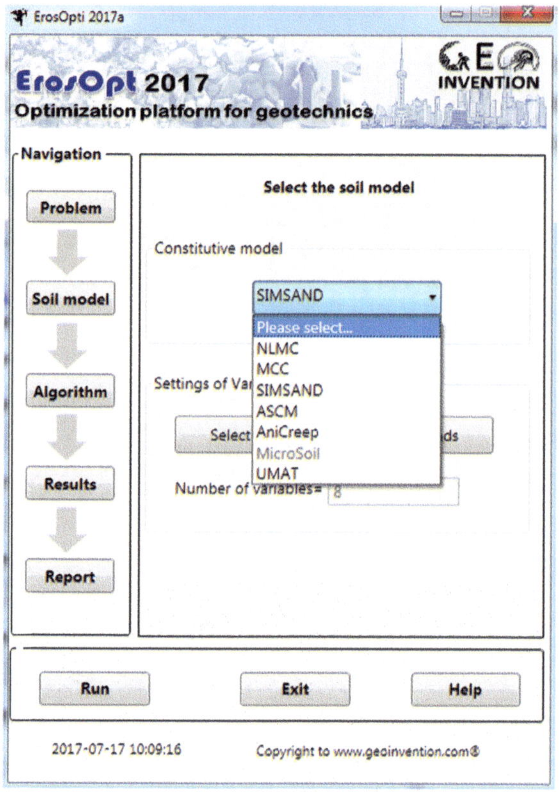

Fig. 10.31 Constitutive models in the ErosOpt

parameters needed to be optimized and set the bounds (see Fig. 10.33), and more settings can be found in the "Advanced". Except the optimized parameters, the other parameters are given fixed values. Figure 10.35 shows the window when the selection is finished. The number of variables will be automatically calculated and is shown in the interface (Fig. 10.34).

10.7.3 Selection of Optimization Algorithms

Click the button "Algorithm"; three optimization algorithms are provided in the ErosOpt, as shown in Fig. 10.36. The user can select one of them and set the settings (see Fig. 10.37) (such as the maximum generation number, the probability of crossover, and the probability of mutation) to conduct the optimization. In terms of the settings for each algorithm, the user can change their values according to the problem or keep the default the values. Note that the default values are recommended.

After setting, the program can be run by clicking the button "Run". During the calculating process, the minimum objective error at every generation with the increase

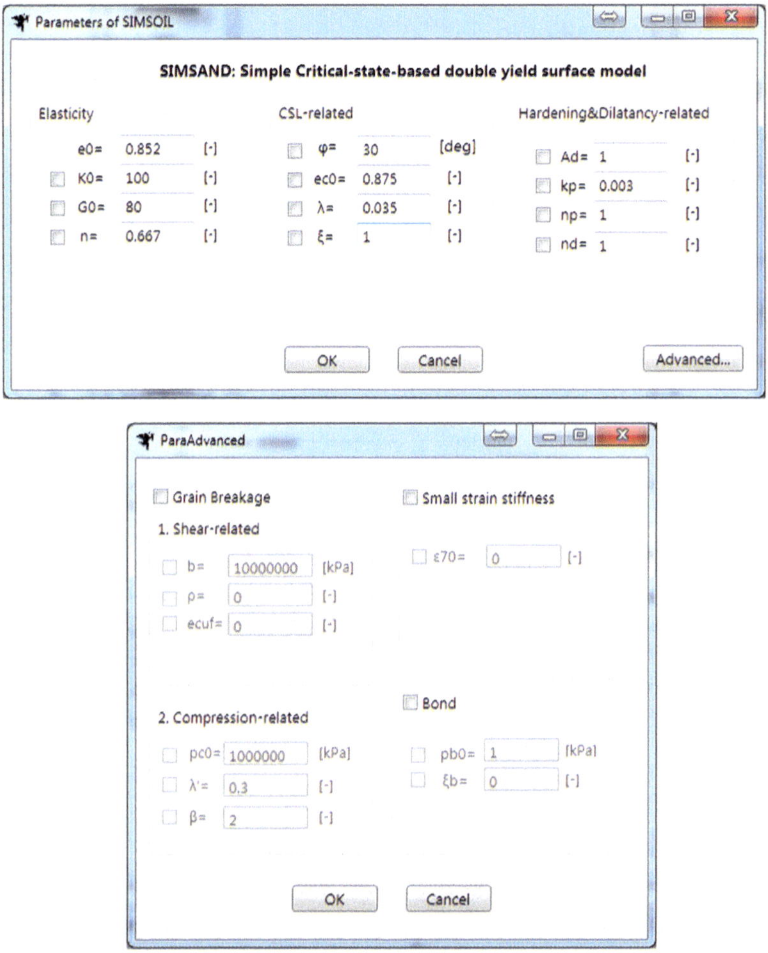

Fig. 10.32 Selection of constitutive model used in the optimization

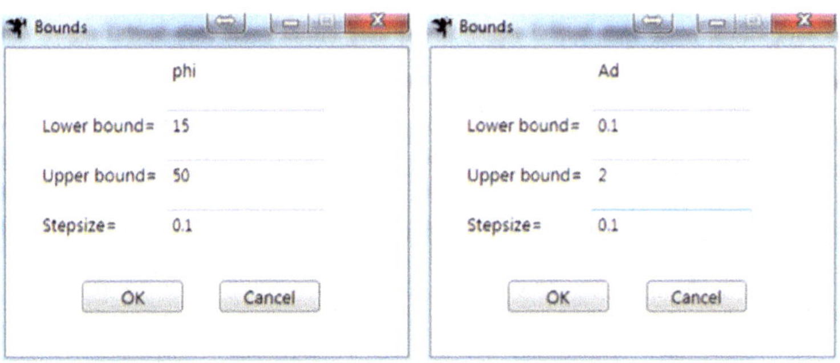

Fig. 10.33 Window for showing the setting of bounds and step size

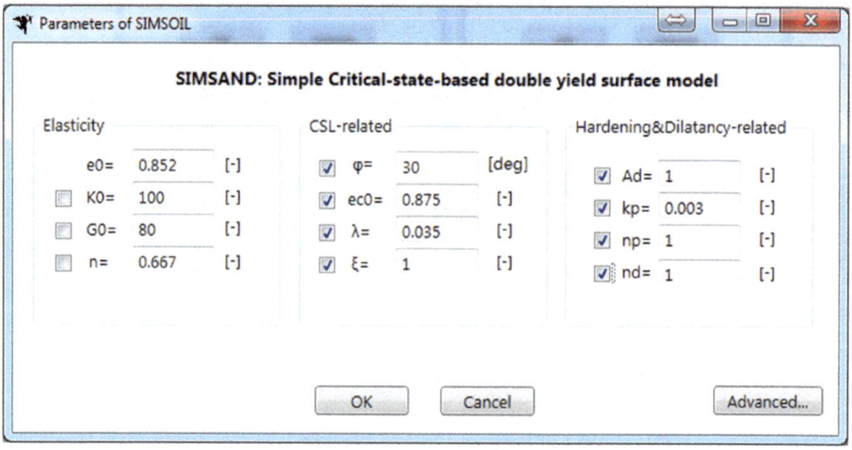

Fig. 10.34 Window after selecting the parameters needed to be optimized

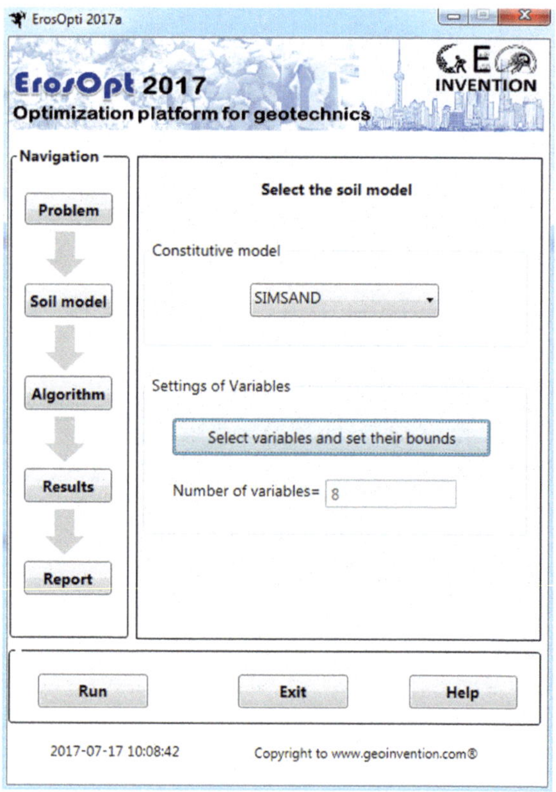

Fig. 10.35 Window after finishing the selection of model and parameters

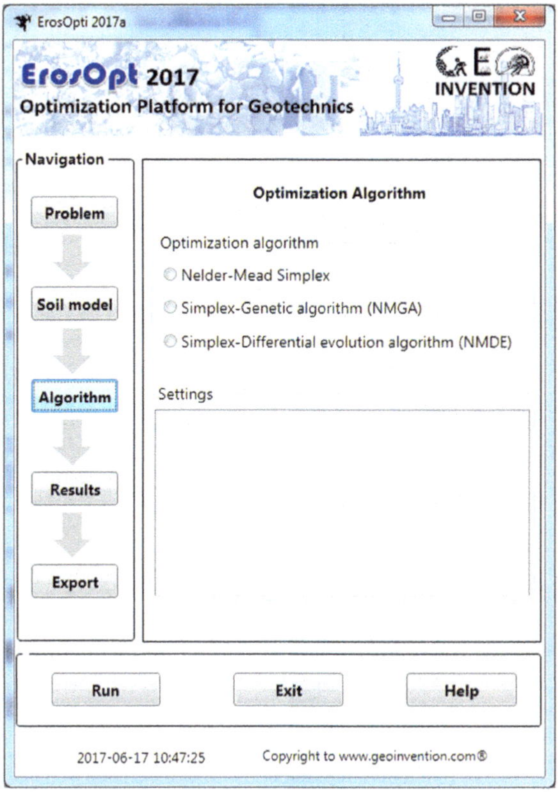

Fig. 10.36 Three optimization algorithms in the ErosOpt

of generation number will be presented in the "Error-generation number" figure (Fig. 10.38). The calculating time is different for different optimization problem, which depends on the selection of test type, soil model, and optimization algorithm.

10.7.4 Results

Click the button "Results", then go to the window for showing the results, as shown in Fig. 10.39. The final optimal parameters corresponding to the selected soil model and the value of error will be presented in the window by clicking the button "Results of optimal parameters", as shown in Fig. 10.40. By clicking the button "Optimal simulations vs. Objectives", the user can obtain the comparison of optimal simulations and objectives, as shown in Fig. 10.41. Then the solutions of the selected optimization problem and the optimal simulations can be exported to Excel files by clicking the buttons "Export optimal solutions" and "Export optimal simulations" (Fig. 10.42).

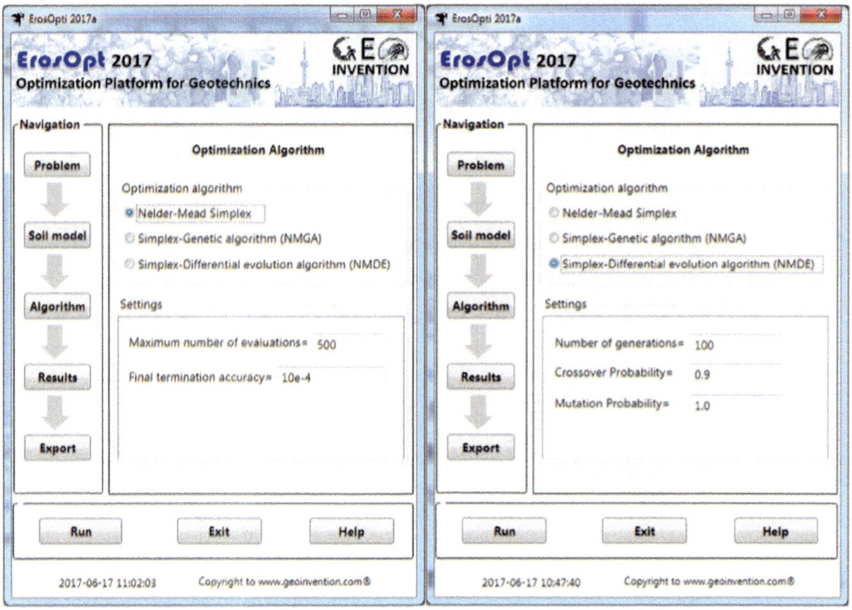

Fig. 10.37 Settings of selected algorithm

Fig. 10.38 Minimum objective error with the increase of generation number

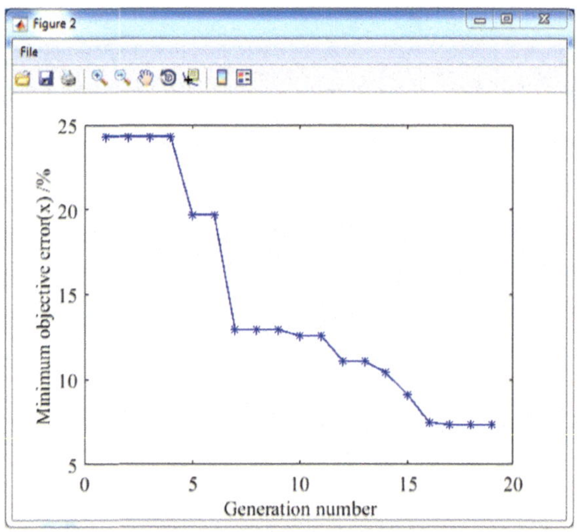

Fig. 10.39 Window of results

Fig. 10.40 Window for showing the optimal results

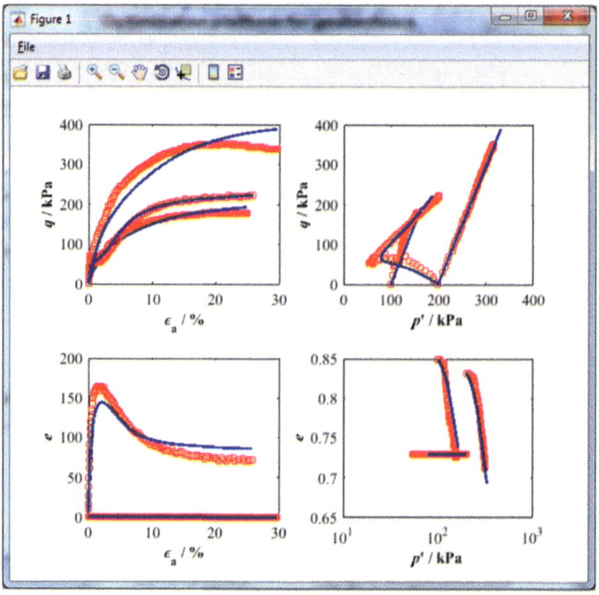

Fig. 10.41 Comparison of optimal simulations and objectives

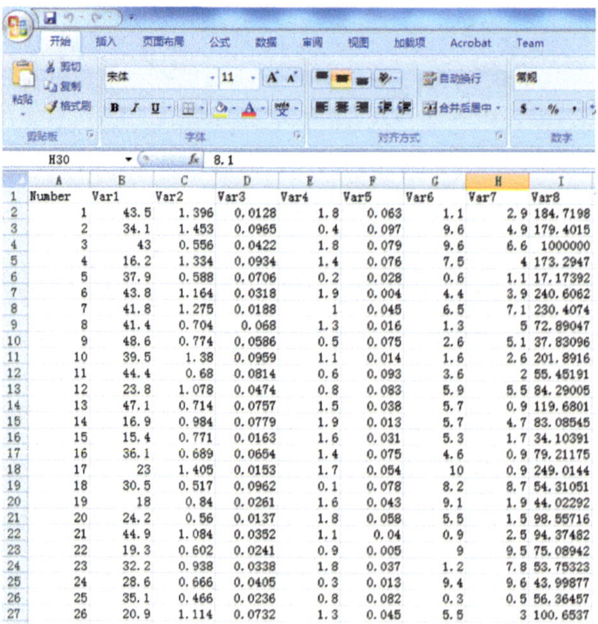

	A	B	C	D	E	F	G	H	I
1	Number	Var1	Var2	Var3	Var4	Var5	Var6	Var7	Var8
2	1	43.5	1.396	0.0128	1.8	0.063	1.1	2.9	184.7198
3	2	34.1	1.453	0.0965	0.4	0.097	9.6	4.9	179.4015
4	3	43	0.556	0.0422	1.8	0.079	9.6	6.6	1000000
5	4	16.2	1.334	0.0934	1.4	0.076	7.5	4	173.2947
6	5	37.9	0.588	0.0706	0.2	0.028	0.6	1.1	17.17392
7	6	43.8	1.164	0.0318	1.9	0.004	4.4	3.9	240.6062
8	7	41.8	1.275	0.0188	1	0.045	6.5	7.1	230.4074
9	8	41.4	0.704	0.068	1.3	0.016	1.3	5	72.89047
10	9	48.6	0.774	0.0586	0.5	0.075	2.6	5.1	37.83096
11	10	39.5	1.38	0.0959	1.1	0.014	1.6	2.6	201.8916
12	11	44.4	0.68	0.0814	0.6	0.093	3.6	2	55.45191
13	12	23.8	1.078	0.0474	0.8	0.083	5.9	5.5	84.29005
14	13	47.1	0.714	0.0757	1.5	0.038	5.7	0.9	119.6801
15	14	16.9	0.984	0.0779	1.9	0.013	5.7	4.7	83.08545
16	15	15.4	0.771	0.0163	1.6	0.031	5.3	1.7	34.10391
17	16	36.1	0.689	0.0654	1.4	0.075	4.6	0.9	79.21175
18	17	23	1.405	0.0153	1.7	0.054	10	0.9	249.0144
19	18	30.5	0.517	0.0962	0.1	0.078	8.2	8.7	54.31051
20	19	18	0.84	0.0261	1.6	0.043	9.1	1.9	44.02292
21	20	24.2	0.56	0.0137	1.8	0.058	5.5	1.5	98.55716
22	21	44.9	1.084	0.0352	1.1	0.04	0.9	2.5	94.37482
23	22	19.3	0.602	0.0241	0.9	0.005	9	9.5	75.08942
24	23	32.2	0.938	0.0338	1.8	0.037	1.2	7.8	53.75323
25	24	28.6	0.666	0.0405	0.3	0.013	9.4	9.6	43.99877
26	25	35.1	0.466	0.0236	0.8	0.082	0.3	0.5	56.36457
27	26	20.9	1.114	0.0732	1.3	0.045	5.5	3	100.6537

Fig. 10.42 Export the optimal solutions to Excel file

10.8 Examples

10.8.1 Parameter Identification of SIMSAND Model Based on Results of Hostun Sand

1. Select the optimization problem, as shown in Fig. 10.43.
2. Import the objective data, as shown in Fig. 10.44.
3. Selection of soil model and the parameters to be optimized, as shown in Figs. 10.45 and 10.46.
4. Selection of algorithm, as shown in Fig. 10.47
5. Run the program, as shown in Fig. 10.48 and the optimization process is shown in Fig. 10.49.
6. Obtain the results, as shown in Fig. 10.50. The comparison of optimal simulations and objectives is shown in Fig. 10.51. Exporting the optimal solutions to Excel file is shown in Fig. 10.52. Exporting the optimal simulation to Excel file is shown in Fig. 10.53.

10.9 Summary

In this paper, the development of ErosOpt, an optimization-based parameter identification tool for geotechnical engineering, was described. The tool also provides support for both research and teaching regarding the practice of optimization methods in the fields of geomechanics and geotechnics. Simple and clear interfaces enable great ease of use for engineers, while the friendly graphical interface help users to view and analyze results. Various constitutive models can be used with an open interface for a user-defined model. The performances of different optimization algorithms can be compared while their results can be discussed. In this research, two selected case studies on typical problems surrounding the identification of soil parameters from both laboratory tests and field measurements were carried out, the outcome demonstrating that ErosOpt is a highly useful tool in engineering practice.

More details can be found in Jin et al. [22].

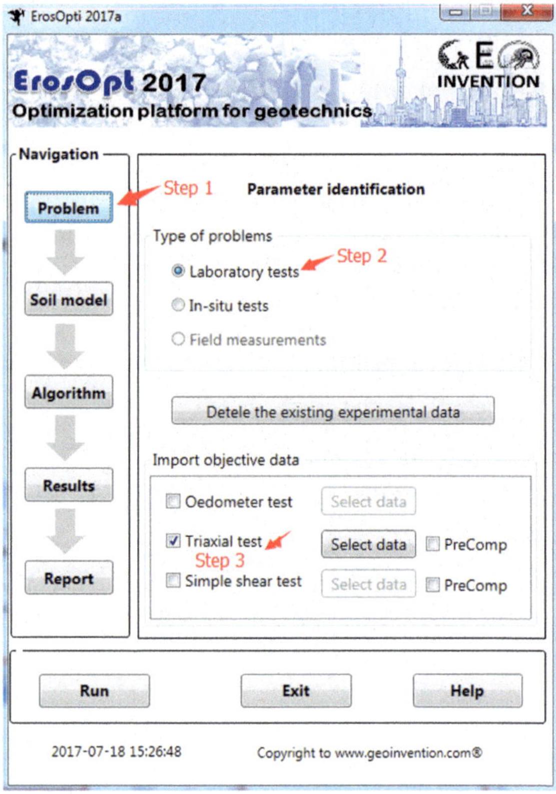

Fig. 10.43 Selection of optimization problem

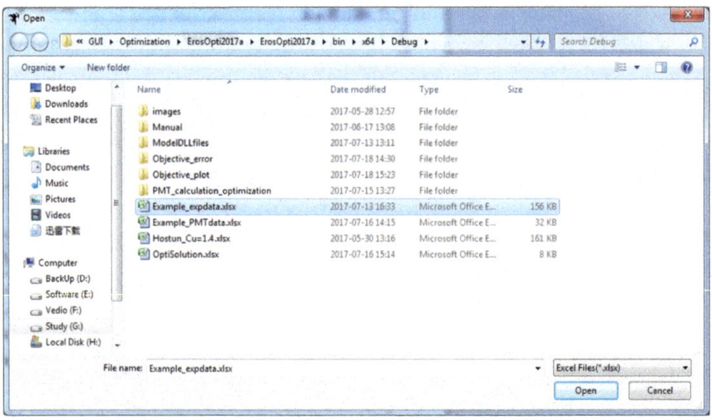

Fig. 10.44 Import objective data

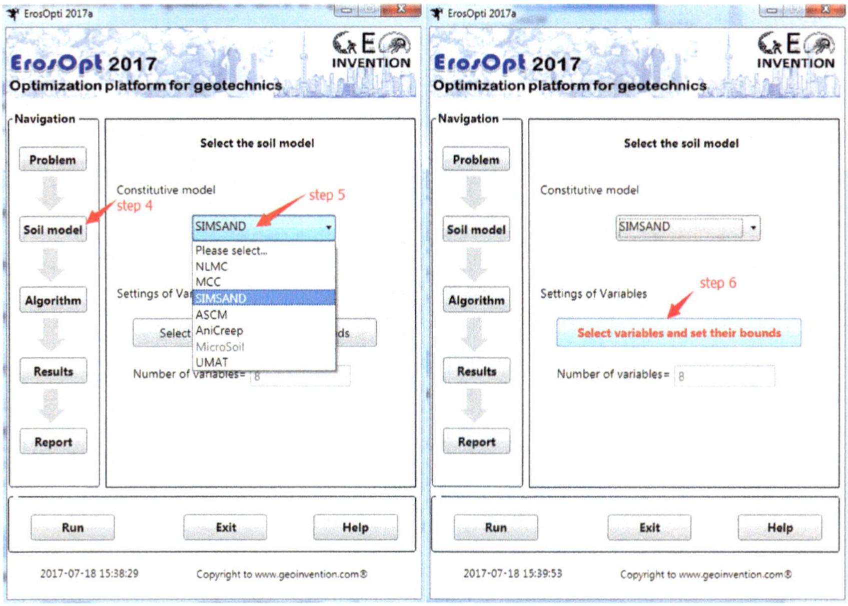

Fig. 10.45 Select the soil model

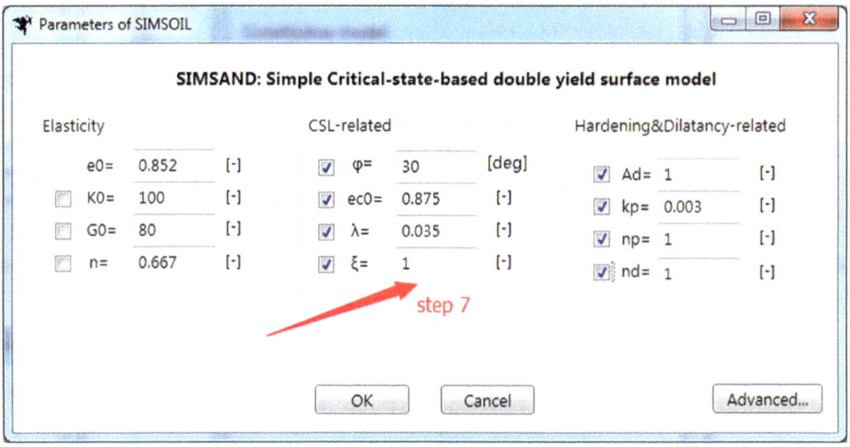

Fig. 10.46 Select the parameters to be optimized

Fig. 10.47 Select the optimization algorithm

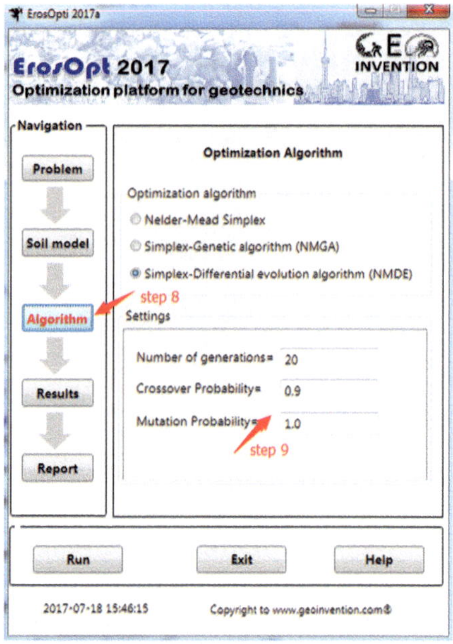

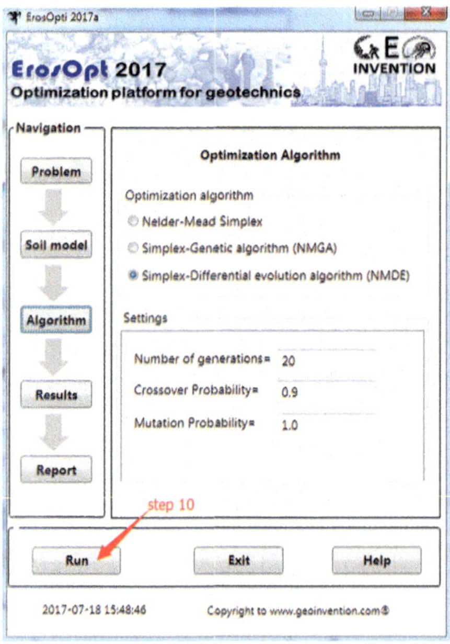

Fig. 10.48 Run the program

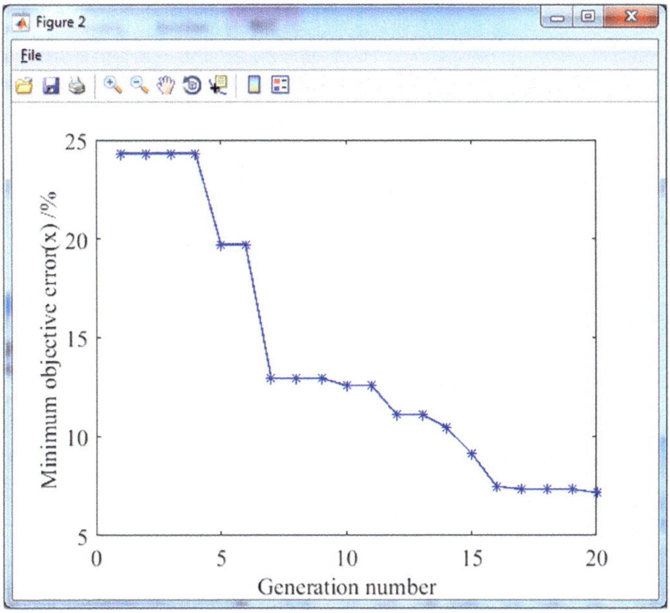

Fig. 10.49 Minimum objective error with the increase of generation number

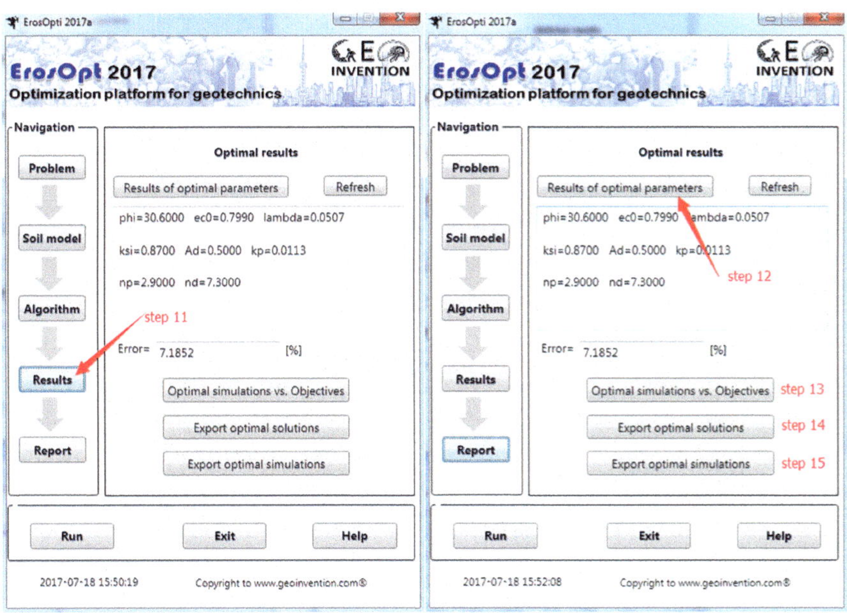

Fig. 10.50 Obtain the optimal parameters and export the optimal solutions

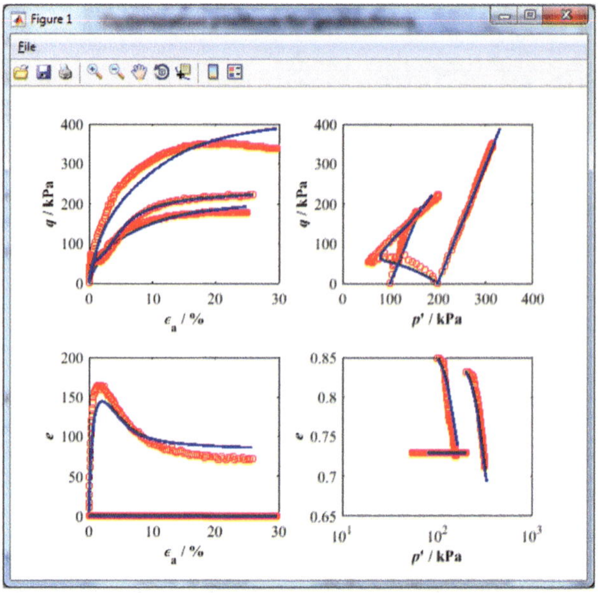

Fig. 10.51 Comparison of optimal simulations and objectives

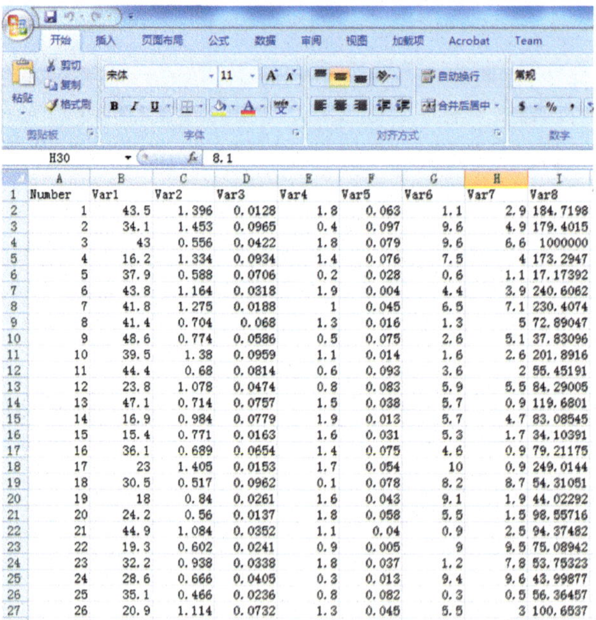

	A	B	C	D	E	F	G	H	I
1	Number	Var1	Var2	Var3	Var4	Var5	Var6	Var7	Var8
2	1	43.5	1.396	0.0128	1.8	0.063	1.1	2.9	184.7198
3	2	34.1	1.453	0.0965	0.4	0.097	9.6	4.9	179.4015
4	3	43	0.556	0.0422	1.8	0.079	9.6	6.6	1000000
5	4	16.2	1.334	0.0934	1.4	0.076	7.5	4	173.2947
6	5	37.9	0.588	0.0706	0.2	0.028	0.6	1.1	17.17392
7	6	43.8	1.164	0.0318	1.9	0.004	4.4	3.9	240.6062
8	7	41.8	1.275	0.0188	1	0.045	6.5	7.1	230.4074
9	8	41.4	0.704	0.068	1.3	0.016	1.3	5	72.89047
10	9	48.6	0.774	0.0586	0.5	0.075	2.6	5.1	37.83096
11	10	39.5	1.38	0.0959	1.1	0.014	1.6	2.6	201.8916
12	11	44.4	0.68	0.0814	0.6	0.093	3.6	2	55.45191
13	12	23.8	1.078	0.0474	0.8	0.083	5.9	5.5	84.29005
14	13	47.1	0.714	0.0757	1.5	0.038	5.7	0.9	119.6801
15	14	16.9	0.984	0.0779	1.9	0.013	5.7	4.7	83.08545
16	15	15.4	0.771	0.0163	1.6	0.031	5.3	1.7	34.10391
17	16	36.1	0.689	0.0654	1.4	0.075	4.6	0.9	79.21175
18	17	23	1.405	0.0153	1.7	0.054	10	0.9	249.0144
19	18	30.5	0.517	0.0962	0.1	0.078	8.2	8.7	54.31051
20	19	18	0.84	0.0261	1.6	0.043	9.1	1.9	44.02292
21	20	24.2	0.56	0.0137	1.8	0.058	5.5	1.5	98.55716
22	21	44.9	1.084	0.0352	1.1	0.04	0.9	2.5	94.37482
23	22	19.3	0.602	0.0241	0.9	0.005	9	9.5	75.08942
24	23	32.2	0.938	0.0338	1.8	0.037	1.2	7.8	53.75323
25	24	28.6	0.666	0.0405	0.3	0.013	9.4	9.6	43.99877
26	25	35.1	0.466	0.0236	0.8	0.082	0.3	0.5	56.36457
27	26	20.9	1.114	0.0732	1.3	0.045	5.5	3	100.6537

Fig. 10.52 Excel file of exported optimal solutions

A	B	C	D	E	F	G	H	I	J	K	L	M	N
Lstep	Time	Eps1	Eps2	Eps3	Eps4	Eps5	Eps6	Sig1	Sig2	Sig3	Sig4	Sig5	Sig6
0	0	0	0	0	0	0	0	200	200	200	0	0	0
1	0.0001	-5.5E-06	2.97E-05	-5.5E-06	0	0	0	200.021	201.534	200.021	0	0	0
3	0.0003	-1.6E-05	8.92E-05	-1.6E-05	0	0	0	200.136	202.175	200.136	0	0	0
6	0.0006	-3.3E-05	0.000178	-3.3E-05	0	0	0	200.312	203.143	200.312	0	0	0
10	0.001	-5.5E-05	0.000297	-5.5E-05	0	0	0	200.55	204.434	200.55	0	0	0
15	0.0015	-9E-05	0.000446	-9E-05	0	0	0	200.075	205.308	200.075	0	0	0
21	0.0021	-0.00012	0.000625	-0.00012	0	0	0	200.456	207.242	200.456	0	0	0
28	0.0028	-0.00017	0.000833	-0.00017	0	0	0	200.147	208.763	200.147	0	0	0
36	0.0036	-0.00021	0.001071	-0.00021	0	0	0	200.694	211.342	200.694	0	0	0
45	0.0045	-0.00027	0.001338	-0.00027	0	0	0	200.588	213.516	200.588	0	0	0
55	0.0055	-0.00033	0.001636	-0.00033	0	0	0	200.545	215.966	200.545	0	0	0
66	0.0066	-0.0004	0.001963	-0.0004	0	0	0	200.651	218.767	200.651	0	0	0
78	0.0078	-0.00048	0.00232	-0.00048	0	0	0	200.104	221.1	200.104	0	0	0
91	0.0091	-0.00056	0.002707	-0.00056	0	0	0	200.464	224.532	200.464	0	0	0
105	0.0105	-0.00065	0.003123	-0.00065	0	0	0	200.119	227.405	200.119	0	0	0
120	0.012	-0.00075	0.003569	-0.00075	0	0	0	200.026	230.672	200.026	0	0	0
136	0.0136	-0.00085	0.004045	-0.00085	0	0	0	200.029	234.194	200.029	0	0	0
153	0.0153	-0.00096	0.004551	-0.00096	0	0	0	200.247	238.103	200.247	0	0	0
171	0.0171	-0.00108	0.005086	-0.00108	0	0	0	200.719	242.394	200.719	0	0	0
190	0.019	-0.0012	0.005651	-0.0012	0	0	0	200.519	246.047	200.519	0	0	0
210	0.021	-0.00134	0.006246	-0.00134	0	0	0	200.421	249.901	200.421	0	0	0
231	0.0231	-0.00148	0.006871	-0.00148	0	0	0	200.041	253.463	200.041	0	0	0
253	0.0253	-0.00163	0.007525	-0.00163	0	0	0	200.6	258.259	200.6	0	0	0
276	0.0276	-0.00179	0.008209	-0.00179	0	0	0	200.486	262.286	200.486	0	0	0
300	0.03	-0.00195	0.008923	-0.00195	0	0	0	200.511	266.522	200.511	0	0	0
325	0.0325	-0.00213	0.009666	-0.00213	0	0	0	200.209	270.39	200.209	0	0	0
351	0.0351	-0.00231	0.01044	-0.00231	0	0	0	200.386	274.87	200.386	0	0	0
378	0.0378	-0.0025	0.011243	-0.0025	0	0	0	200.75	279.608	200.75	0	0	0
406	0.0406	-0.0027	0.012076	-0.0027	0	0	0	200.425	283.481	200.425	0	0	0
435	0.0435	-0.00292	0.012938	-0.00292	0	0	0	200.259	287.55	200.259	0	0	0
465	0.0465	-0.00313	0.013831	-0.00313	0	0	0	200.272	291.844	200.272	0	0	0
496	0.0496	-0.00336	0.014753	-0.00336	0	0	0	200.87	296.907	200.87	0	0	0
528	0.0528	-0.0036	0.015704	-0.0036	0	0	0	200.696	300.926	200.696	0	0	0
561	0.0561	-0.00385	0.016686	-0.00385	0	0	0	200.091	304.217	200.091	0	0	0
595	0.0595	-0.0041	0.017697	-0.0041	0	0	0	200.53	309.128	200.53	0	0	0
630	0.063	-0.00437	0.018738	-0.00437	0	0	0	200.323	313.008	200.323	0	0	0
666	0.0666	-0.00464	0.019809	-0.00464	0	0	0	200.337	317.161	200.337	0	0	0
703	0.0703	-0.00493	0.020909	-0.00493	0	0	0	200.594	321.627	200.594	0	0	0
741	0.0741	-0.00523	0.02204	-0.00523	0	0	0	200.134	324.862	200.134	0	0	0
780	0.078	-0.00553	0.0232	-0.00553	0	0	0	200.893	330.16	200.893	0	0	0
[illegible]	[illegible]	[illegible]	[illegible]	[illegible]	[illegible]	[illegible]	[illegible]	[illegible]	[illegible]	[illegible]	[illegible]	[illegible]	[illegible]

Fig. 10.53 Excel file of exported optimal simulations

References

1. Jin Y-F, Yin Z-Y, Shen S-L, Hicher P-Y (2016) Selection of sand models and identification of parameters using an enhanced genetic algorithm. Int J Numer Anal Methods Geomech 40(8):1219–1240
2. Roscoe KH, Burland J (1968) On the generalized stress-strain behaviour of wet clay. Engineering Plasticity. Cambridge University Press, Cambridge, UK, pp 535–609
3. Jin Y-F, Yin Z-Y, Shen S-L, Hicher P-Y (2016) Investigation into MOGA for identifying parameters of a critical-state-based sand model and parameters correlation by factor analysis. Acta Geotech 11(5):1131–1145
4. Yang J, Yin Z-Y, Huang H-W, Jin Y-F, Zhang D-M (2017) A bounding surface plasticity model of structured clays using disturbed state concept based hardening variables. Chin J Geotech Eng 39:554–561
5. Yin ZY, Chang CS, Karstunen M, Hicher PY (2010) An anisotropic elastic-viscoplastic model for soft clays. Int J Solids Struct 47(5):665–677
6. Yin Z-Y, Yin J-H, Huang H-W (2015) Rate-Dependent and long-term yield stress and strength of soft Wenzhou Marine clay: experiments and modeling. Mar Georesour Geotechnol 33(1):79–91
7. Yin ZY, Karstunen M, Chang CS, Koskinen M, Lojander M (2011) Modeling time-dependent behavior of soft sensitive clay. J Geotech Geoenviron Eng 137(11):1103–1113
8. Ye L, Jin Y-F, Shen S-L, Sun P-P, Zhou C (2016) An efficient parameter identification procedure for soft sensitive clays. J Zhejiang University SCIENCE A 17(1):76–88
9. Jin Y-F, Yin Z-Y, Shen S-L, Zhang D-M (2017) A new hybrid real-coded genetic algorithm and its application to parameters identification of soils. Inverse Prob Sci Eng 25(9):1343–1366

10. Jin Y-F, Wu Z-X, Yin Z-Y, Shen JS (2017) Estimation of critical state-related formula in advanced constitutive modeling of granular material. Acta Geotech 12:1329
11. Yin Z-Y, Jin Y-F, Shen JS, Hicher P-Y (2017) Optimization techniques for identifying soil parameters in geotechnical engineering: comparative study and enhancement. Int J Numer Anal Methods Geomech 42:70
12. Yin Z-Y, Jin Y-F, Shen S-L, Huang H-W (2016) An efficient optimization method for identifying parameters of soft structured clay by an enhanced genetic algorithm and elastic–viscoplastic model. Acta Geotech 12:1–19
13. Graham J, Houlsby G (1983) Anisotropic elasticity of a natural clay. Geotechnique 33(2):165–180
14. Sheng D, Sloan S, Yu H (2000) Aspects of finite element implementation of critical state models. Comput Mech 26(2):185–196
15. Yao Y, Hou W, Zhou A (2009) UH model: three-dimensional unified hardening model for overconsolidated clays. Geotechnique 59(5):451–469
16. Yao YP, Sun DA (2000) Application of Lade's criterion to Cam-Clay model. J Eng Mech 126(1):112–119
17. Yao Y, Lu D, Zhou A, Zou B (2004) Generalized non-linear strength theory and transformed stress space. Sci China Ser E: Technol Sci 47(6):691–709
18. Hu W, Yin ZY, Dano C, Hicher PY (2011) A constitutive model for granular materials considering grain breakage. Science China-Technological Sci 54(8):2188–2196
19. Yin Z-Y, Hicher P-Y, Dano C, Jin Y-F (2016) Modeling mechanical behavior of very coarse granular materials. J Eng Mech 143:C4016006
20. Dafalias YF (1986) Bounding surface plasticity I: Mathematical foundation and hypoplasticity. J Eng Mech 112(9):966–987
21. Dafalias YF, Herrmann LR (1986) Bounding surface plasticity II: application to isotropic cohesive soils. J Eng Mech 112(12):1263–1291
22. Jin Y-F, Yin Z-Y (2018) ErosLab: a modelling tool for soil tests. Adv Eng Softw 121:84–97

Appendix A: ANICREEP

Copyright (c) 2017, geoinvention "www.geoinvention.com"
*****************************Main program ***********************************

```
!  -------------------------------------------------------------------------------------------
       SUBROUTINE UMAT(STRESS,STATEV,DDSDDE,SSE,SPD,SCD,
     1 RPL,DDSDDT,DRPLDE,DRPLDT,
     2 STRAN,DSTRAN,TIME,DTIME,TEMP,DTEMP,PREDEF,DPRED,CMNAME,
     3 NDI,NSHR,NTENS,NSTATV,PROPS,NPROPS,COORDS,DROT,PNEWDT,
     4 CELENT,DFGRD0,DFGRD1,NOEL,NPT,LAYER,KSPT,KSTEP,KINC)
C
       INCLUDE 'ABA_PARAM.INC'
C

       CHARACTER*80 CMNAME
       DIMENSION STRESS(6),STATEV(NSTATV),
     1 DDSDDE(NTENS,NTENS),
     2 DDSDDT(NTENS),DRPLDE(NTENS),
     3 STRAN(NTENS),DSTRAN(NTENS),TIME(2),PREDEF(1),DPRED(1),
     4 PROPS(NPROPS),COORDS(3),DROT(3,3),DFGRD0(3,3),DFGRD1(3,3)


      Double Precision  StVar(NSTATV),StVar_0(NSTATV)
      Double Precision  StVar_dds(NSTATV)
      Double Precision  Sig0(6), Sig(6),Sig_ini(6),Sig_new(6)
      Double Precision  Stress_star(6),sig_star(6),stress_old(6)
      Double Precision  dsig(6)
      Double Precision  Eps0(6),dEps(6),dEps_dds(6)
      Double Precision  dstran_star_dt(6),deps_ini(6)
      Double Precision  D(6,6),DDS(6,6),DD(6,6),DDE(6,6)
      Double Precision  DDSDD(6,6)
      Double Precision  Sig0_mod(6),sigmod_d(6),alpha_d(6)
      Double Precision  pdash,term(6),POP,eta_K0,dMc,alpha0
      Double Precision  dM_ratio,phi,dK0nc,q,g_theta,dM,pm0,St,x0,Pmi
      Double Precision  theta
      integer           ntens,ii,jj,kk
      integer           i,j

      PARAMETER(ZERO=0.D0,ONE=1.D0,TWO=2.D0,THREE=3.D0,SIX=6.D0,
     1          ENUMAX=.4999D0,NEWTON=10.D0,TOLER=1.0D-6)
      parameter(theta = 1.0D-8)
```


Z. Yin and Y. Jin, *Practice of Optimisation Theory in Geotechnical Engineering*,
https://doi.org/10.1007/978-981-13-3408-5

```
! ---------------------------------------------------------------------------------
! Expected contents of Props(1..50)
!
!  1 : dny           Poisson's ratio
!  2 : dkappa        slope of post yield compression line from e-ln_p'-diagram
!  3 : dlambda       slope of swelling line from e-ln_p'-diagram
!  4 : e0            initial voil ratio
!  5 : dM            critical state dM value (in triaxial compression)
!  6 : Cae           secondary compression coefficient
!  7 : tau_in        reference time
!  8 : St            soil sensitivity from oedometer tests
!  9 : a             absolute effectiveness of destructurational hardening
! 10 : b             relative effectiveness of destructurational hardening
! 11 : POP           pre-overburden pressure like in PLAXIS
! ---------------------------------------------------------------------------------
! ---------------------------------------------------------------------------------
! Expected contents of StVar(10)
!
!  1-6  : alpha(1:6)
!  7    : alpha_scalar
!  8    : pmi            pmi=pm0/(1+xx)
!  9    : xx
!  10   : ee
! ---------------------------------------------------------------------------------

! ---------------------------------------------------------------------------------
! Read STRESS STRAIN
! ---------------------------------------------------------------------------------
      Sig0 = STRESS
      Eps0 = STRAN
! ---------------------------------------------------------------------------------
! Read STRESS STRAIN
! ---------------------------------------------------------------------------------
      Sig0(1:4)   = STRESS(1:4)
      Eps0(1:4)   = STRAN(1:4)
      dEps(1:4)   = DSTRAN(1:4)
      Sig0(5:6)   = zero
      Eps0(5:6)   = zero
      dEps(5:6)   = zero
      if( NSHR .GT. 1 ) then
          Sig0(5:6)   = STRESS(5:6)
          Eps0(5:6)   = STRAN(5:6)
          dEps(5:6)   = DSTRAN(5:6)
      end if
      nStat = NSTATV
! ---------------------------------------------------------------------------------
! Get Material Constants
! ---------------------------------------------------------------------------------
        dny      = Props(1)
        dkappa   = Props(2)
        dlambda  = Props(3)
        e0       = Props(4)
        dMc      = Props(5)
        OCR      = Props(6)
        POP      = Props(7)
        Cae      = Props(8)
        tau_in   = Props(9)
        St       = Props(10)
        a        = Props(11)
        b        = Props(12)
        anisotropy= Props(13)
        E_ratio   = Props(14)
! ---------------------------------------------------------------------------------
```

```
        eta_K0     = 3.*dMc/(6.-dMc)
        alpha0     = eta_K0 - (dMc**2-eta_K0**2)/3.
        dM_ratio   = 3./(3.+dMc)
! ------------------------------------------------------------------------------------------
! Read state variables
! ------------------------------------------------------------------------------------------
       STVAR = Statev
! ------------------------------------------------------------------------------------------
! Initialize state variables
! ------------------------------------------------------------------------------------------
       IF (KSTEP == 1 .AND. TIME(2) == 0.) THEN
c              write(*,*)'Step1'
           ! get dK0nc value
           phi      = asin(3.*dMc/(6.+dMc))
           dK0nc    = 1.-sin(phi)
           if (anisotropy .lt. 0.1) then     !for MCC creep model
              etaK0=(sqrt(9.+4.*dMc*dMc)-3.)/2.
              dK0nc=(3.-etaK0)/(2.*etaK0+3.)
          end if
             !adjust Sig0(1:3) due to POP
           POP=abs(POP)*(-1.)
           if (POP .eq. 0.) POP = -0.1
           Sig0_mod(1)  = POP*dK0nc
           Sig0_mod(2)  = POP
           Sig0_mod(3)  = POP*dK0nc
           Sig0_mod(4)  = 0.
           Sig0_mod(5)  = 0.
           Sig0_mod(6)  = 0.

! change sign due to EVP-MCC formulation (compression positive)
          Sig0_mod = (-1.)*Sig0_mod

          !! calculate stress invariants
           pdash          =  ( Sig0_mod(1) + Sig0_mod(2) + Sig0_mod(3) ) /3.
           sigmod_d(1) =   Sig0_mod(1) - pdash
           sigmod_d(2) =   Sig0_mod(2) - pdash
           sigmod_d(3) =   Sig0_mod(3) - pdash
           sigmod_d(4) =   Sqrt(2.) * Sig0_mod(4)
           sigmod_d(5) =   Sqrt(2.) * Sig0_mod(5)
           sigmod_d(6) =   Sqrt(2.) * Sig0_mod(6)
           q             =  Sqrt(3./2.*( sigmod_d(1)**2. + sigmod_d(2)**2.
     &                    + sigmod_d(3)**2. + sigmod_d(4)**2.
     &                    + sigmod_d(5)**2. + sigmod_d(6)**2. ))

! checking input variables
           if ((pdash==0.).or.(dMc**2.==alpha0**2.)) then  ! pm not calculable
              Write(*,*)'ERROR: pm not calculable - division by zero'
             Stop
           end if


! pre-set state variables
           StVar(1)     = -(alpha0/3.)+1.          ! alpha_x
           StVar(2)     = (2.*alpha0/3.)+1.        ! alpha_y
           StVar(3)     = -(alpha0/3.)+1.          ! alpha_z
           StVar(4)     = 0.                       ! alpha_xy
           StVar(5)     = 0.                       ! alpha_yz
           StVar(6)     = 0.                       ! alpha_zx
           StVar(7)     = alpha0                   ! alpha_scalar
```

```
! determine size of the initial yield curve
            alpha_d(1)  =   StVar(1) - 1.
            alpha_d(2)  =   StVar(2) - 1.
            alpha_d(3)  =   StVar(3) - 1.
            alpha_d(4)  =   StVar(4) * Sqrt(2.)
            alpha_d(5)  =   StVar(5) * Sqrt(2.)
            alpha_d(6)  =   StVar(6) * Sqrt(2.)
! determine size of the initial yield curve
            term        =  sigmod_d-pdash*alpha_d
          call Lode_angle(term,dM_ratio,g_theta)
c                g_theta=1.
!   call Lode_angle(Sig0_mod,dM_ratio,g_theta)
            dM  = dMc*g_theta
            pm0 = ((q-alpha0*pdash)**2./(pdash*(dM**2.-alpha0**2.)))+pdash
!           write(*,*) pm0
            pmi = pm0/(1.+x0)
            StVar(8) = pmi   ! pmi
            StVar(9) = St-1.
            StVar(10)= e0
      END IF
! ---------------------------------------------------------------------------------
! Calculate elastic matrix [D]
! ---------------------------------------------------------------------------------
      CALL MATRIXDE_ANI(SIG0,PROPS,STVAR,D)
      StVar_0=StVar
      sig_ini = Sig0
      DDS=D
C     write(*,21)
! ---------------------------------------------------------------------------------

! ---------------------------------------------------------------------------------
    dEps   = DSTRAN
    deps_dds = dEps
c   write(*,21) dEps
     CALL EVPSCLAY1S   (dTime, Props, Sig0, dEps,
    &               D, Sig, StVar, nStat)
c       write(*,21) sig
! ---------------------------------------------------------------------------------
! UPDATE STRESS STRAIN AND STATE VARIABLES
! ---------------------------------------------------------------------------------
      STRESS(1:4) = Sig(1:4)
      if( NSHR .GT. 1 ) then
          STRESS(5:6) = Sig(5:6)
      end if
      STATEV = STVAR
```

```
! -------------------------------------------------------------------------------------
! Give the DDSDDE BY NUMERICAL DISTURBED METHOD
! -------------------------------------------------------------------------------------
        do jj=1,ntens
            stress_old          = sig_ini
            sig_star            = (-1.0)*sig
            StVar_dds           = StVar_0
            DDE                 = DDS
            dstran_star_dt      = (-1.0)*deps_dds
            dstran_star_dt(jj) = dstran_star_dt(jj)+theta
            deps_ini            = (-1.0)*dstran_star_dt
c       write(*,21) StVar_dds
       CALL EVPSCLAY1S (dTime, Props, stress_old, deps_ini,
     &                  DDE, Stress_star, StVar_dds, nStat)
        stress_star = -1.0*stress_star
c       write(*,21) stress_star
c ... stiffness components in column jj (kk = row index, jj = column index)
c         write(*,21) sig_star
         do kk=1,NTENS
            dsig(kk) = stress_star(kk)-sig_star(kk)
            DD(kk,jj)= dsig(kk)/theta
         enddo !kk
c        write(*,21) dsig
c        write(*,*)'a'
        enddo !jj
c Consistent tangent stiffness
      do j=1,NTENS
        do i=1,NTENS
          if((i.le.3).and.(j.le.3)) then
             DDSDDE(i,j) = DD(i,j)
          else
             DDSDDE(i,j) = DD(i,j)
          end if
        end do
      enddo


 c Consistent tangent stiffness
       do j=1,NTENS
         do i=1,NTENS
           if((i.le.3).and.(j.le.3)) then
              DDSDDE(i,j) = DD(i,j)
           else
              DDSDDE(i,j) = DD(i,j)
           end if
         end do
       enddo
 c     c Consistent tangent stiffness
       do j=1,NTENS
         do i=1,NTENS
           if (DDSDDE(i,j)<1.0) then
              DDSDDE(i,j) = 0
           end if
         end do
       enddo

 21    format(6(1x,1p,e12.5))

       Return
       END SUBROUTINE UMAT
```

****************************Subroutines**************************************

```
      Subroutine EVPSCLAY1S (dTime, Props, Sig0, dEps,
     &              D, Sig, StVar, nStat)

      INCLUDE 'ABA_PARAM.INC'
c      Implicit Double Precision (A-H, O-Z)
      Double Precision DDSDDE(6,6)

      Double Precision   StVar(nStat)
      Double Precision   Sig0(6), dsig(6), Sig(6), sig_d(6)
      Double Precision   dEps(6), dEps_vp(6)
      Double Precision   D(6,6)
      Double Precision   Props(50)
      Double Precision   hh(6,6),hh1(6,6),hh2(6,6),hh3(6,6),hh4(6,6),
     &                   hh5(6,6),hh6(6,6),hh7(6),hh8(6,6),htte(6)
      Double Precision   dF_dSig(6),VID(6),dGdSig(6,6)
      Double Precision   ht(6,6),hd(6,6),ht1(6),ht2(6),hdd(6,6),ht3(6)
      Double Precision  dEps0_trial(6),dEps0_vp(6),Sig_de(6)
      Double Precision  Sig_triale(6),Sig0_trial(6),GG(6),QQ(6)
      Double Precision  dddSig_trial(6)


      DATA VID/1.d0, 1.d0, 1.d0, 0.d0, 0.d0, 0.d0/

      Double Precision      alpha(6)         ! current alpha
      Double Precision      alpha_d(6)       ! deviatoric fabric tensor
      Double Precision      term(6)          ! auxilary vector for evaluation
      Double Precision      Sig_trial(6)     ! trial stress state
      Double Precision      dSig_trial(6)    ! trial stress increment
      Double Precision      dEps_trial(6)    ! target strain increment
      Double Precision      dEpsDev(6)       ! plastic deviatoric strain increment vector
      Double Precision      dad(6)           ! change of alpha_d
      DOUBLE PRECISION      Sig0_d(6)
! ------------------------------------------------------------------------------------------

! ------------------------------------------------------------------------------------------
! Get Material Constants
! ------------------------------------------------------------------------------------------
    dny      = Props(1)
    dkappa   = Props(2)
    dlambda  = Props(3)
    e0       = Props(4)
    dMc      = Props(5)
    OCR      = Props(6)
    POP      = Props(7)
    Cae      = Props(8)
    tau_in   = Props(9)
    St       = Props(10)
    a        = Props(11)
    b        = Props(12)
    anisotropy= Props(13)
    E_ratio  = Props(14)
    I_undrain=Props(15)   ! =0 for drained, =1 for undrained
    thet         = 1.0
c
    eta_K0     = 3.*dMc/(6.-dMc)
    alpha0     = eta_K0 - (dMc**2-eta_K0**2)/3.
    beta       = 3.*(4.*dMc**2-4.*eta_K0**2-3.*eta_K0)/
     &           ( 8.*(eta_K0**2+2.*eta_K0-dMc**2) )
    dw         = (1.+e0)/(dlambda-dkappa)*log((10.*dMc**2-2.*alpha0*beta)/
     &                                  (dMc**2-2.*alpha0*beta))
    x0         = St-1.
```

```
     dnn        = (dlambda-dkappa)/Cae
     gamma      = Cae*(dMc**2-alpha0**2)/(tau_in*(1.+e0)*(dMc**2-
      &                                            eta_K0**2))
     dM_ratio   = 3./(3.+dMc)
     power      = 0.
       if (anisotropy .lt. 0.1) then
              alpha0 = 0.0
              dw     = 0.0
              beta   = 0.0
       end if
! -----------------------------------------------------------------------------------
! Change sign due to SCLAY formulation (compression positive)
! -----------------------------------------------------------------------------------
       Sig0      = -1.*Sig0
       dEps      = -1.*dEps
! -----------------------------------------------------------------------------------
! Start strain increament and Get the state varaibles
! -----------------------------------------------------------------------------------
     dEps_trial   = dEps
     dEps_vp      = 0.         ! reset plastic strain increment
     dSig_trial   = 0.

     ! get state variables
     alpha_scalar= StVar(7)
     xx           = StVar(9)
     ee           = StVar(10)
     pmi          = StVar(8)
     alpha(1:6)   = StVar(1:6)
     pm           = (1.+xx) * pmi

     ! calculate generalized alpha
     alpha_d(1)     = alpha(1) - 1.
     alpha_d(2)     = alpha(2) - 1.
     alpha_d(3)     = alpha(3) - 1.
     alpha_d(4)     = alpha(4) * Sqrt(2.)
     alpha_d(5)     = alpha(5) * Sqrt(2.)
     alpha_d(6)     = alpha(6) * Sqrt(2.)

! -----------------------------------------------------------------------------------
! Check yield criterion
! -----------------------------------------------------------------------------------
21     format(6(1x,1p,e12.5))
c       write(*,*)'D'
c      write(*,21) D
       ! calculate new trial stress increment dSig_trial=dEps_trial*D
       Call MatVec( D, 6, dEps_trial, 6, dSig_trial)

     pdash0         = ( Sig0(1) + Sig0(2) + Sig0(3) ) /3.
     Sig0_d(1)      = Sig0(1) - pdash0
     Sig0_d(2)      = Sig0(2) - pdash0
     Sig0_d(3)      = Sig0(3) - pdash0
     Sig0_d(4)      = Sqrt(2.) * Sig0(4)
     Sig0_d(5)      = Sqrt(2.) * Sig0(5)
     Sig0_d(6)      = Sqrt(2.) * Sig0(6)
     term           = Sig0_d - pdash0*alpha_d ! calculating auxilary vector
       ! Lode angle
         call Lode_angle(term,dM_ratio,g_theta)
!        call Lode_TS (Sig0,Props,dM_ratio)

     dM             = dMc*g_theta
     Sig_triale     = Sig0+dSig_trial

     ! calculate stress invariants
     ppdash         = ( Sig_triale(1) + Sig_triale(2) + Sig_triale(3) ) /3.

     ! calculate generalized deviatoric stress vector sig_d
       sig_de(1)      =   Sig_triale(1) - ppdash
       sig_de(2)      =   Sig_triale(2) - ppdash
```

```
     sig_de(3)      =   Sig_triale(3) - ppdash
     sig_de(4)      =   Sqrt(2.) * Sig_triale(4)
     sig_de(5)      =   Sqrt(2.) * Sig_triale(5)
     sig_de(6)      =   Sqrt(2.) * Sig_triale(6)
     term           =   sig_de - ppdash*alpha_d ! calculating auxilary vector

   if (abs(ppdash) .lt. 5.) Then         !we assume creep occurs only when p'>5 Pa
       Sig_trial    =   Sig_triale
   end if
! ----------------------------------------------------------------------------------------
! Begin viscoplastic stress correction: associated flow
! ----------------------------------------------------------------------------------------

     If (abs(ppdash) .ge. 5.) Then

   ! initial conditions
   Sig_trial      = Sig0
   Sig0_trial     = Sig0
   dEps0_trial    = dEps_trial
   dSig_trial     = 0.
   dEps_vp        = 0.
   pm0            = pm
   alpha_scalar0 = alpha_scalar
   pmi0           = pmi
   xx0            = xx
   ! set newton-raphson iteration number
   inmax          = 1000
   inewt          = 0
! ----------------------------------------------------------------------------------------
! Newton-Raphson iteration
! ----------------------------------------------------------------------------------------


          ! start iteration
   10     continue

              inewt    = inewt+1
   !      write(*,*)'Number of iteration',inewt

          ! calculate stress invariants for time=t_inewt
              pdash    = ( Sig_trial(1) + Sig_trial(2) + Sig_trial(3) ) /3.
   !      write(*,*)'pdash',pdash

          ! calculate generalized deviatoric stress vector sig_d
            sig_d(1)= Sig_trial(1)  - pdash
            sig_d(2)= Sig_trial(2)  - pdash
            sig_d(3)= Sig_trial(3)  - pdash
            sig_d(4)= Sqrt(2.)  * Sig_trial(4)
            sig_d(5)= Sqrt(2.)  * Sig_trial(5)
            sig_d(6)= Sqrt(2.)  * Sig_trial(6)
              term  = sig_d - pdash*alpha_d      ! calculating auxilary vector

   c-------------- Evaluation of matrix H ---
            ! initial matrix
            hh   = 0.
            hh1  = 0.
            hh2  = 0.
            hh3  = 0.
            hh4  = 0.
            hh5  = 0.
            hh6  = 0.
            hh7  = 0.
            hh8  = 0.

            ! dS(ij)/dSig(kl) named as hh1
            hh1(1,1)  = 2./3.
            hh1(1,2)  = -1./3.
            hh1(1,3)  = -1./3.

            hh1(2,1)  = -1./3.
            hh1(2,2)  = 2./3.
            hh1(2,3)  = -1./3.
```

```
        hh1(2,1) = -1./3.
        hh1(2,2) = 2./3.
        hh1(2,3) = -1./3.

        hh1(3,1) = -1./3.
        hh1(3,2) = -1./3.
        hh1(3,3) = 2./3.

        hh1(4,4) = Sqrt(2.)
        hh1(5,5) = Sqrt(2.)
        hh1(6,6) = Sqrt(2.)

      ! delta(ij)*delta(kl)
        do i = 1,3
            do j = 1,3
          hh2(i,j) = 1.
          end do
        end do


    ! s(ij)*delta(kl)+s(kl)*delta(ij)
    hh3(1,1) = 2.*Sig_d(1)
    hh3(1,2) = Sig_d(1)+Sig_d(2)
    hh3(1,3) = Sig_d(1)+Sig_d(3)
    hh3(1,4) = Sig_d(4)
    hh3(1,5) = Sig_d(5)
    hh3(1,6) = Sig_d(6)
    hh3(2,1) = hh3(1,2)
    hh3(2,2) = 2.*Sig_d(2)
    hh3(2,3) = Sig_d(2)+Sig_d(3)
    hh3(2,4) = Sig_d(4)
    hh3(2,5) = Sig_d(5)
    hh3(2,6) = Sig_d(6)
    hh3(3,1) = hh3(1,3)
    hh3(3,2) = hh3(2,3)
    hh3(3,3) = 2.*Sig_d(3)
    hh3(3,4) = Sig_d(4)
    hh3(3,5) = Sig_d(5)
    hh3(3,6) = Sig_d(6)
    hh3(4,1) = hh3(1,4)
    hh3(4,2) = hh3(2,4)
    hh3(4,3) = hh3(3,4)
    hh3(5,1) = hh3(1,5)
    hh3(5,2) = hh3(2,5)
    hh3(5,3) = hh3(3,5)
    hh3(6,1) = hh3(1,6)
    hh3(6,2) = hh3(2,6)
    hh3(6,3) = hh3(3,6)



   ! d[dF/dSig(ij)]/dSig(kl)
     hh4        = 1./(dM**2.-alpha_scalar**2) * (3./pdash*hh1+
    &  (DInProd(Sig_d,Sig_d,6))/3./pdash**3*hh2-1./pdash**2*hh3)

   ! dF/dSig(ij)
     dF_dSig   = 1./(dM**2.-alpha_scalar**2)/pdash * ( 3.*term-
    &  ((DInProd(term,alpha_d,6))+1./2./pdash*
    &    (DInProd(term,term,6)))*VID ) + 1./3.*VID

   ! [dPhi/dSig(kl)]*[dF/dSig(ij)]
   pmd        = 3./2.*(DInProd(term,term,6))/(dM**2.-alpha_scalar**2)
    &  /pdash + pdash
   PhiF       = (pmd/pm)**dnn                     ! for non-linear creep
!  write(*,*)'PhiF',PhiF
   if (PhiF .lt. 0.) PhiF = 0.

     ! QQ^n (constant value during the iteration)
   if (inewt .eq. 1) then
     dEps0_vp = dTime*gamma*PhiF*dF_dSig
```

```
        hdd=d
        call brinv(hdd,6,L)
        call MatVec (hdd,6,Sig0_trial,6,htte)
        QQ = dEps0_trial-(1-thet)*dEps0_vp+htte
    end if

    ! GG^i for iteration
      call MatVec (hdd,6,Sig_trial,6,htte)
    GG = htte+thet*dTime*gamma*PhiF*dF_dSig
!   write(*,*)'GG',GG
    ! hh5
      do i = 1,6
        do j = 1,6
         hh5(i,j) = dnn/pm*(pmd/pm)**(dnn-1.)*dF_dSig(i)*dF_dSig(j)
        end do
        end do
!        write(*,*)'hh5',hh5

    ! [H] = dEps_vpRate(ij)/dSig(kl)
      hh      = gamma*(hh5+PhiF*hh4)
!        write(*,*)'hh',hh

    ! dGdSig
    dGdSig = hdd+dTime*thet*hh
    call brinv(dGdSig,6,L)
!   write(*,*)'dGdSig',dGdSig
! -------------------------------------------------------------------------------------------
! Calculate DDSDDE: dSig/dEsp
! -------------------------------------------------------------------------------------------
c   DDSDDE = dGdSig

    GG = QQ-GG
    call MatVec (dGdSig,6,GG,6,dSig_trial)    ! dSig
!   write(*,*)'dSig_trial',dSig_trial

      ! update stress strain
      Sig_trial     = Sig_trial+dSig_trial
      dddSig_trial = Sig_trial-Sig0_trial
      call MatVec (hdd,6,dddSig_trial,6,ht3)
      dEps_vp       = dEps_trial- ht3


      ! calculate volumic plastic strain and deviatoric plastic strain
      dEpsVol        = dEps_vp(1) + dEps_vp(2) + dEps_vp(3)
      dEpsVolMac     = dEpsVol
      if (dEpsVolMac.lt.0.) dEpsVolMac = 0. !Macauley brackets

      dEpsDev(1)     = (2.*dEps_vp(1) - dEps_vp(2) - dEps_vp(3))/3.
      dEpsDev(2)     = (2.*dEps_vp(2) - dEps_vp(1) - dEps_vp(3))/3.
      dEpsDev(3)     = (2.*dEps_vp(3) - dEps_vp(2) - dEps_vp(1))/3.
      dEpsDev(4)     = Sqrt(2.)*dEps_vp(4)
      dEpsDev(5)     = Sqrt(2.)*dEps_vp(5)
      dEpsDev(6)     = Sqrt(2.)*dEps_vp(6)
      dEpsDevScalar = Sqrt(2./3.*(DInProd(dEpsDev,dEpsDev,6)))


      ! update state variables
      dpmi = ((1.+e0)*pmi*dEpsVol)/(dlambda-dkappa)
      dxx  = -(a*xx*( Abs(dEpsVol) + b*Abs(dEpsDevScalar) ))
      pmi  = pmi0+dpmi
      xx   = xx0+dxx
      pm   = pmi * (1. + xx)

!   write(*,*)'GG',GG

      ! set convergence criterion
    verr=sqrt(dSig_trial(1)**2+dSig_trial(2)**2+dSig_trial(3)**2+
     &    dSig_trial(4)**2+dSig_trial(5)**2+dSig_trial(6)**2)

    if (inewt .eq. inmax) then
!       write(*,*)'can not convergence'
        stop 'can not convergence'
    endif
```

```
    if (verr .gt. 1.d-4) goto 10
! ------------------------------------------------------------------------------
! End of Newton-Raphson iteration
! ------------------------------------------------------------------------------
!      write(*,*)'DDSDDE',DDSDDE
! ------------------------------------------------------------------------------
! Update State Variables
! ------------------------------------------------------------------------------
        dad             = dw*(((3.*sig_d)/(4.*pdash)-alpha_d)*dEpsVolMac
     &     + beta*(sig_d/(3.*pdash)-alpha_d)*dEpsDevScalar)
        alpha_d         = alpha_d+dad
        alpha_scalar    = Sqrt(3./2.*(DInProd(alpha_d,alpha_d,6)))   ! get alpha explicitly

      StVar(1)  = (alpha_d(1)) + 1.
      StVar(2)  = (alpha_d(2)) + 1.
      StVar(3)  = (alpha_d(3)) + 1.
      StVar(4)  = (alpha_d(4)) / Sqrt(2.)
      StVar(5)  = (alpha_d(5)) / Sqrt(2.)
      StVar(6)  = (alpha_d(6)) / Sqrt(2.)
      StVar(7)  = alpha_scalar
      StVar(8)  = pmi
      StVar(9)  = xx


      endif                        ! end of viscoplasticity
! ------------------------------------------------------------------------------
! End of viscoplasticity
! ------------------------------------------------------------------------------



      ! re-change sign due to SCLAY formulation (compression positive)
      Sig  = (-1.)*Sig_trial
      dEps = (-1.)*dEps

      Return
      End Subroutine EVPSCLAY1S
!---------------------- some subroutines for the calculation --------------------------



!***********************************************************************
      Subroutine MatVec(xMat,IM,Vec,N,VecR)
C***********************************************************************
C
C     Calculate VecR = xMat*Vec
C
C I   xMat  : (Square) Matrix (IM,*)
C I   Vec   : Vector
C I   N     : Number of rows/colums
C O   VecR  : Resulting vector
C
C***********************************************************************
      INCLUDE 'ABA_PARAM.INC'
c     Implicit Double Precision (A-H,O-Z)
      Dimension xMat(IM,*),Vec(*),VecR(*)
C***********************************************************************
      Do I=1,N
        X=0
        Do J=1,N
          X=X+xMat(I,J)*Vec(J)
        End Do
        VecR(I)=X
      End Do
      Return
      End     ! Subroutine MatVec
C
```

```
C***********************************************************************
      Double Precision Function DInProd(A,B,N)
C***********************************************************************
C
C     Returns the Inproduct of two vectors
C
C I   A,B   : Two vectors
C I   N     : Used length of vectors
C***********************************************************************
      INCLUDE 'ABA_PARAM.INC'
c      Implicit Double Precision (A-H,O-Z)
      Dimension A(*),B(*)
C***********************************************************************
        X = 0
      Do I=1,N
        X = X + A(I)*B(I)
      End Do
        DInProd = X
      Return
      End      ! Function DInProd
C
C***********************************************************************

C===========================================================================
C
C            GAUSS-JORDAN ALGORYTHM FOR MATRIX INVERSE
C
C===========================================================================
C
      SUBROUTINE BRINV(A,N,L)
      INCLUDE 'ABA_PARAM.INC'
      DIMENSION A(N,N),IS(N),JS(N)
      DOUBLE PRECISION A,T,D
        L=1
      DO 100 K=1,N
          D=0.0
      DO 10 I=K,N
        DO 10 J=K,N
        IF (ABS(A(I,J)).GT.D) THEN
          D=ABS(A(I,J))
          IS(K)=I
          JS(K)=J
        END IF
10    CONTINUE
        IF (D+1.0.EQ.1.0) THEN
            L=0
        WRITE(*,20)
        RETURN
        END IF
20    FORMAT(1X,'ERR**NOT INV')
      DO 30 J=1,N
        T=A(K,J)
        A(K,J)=A(IS(K),J)
        A(IS(K),J)=T
30    CONTINUE
      DO 40 I=1,N
        T=A(I,K)
        A(I,K)=A(I,JS(K))
        A(I,JS(K))=T
40    CONTINUE
        A(K,K)=1/A(K,K)
      DO 50 J=1,N
```

```
            IF (J.NE.K) THEN
              A(K,J)=A(K,J)*A(K,K)
            END IF
50      CONTINUE
        DO 70 I=1,N
            IF (I.NE.K) THEN
            DO 60 J=1,N
                IF (J.NE.K) THEN
                  A(I,J)=A(I,J)-A(I,K)*A(K,J)
                END IF
60          CONTINUE
            END IF
70      CONTINUE
        DO 80 I=1,N
            IF (I.NE.K) THEN
              A(I,K)=-A(I,K)*A(K,K)
            END IF
80      CONTINUE
100       CONTINUE
          DO 130 K=N,1,-1
          DO 110 J=1,N
            T=A(K,J)
            A(K,J)=A(JS(K),J)
            A(JS(K),J)=T
110       CONTINUE
        DO 120 I=1,N
            T=A(I,K)
            A(I,K)=A(I,IS(K))
            A(I,IS(K))=T
120     CONTINUE
130         CONTINUE
          RETURN
          END


      SUBROUTINE MATRIXDE_ANI(SIG,PROPS,STVAR,D)
c======================================================================================
c
c                     Calculation of Anisotropic Elastic Matrix D for SIG=D*EPS
c
c          Remark: E_ratio = Eh / Ev ( = 1 for isotropic elasticity)
c                  E_para can be dermined from kapa-line ( = 1 for isotropic elasticity)
c======================================================================================
      INCLUDE 'ABA_PARAM.INC'
c     IMPLICIT DOUBLE PRECISION(A-H,O-Z)
      DIMENSION SIG(6),PROPS(*),STVAR(*),D(6,6)
c
      ! get dny value
      xNu     = props(1)
      dkappa  = props(2)
      ee      = props(4)    ! keep constant e as the definition of compression curve from experiment 11
      E_ratio = props(14)
      if (E_ratio .gt. 1.) E_ratio = 1.
      E_para  = (3./(2.*E_ratio+1.))**1.5          ! to keep kappa-line the same
c     write(*,*) "E_ratio",E_ratio

    ! determine shear modulus from Youngs' modulus
    pdash = ( Sig(1) + Sig(2) + Sig(3) ) /3.          ! global mean stress
    if ((pdash.gt.-10.).and.(pdash.le.0.)) pdash =-10.
    if ((pdash.lt. 10.).and.(pdash.gt.0.)) pdash = 10.

        dKdash= abs((1.+ee)*pdash/dkappa)                      !..compression modulus
        Ev = dKdash*(1.-4.*xNu+2./E_ratio-2.*xNu/dsqrt(E_ratio))
!       E     = 3.*(1.-(2.*xNu))*dKdash*E_para                 ! (Young's) from dkappa
        xNu = xNu*dsqrt(E_ratio)

    ! compose liner elastic material stiffness matrix
```

```
    ! compose liner elastic material stiffness matrix
        D      = 0.                                      ! initial Matrix
        F1     = Ev/(1.+xNu)/(1.-2.*xNu)
        alph   = sqrt(E_ratio)
        D(1,1) = F1*(1.-xNu)*alph**2.
        D(2,2) = F1*(1.-xNu)
        D(3,3) = F1*(1.-xNu)*alph**2.
        D(4,4) = F1*(1.-2.*xNu)*alph
        D(5,5) = F1*(1.-2.*xNu)*alph
        D(6,6) = F1*(1.-2.*xNu)*alph**2.
        D(1,2) = F1*xNu*alph
        D(1,3) = F1*xNu*alph**2.
        D(2,3) = F1*xNu*alph
        D(2,1) = D(1,2)
        D(3,1) = D(1,3)
        D(3,2) = D(2,3)          ! for plaxis, term 2 is the vertical tensor          ! for
c-------------------------------------------------------------------------------------------
      END


      SUBROUTINE Lode_angle(Sig,dM_ratio,g_theta)
C=========================================================================
c
c                 calculate the crrection of dM by Lode angle
c                         [Sheng D., Sloan S.W. and Yu H.S. 2000]
c                          - Aspects of finite element implementation of critical state models
c
C=========================================================================
      INCLUDE 'ABA_PARAM.INC'
c     IMPLICIT DOUBLE PRECISION(A-H,O-Z)
      DIMENSION SIG(6)
      ! compression positive

        dJ2 = ((Sig(1)-Sig(2))**2+(Sig(1)-Sig(3))**2+(Sig(2)
     &      -Sig(3))**2)/6+Sig(4)**2+Sig(5)**2+Sig(6)**2
        pdash = (Sig(1)+Sig(2)+Sig(3))/3.
        s1    = Sig(1) - pdash
        s2    = Sig(2) - pdash
        s3    = Sig(3) - pdash
        s4    = Sqrt(2.) * Sig(4)
        s5    = Sqrt(2.) * Sig(5)
        s6    = Sqrt(2.) * Sig(6)
        dJ3   = s1*s2*s3+2*s4*s5*s6-s1*s6**2-s2*s5**2-s3*s4**2

     if (dJ2 .lt. 1.d-6) then
         g_theta = (2.*dM_ratio**4/(1.+dM_ratio**4))**0.25
        else
         sin3fi  = -3.*3.**0.5*dJ3/2./dJ2**1.5
         g_theta=(2.*dM_ratio**4/(1.+dM_ratio**4+(1.-dM_ratio**4)
     &                 *sin3fi))**0.25
     end if

    end
C
c
C***********************************************************************
C                             End Of program
C***********************************************************************
```

Appendix B: SCLAY1-S-SS

(Copyright (c) 2017, geoinvention www.geoinvention.com)

************************* Main program *************************

```
! ---------------------------------------------------------------------------------
! Subroutines:
!     UMAT                                        : Link GCAM to ABAQUS
! ---------------------------------------------------------------------------------
      SUBROUTINE UMAT(STRESS,STATEV,DDSDDE,SSE,SPD,SCD,
     1 RPL,DDSDDT,DRPLDE,DRPLDT,
     2 STRAN,DSTRAN,TIME,DTIME,TEMP,DTEMP,PREDEF,DPRED,CMNAME,
     3 NDI,NSHR,NTENS,NSTATV,PROPS,NPROPS,COORDS,DROT,PNEWDT,
     4 CELENT,DFGRD0,DFGRD1,NOEL,NPT,LAYER,KSPT,KSTEP,KINC)
C
      INCLUDE 'ABA_PARAM.INC'
C
      CHARACTER*80 CMNAME
      Dimension STRESS(NTENS),STATEV(NSTATV),
     1 DDSDDE(NTENS,NTENS),
     2 DDSDDT(NTENS),DRPLDE(NTENS),
     3 STRAN(NTENS),DSTRAN(NTENS),TIME(2),PREDEF(1),DPRED(1),
     4 PROPS(NPROPS),COORDS(3),DROT(3,3),DFGRD0(3,3),DFGRD1(3,3)
C------------------------------------------------------------------------------------------
C                          DEFINE THE VARIABLES
C------------------------------------------------------------------------------------------
!    Implicit Double Precision (A-H, O-Z)
      Double precision    cm(NPROPS),CC(6,6)
      Double precision    Sig(6),eps(6),dSig(6),dEps(6),Epsp(6)
      Double precision    hsv(NSTATV),deps0(6),depsp(6),Sig0(6)
      Double precision    Sig_temp(6),Eps0(6),Sig0_mod(6)
      Double precision    Sig_new(6),sij_mod(6),deps_temp(6)
      Integer             crit,LSTEP, ID_step
      Integer             Nstep,niter,ISTEP,k_input
      Logical             :: converged       ! converged for iteration
      Parameter( zero = 0.D0, one = 1.D0, two = 2.D0, three = 3.D0,
     1  third = one/three, half = 0.5D0, twoThirds = two/three,
     2  threeHalfs = 1.5D0 )
C
      Parameter(theta = 1.0D-8)
C------------------------------------------------------------------------------------------
C                           READ model parameters
C------------------------------------------------------------------------------------------
```


Z. Yin and Y. Jin, *Practice of Optimisation Theory in Geotechnical Engineering*,
https://doi.org/10.1007/978-981-13-3408-5

```
      cm        = props
! -----------------------------------------------------------------------
! Get Material Constants
! -----------------------------------------------------------------------
      pois      = cm(1)     ! Poisson's ratio =0.15~0.35
      dkappa    = cm(2)     ! slope of loading-reloading curve in e-ln(p) [typical value: 0.02~0.001]
      dlambda   = cm(3)     ! slope of post-yield compression curve in e-ln(p) [typical value: 0.2~0.05]
      e0        = cm(4)     ! initial void ratio [typical value: 0.5~3]
      dMc       = cm(5)     ! slope of critical state line in q-p plane [typical value: 0.8~1.5]
      POP       = cm(6)     ! preconsolidation pressure [typical value: 1.~1000.]
      alpha_K0  = cm(7)     ! initial inclination of yield surface [typical value: 0.~0.5]
      aa        = cm(8)     ! distance betwwen ICL and CSL for infinite epsv [typical value: 0.75]
      bb        = cm(9)     ! distance betwwen ICL and CSL for infinite epsd [typical value: 0.~0.3333]
      dw        = cm(10)    ! rotation effect for plastic strains [typical value: 0.~50.]
      dw_d      = cm(11)    ! rotation effect for plastic deviatoric strain [typical value: 0.~0.8]
      x0        = cm(12)    ! bonding ratio =St-1. [typical value: 0.~50]
      dksi_v    = cm(13)    ! debonding effect due to plastic volumetric strain [typical value: 0.~20]
      dksi_d    = cm(14)    ! [typical value: 0.~0.5]
      pb0       = cm(15)    ! bonding amount for CSL = cohesion/Mc [typical value: 0.~100.]
      dksi_s    = cm(16)    ! debonding effect for cohesion due to plastic diviatoric strain [typical va
      E_ratio   = cm(17)    ! Anisotropy of elastic modulus E_ratio=Eh/Ev, <=1 [typical value: 1.]
      dLode     = cm(18)    ! dLode = 1 (if dLode>=0.5) for TS; dLode = 0 (if dLode<=0.5) for Sheng
      dM_ratio  = cm(19)
      eps70     = cm(20)    ! for small strain stiffness, value of eps_d at 70%*Gmax
! Initialisation section
c      write(*,21) cm
c     stop
! -----------------------------------------------------------------------
c**********************************************************************c
21    format(6(1x,1p,e12.5))
! Initialize state variables
! -----------------------------------------------------------------------

! -----------------------------------------------------------------------
! Read STRESS STRAIN
! -----------------------------------------------------------------------
      Sig0(1:4)   = STRESS(1:4)
      Eps0(1:3)   = STRAN(1:3)
      Eps0(4)     = STRAN(4)*0.5    ! change engineering strain to normal strain
      dEps(1:3)   = DSTRAN(1:3)
      dEps(4)     = DSTRAN(4)*0.5   ! change engineering strain to normal strain
      Sig0(5:6)   = zero
      Eps0(5:6)   = zero
      dEps(5:6)   = zero
      if( NSHR .GT. 1 ) then
          Sig0(5:6)   = STRESS(5:6)
          Eps0(5:6)   = STRAN(5:6)*0.5    ! change engineering strain to normal strain
          dEps(5:6)   = DSTRAN(5:6)*0.5   ! change engineering strain to normal strain
      end if
c change sign due to EVP-MCC formulation (compression positive)
! -----------------------------------------------------------------------
! Read state variables
! -----------------------------------------------------------------------
      hsv=STATEV
! -----------------------------------------------------------------------
! Initialize state variables
! -----------------------------------------------------------------------
      IF (KSTEP == 1 .AND. TIME(2) == 0.) THEN
      hsv = 0.
      Sig_temp=( -1.)*Sig0
    ! get K0nc value
      eta_K0 = 3.*dMc/(6.-dMc)
      alpha_K0 = eta_K0 - (dMc**2-eta_K0**2)/3.
      dw_d    = 3.*(4.*dMc**2-4.*eta_K0**2-3.*eta_K0)/
     &          ( 8.*(eta_K0**2+2.*eta_K0-dMc**2) )
      dw      = (1.+e0)/(dlambda-dkappa)*log((10.*dMc**2-2.*alpha_K0*dw_d)/
     &          (dMc**2-2.*alpha_K0*dw_d))
      phi     = asin(3.*dMc/(6.+dMc))
      dK0nc = 1.-sin(phi)
c     dK0nc = 1.
      POP     = abs(POP)
      if (POP.le.0.) POP = 0.1
```

```
      ! adjust Sig0(1:3) due to POP    Sig_temp(1)+
      Sig0_mod(1)= Sig_temp(1)+POP*dK0nc
      Sig0_mod(2)= Sig_temp(2)+POP
      Sig0_mod(3)= Sig_temp(3)+POP*dK0nc
      Sig0_mod(4)= 0.
      Sig0_mod(5)= 0.
      Sig0_mod(6)= 0.

    ! calculate stress invariants and lode angle effect
      call stress_invariant(Sig0_mod, dI1, dI2, dI3, p, q)
      if ((p==0.).or.(dMc**2.==alpha_K0**2.)) then  ! pm not calculable
          Write(*,*)'ERROR: pm not calculable - division by zero'
c         Stop
      end if

    ! pre-set state variables
      hsv(1) = -(alpha_K0/3.)+1.         ! alpha_x
      hsv(2) = (2.*alpha_K0/3.)+1.       ! alpha_y    ! y-direction is vertical
      hsv(3) = -(alpha_K0/3.)+1.         ! alpha_z
      hsv(4) = 0.                        ! alpha_xy
      hsv(5) = 0.                        ! alpha_yz
      hsv(6) = 0.                        ! alpha_zx
      hsv(7) = alpha_K0                  ! alpha_scalar
c-----------------------------------------------------------------------------------------------
      alpha_d1 = hsv(1) - 1.
      alpha_d2 = hsv(2) - 1.
      alpha_d3 = hsv(3) - 1.
      alpha_d4 = hsv(4)
      alpha_d5 = hsv(5)
      alpha_d6 = hsv(6)

      sij_mod(1)=Sig0_mod(1)-(p+pb0)-(p+pb0)*alpha_d1
      sij_mod(2)=Sig0_mod(2)-(p+pb0)-(p+pb0)*alpha_d2
      sij_mod(3)=Sig0_mod(3)-(p+pb0)-(p+pb0)*alpha_d3
      sij_mod(4)=Sig0_mod(4)-(p+pb0)*alpha_d4
      sij_mod(5)=Sig0_mod(5)-(p+pb0)*alpha_d5
      sij_mod(6)=Sig0_mod(6)-(p+pb0)*alpha_d6

      if (dLode .gt. 0.5) then
        call Lode_TS (Sig0_mod,cm,dMratio)
      else
        call Lode_sheng(sij_mod,cm,dMratio)
      end if
      dM = dMc*dMratio
c-----------------------------------------------------------------------------------------------
    ! determine size of the initial yield curve(crosspoint of p-axial and outer elipse)
      pmd0=((q-(p+pb0)*alpha_K0)**2+
     &  (dM**2-alpha_K0**2)*(p+pb0)*p)/(dM**2-alpha_K0**2)/(p+pb0)
c      write(*,*) 'pmd0=',pmd0
c      stop

      hsv(8) = pmd0/(1.+x0)      ! pmi
      hsv(9) = x0                ! x
      hsv(10)= pmd0              ! pmd Size of dynamic yield surface
      hsv(11)= e0                ! e_current (current void ratio)
      hsv(12)= pb0               ! cohesion in p-axe, positive value for negative position
      hsv(13:18)= 0.             ! plestic strain
      hsv(19)   = 0.             ! peeq
! initialization for small strain stiffness
      do 11 i = 1, 6
      hsv(20+i) = 0.    ! initial dstrain deps0(i)
      hsv(26+i) = 0.    ! initial strain eps0(i)
      hsv(32+i) = 0.    ! initial reverse strain point epsR(i)
11    continue          ! Next i
      hsv(39)= 0.       ! ID_reverse = 0
      hsv(40)= 0.       ! eps_d_m=0.
      hsv(41)= 1.       ! costheta=1. no reverse, angle between previous and current loadings
c
      end if
c       write(*,21) hsv
c       stop
! ---------------------------------------------------------------------------------------
```

```
! ------------------------------------------------------------------------------------
! Strain Load
! ------------------------------------------------------------------------------------
         Call GCAM_model(cm,dEps,Sig0,Sig_new,hsv,Eps0)
! ------------------------------------------------------------------------------------
! ------------------------------------------------------------------------------------
! UPDATE STRESS STRAIN AND STATE VARIABLES
! ------------------------------------------------------------------------------------
       STRESS(1:4) = Sig_new(1:4)
       if( NSHR .GT. 1 ) then
           STRESS(5:6) = Sig_new(5:6)
       end if
       STATEV = hsv
!-------------------------------------------------------------------------------------
! Give the DDSDDE BY NUMERICAL DISTURBED METHOD
!-------------------------------------------------------------------------------------
       Sig_new =-1.*Sig_new
       if (eps70 .gt. 0.) then
            Call MATRIXDE_ANI_SS(Sig_new,cm,hsv,CC)
       else
            Call MATRIXDE_ANI(Sig_new,cm,hsv,CC)
       end if
       Do i=1,NTENS
          Do j=1,NTENS
             DDSDDE(i,j)=CC(i,j)
          End Do
       End Do
       DDSDDE(4,4)=CC(4,4)*0.5    ! change to engineering strain for abaqus-umat
       if( NSHR .GT. 1 ) then
           DDSDDE(5,5)=CC(5,5)*0.5
           DDSDDE(6,6)=CC(6,6)*0.5
       end if
       end SUBROUTINE UMAT

!=====================================================================================
! Subroutines for the soil model Anisotropic Modified Cam Clay with small strain=========
!=====================================================================================
       Subroutine GCAM_model (cm,dEps,Sig,Sig_new,hsv,Eps0)  ! using normal strain,  not engineering str
       INCLUDE 'ABA_PARAM.INC'
       Character*80 cmname
c       Implicit Double Precision (A-H, O-Z)
       Double Precision  cm(*),Sig(6),hsv(*),dEps(6),dSig(6),Eps0(6)
       Double Precision  alpha_d(6),alpha(6),Sig0_d(6),s(6),CC(6,6)
       Double Precision  s_alph(6),alpha0,alpha1,deps0(6),eps(6)
       Double Precision  dfds(6),dfdsig(6),dEps_e(6),dEps_vp(6),dad(6)
       Double Precision  dGdSig(6),dSigdLam(6),depsp(6),Sig0(6)
       Double Precision  dfdalphad(6),Sig_new(6),depse(6)
       Double Precision  dalphaddepsv(6),dalphaddepsd(6),tta(6)
       Double Precision  Sig0_mod(6),s0(6)
       Logical            :: converged
c
       pois       = cm(1)    ! Poisson's ratio =0.15~0.35
       dkappa     = cm(2)    ! slope of loading-reloading curve in e-ln(p) [typical value: 0.02~0.001]
       dlambda    = cm(3)    ! slope of post-yield compression curve in e-ln(p) [typical value: 0.2~0.05]
       e0         = cm(4)    ! initial void ratio [typical value: 0.5~3]
       dMc        = cm(5)    ! slope of critical state line in q-p plane [typical value: 0.8~1.5]
       POP        = cm(6)    ! preconsolidation pressure [typical value: 1.~1000.]
       alpha_K0   = cm(7)    ! initial inclination of yield surface [typical value: 0.~0.5]
       aa         = cm(8)    ! distance betwwen ICL and CSL for infinite epsv [typical value: 0.75]
       bb         = cm(9)    ! distance betwwen ICL and CSL for infinite epsd [typical value: 0.~0.3333]
       dw         = cm(10)   ! rotation effect for plastic strains [typical value: 0.~50.]
       dw_d       = cm(11)   ! rotation effect for plastic deviatoric strain [typical value: 0.~0.8]
       x0         = cm(12)   ! bonding ratio =St-1. [typical value: 0.~50]
       dksi_v     = cm(13)   ! debonding effect due to plastic volumetric strain [typical value: 0.~20]
       dksi_d     = cm(14)   ! [typical value: 0.~0.5]
       pb0        = cm(15)   ! bonding amount for CSL = cohesion/Mc [typical value: 0.~100.]
       dksi_s     = cm(16)   ! debonding effect for cohesion due to plastic diviatoric strain [typical va
       E_ratio    = cm(17)   ! Anisotropy of elastic modulus E_ratio=Eh/Ev, <=1 [typical value: 1.]
       dLode      = cm(18)   ! dLode = 1 (if dLode>=0.5) for TS; dLode = 0 (if dLode<=0.5) for Sheng
       dM_ratio   = cm(19)   ! dM_ratio=Me/Mc only for the case of Sheng
       eps70      = cm(20)   ! for small strain stiffness, value of eps_d at 70%*Gmax

21     format(6(1x,1p,e12.5))
! ------------------------------------------------------------------------------------
! Change sign due to SCLAY formulation (compression positive)
! ------------------------------------------------------------------------------------
```

```
! ------------------------------------------------------------------------------------------
      Sig        = -1.*Sig
      dEps       = -1.*dEps
c-------------------------------------------------------------------------------------------
      deps0= deps
      sig0 = sig
c*************************small strain ****************************************************
      if (eps70 .gt. 0.) then
           Call smallstrain_sim (hsv,Eps0)
c           Call smallstrain (hsv,deps0)
           Call MATRIXDE_ANI_SS(Sig,cm,hsv,CC)
      else
           Call MATRIXDE_ANI(Sig,cm,hsv,CC)
      end if
c******************************************************************************************
      eta_K0 = 3.*dMc/(6.-dMc)
      alpha_K0 = eta_K0 - (dMc**2-eta_K0**2)/3.
      dw_d    = 3.*(4.*dMc**2-4.*eta_K0**2-3.*eta_K0)/
     &         ( 8.*(eta_K0**2+2.*eta_K0-dMc**2) )
      dw      = (1.+e0)/(dlambda-dkappa)*log((10.*dMc**2-2.*alpha0*beta)/
     &          (dMc**2-2.*alpha0*beta))
c-------------------------------------------------------------------------------------------
      ! update state variables: current void ratio
      Call strain_invariant(deps0, dEpsv, dEpsd)
      call stress_invariant(Sig, dI1, dI2, dI3, p, q)
      hsv(11)= hsv(11)-dEpsV*(1.+hsv(11))  !update the void ration
!-------------cutting plane method,iteration for real depsp---------------------------------
      ! reset
      depsp=0.                  ! reset plastic strain increment
      converged=.false.         ! do at least one iteration for each subincrement

      ! loop until convergence is reached
      inewt=0
      Do While (.not.converged)
         inewt=inewt+1
c     WRITE (*,*) '---------------------'
c     WRITE (*,*) 'Number of iteration: ',inewt
        converged =.true. ! gets false again during iteration when plasticity occurs
        deps       = deps-depsp  ! new trial strain increment
      if (inewt .gt. 1000) then
!          write(*,*) 'inewt >1000 unconvergence',inewt
          Call MatVec(CC,6,deps,6,dsig)
          sig  = sig0 +dsig
          goto 101
      end if
        depsp      = 0.          ! reset
c
         Call MatVec(CC,6,deps,6,dsig)
         sig  = sig0 +dsig
c         write(*,21) dsig
c-------------------------------------------------------------------------------------------
    !get initial hardening parameters
       alpha(1:6)   = hsv(1:6)
       alpha0       = hsv(7)      ! alpha_scalar
       pmi          = hsv(8)
       xx           = hsv(9)
       pmd          = (1.+xx)*pmi
       ee           = hsv(11)
       pb           = hsv(12)
!------------------------------------ check yield criterion------------------------
      Call stress_invariant(Sig, dI1, dI2, dI3, pp, qq)

    ! calculate generalized alpha_ij
      alpha_d(1) = alpha(1) - 1.
      alpha_d(2) = alpha(2) - 1.
      alpha_d(3) = alpha(3) - 1.
      alpha_d(4) = alpha(4)
      alpha_d(5) = alpha(5)
      alpha_d(6) = alpha(6)
      alpha1     = 1.5*(alpha_d(1)**2+alpha_d(2)**2+
     & alpha_d(3)**2+2.*alpha_d(4)**2+2.*alpha_d(5)**2+2.*alpha_d(6)**2)
```

```
      s0(1)       = Sig(1)-pp
      s0(2)       = Sig(2)-pp
      s0(3)       = Sig(3)-pp
      s0(4)       = Sig(4)
      s0(5)       = Sig(5)
      s0(6)       = Sig(6)
      s_alph      = s0-alpha_d*(pp+pb)  ! rij=sij-alphaij*(p+pb)

      sa= 1.5*( s_alph(1)*s_alph(1)+s_alph(2)*s_alph(2)+s_alph(3)*s_alph(3)
     & +2.*s_alph(4)*s_alph(4)+2.*s_alph(5)*s_alph(5)+
     &  2.*s_alph(6)*s_alph(6))

      salpha2     = sqrt(sa)
c-----------------------------------------------------------------------------------
      if (dLode .gt. 0.5) then              !when dlode <30 degree
         call Lode_TS (Sig,cm,dMratio)
      else
         call Lode_sheng(s_alph,cm,dMratio)       !when dlode >30 degree
c        write(*,*) 'dMratio=',dMratio
      end if
      dM = dMc*dMratio
c        write(*,*) 'dM=',dM
!---------------------------------------------------------------------------------------
      yield=sa+(dM**2-alpha0**2)*(pp+pb)*(pp-pmd)       !yield function
c        write (*,*) yield
      if (yield.gt.1d-7) then


      !dfdsig
      ds_alphalph=-3.*(s_alph(1)*alpha_d(1)+s_alph(2)*alpha_d(2)
     & +s_alph(3)*alpha_d(3)+2.*s_alph(4)*alpha_d(4)+
     & 2.*s_alph(5)*alpha_d(5)+2.*s_alph(6)*alpha_d(6))
      dfdp        = ds_alphalph+(dM**2-alpha0**2)*(2.*pp+pb-pmd)
      dfds        = 3.*(s0-(pp+pb)*alpha_d)
      dfdsig(1)   = dfdp*1./3.+dfds(1)
      dfdsig(2)   = dfdp*1./3.+dfds(2)
      dfdsig(3)   = dfdp*1./3.+dfds(3)
      dfdsig(4)   =            dfds(4)
      dfdsig(5)   =            dfds(5)
      dfdsig(6)   =            dfds(6)

    !dgdsig
      dgdsig=dfdsig
!-------------------------------------------------------------------------------
    !dd
        Call MatVec( CC, 6, dGdSig, 6, dSigdLam)
        dSigXdLam    = (-1.)*dSigdLam(1)
        dSigYdLam    = (-1.)*dSigdLam(2)
        dSigZdLam    = (-1.)*dSigdLam(3)
        dTauXYdLam   = (-1.)*dSigdLam(4)
        dTauYZdLam   = (-1.)*dSigdLam(5)
        dTauZXdLam   = (-1.)*dSigdLam(6)

     dd =dfdsig(1)*dSigXdLam+dfdsig(2)*dSigYdLam+dfdsig(3)*dSigZdLam
     & +2.*dfdsig(4)*dTauXYdLam+2.*dfdsig(5)*dTauYZdLam
     & +2.*dfdsig(6)*dTauZXdLam
!-----------hh1   ------pmi--------------------------------------------------
      dGdp = dfdp
      tt = dfds(1)**2+dfds(2)**2+dfds(3)**2
     & +2.*(dfds(4)**2+dfds(5)**2+dfds(6)**2)
      dGdq = Sqrt(2./3.*tt) !auxilary scalar

      dfdpmi = -1.*(1.+xx)*(dM**2-alpha0**2)*(pp+pb)
      dpmidepsv=((1.+e0)*pmi)/(dlambda-dkappa)
      dhh1=dfdpmi*dpmidepsv*dGdp
!------------hh2  ------xx----------------------------------------------------
      dfdxx     = -(dM**2-alpha0**2)*(pp+pb)*pmi
      dxxdepsv = -1.*xx*dksi_v
      dxxdepsd = -1.*xx*dksi_v*dksi_d                 !-1.*xx*dksi*dksi_d
      dhh2=dfdxx*(dxxdepsv*abs(dGdp)+dxxdepsd*dGdq)
```

```
!-------------hh3  ------pb-------------------------------------------------
      dfdpb    = ds_alphalph+(dM**2-alpha0**2)*(pp-pmd)
      dpbdepsd = -1.*pb*dksi_s
      dhh3=dfdpb*dpbdepsd*dGdq
!-------------hh4  ------alpha----------------------------------------------
      dfdalphad=-3.*s_alph*(pp+pb)-3.*alpha_d*(pp+pb)*(pp-pmd)

      dalphaddepsv = dw*(aa*s0/(pp+pb)-alpha_d)
      dalphaddepsd = dw*dw_d*(bb*s0/(pp+pb)-alpha_d)
      dGdp_mac     = dGdp

      if (dGdp .lt. 0.) dGdp_mac = 0.
       tta = dalphaddepsv*dGdp_mac+dalphaddepsd*dGdq
       dhh4 = dfdalphad(1)*tta(1)+dfdalphad(2)*tta(2)+dfdalphad(3)*tta(3)
     & +2.*(dfdalphad(4)*tta(4)+dfdalphad(5)*tta(5)+dfdalphad(6)*tta(6))
!---------------------------------------------------------------------------tat
       dhh=dhh1+dhh2+dhh3+dhh4
       dfdlam=dd+dhh

!---------------------------------------------------------tatal lamda
      xlam=-yield/dfdlam
      if (xlam .lt. 0.)  xlam = 0.
!-----------------------------------------------calculate plastic strains
      depsp = xLam * dGdSig
      hsv(13:18)=hsv(13:18)+depsp
c-----------------------------------------------------------------------------------------
      Call strain_invariant(depsp, dEpsp_v, dEpsp_d)
      hsv(19) = hsv(19)+dEpsp_d
      dEpsDev1=(2.*depsp(1)-depsp(2)-depsp(3))/3.
      dEpsDev2=(2.*depsp(2)-depsp(1)-depsp(3))/3.
      dEpsDev3=(2.*depsp(3)-depsp(2)-depsp(1))/3.
      dEpsDev4=depsp(4)
      dEpsDev5=depsp(5)
      dEpsDev6=depsp(6)
      tt4=dEpsDev1*dEpsDev1+dEpsDev2*dEpsDev2+dEpsDev3*dEpsDev3+
     &   2.*(dEpsDev4*dEpsDev4+dEpsDev5*dEpsDev5+dEpsDev6*dEpsDev6)
        dEpsDevScalar = Sqrt(2./3.*tt4)
c
     if (dEpsVolMac .lt.0.) dEpsVolMac = 0.                !Macauley brackets
!--------------------------------------------- update hardening parameters and state variables
     dpmi     = ((1.+e0)*pmi*dEpsp_v)/(dlambda-dkappa)
     dxx      = -1.*xx*dksi_v*(dEpsVolMac+dksi_d*dEpsDevScalar)          !+dksi_d*dEpsp_d)
     dpb      = -1.*pb*dksi_s*dEpsDevScalar
     pmi      = pmi+dpmi
     xx       = xx+dxx
     pb       = pb+dpb
     pmd      = pmi*(1.+xx)
     dad      = dw*((aa*s0/(pp+pb)-alpha_d)*dEpsVolMac+
     &          dw_d*(bb*s0/(pp+pb)-alpha_d)*dEpsDevScalar)
c
     alpha_d = alpha_d+dad
!  get alpha explicitly
     alpha0   = sqrt(3./2.*(alpha_d(1)*alpha_d(1)
    &          +alpha_d(2)*alpha_d(2)+alpha_d(3)*alpha_d(3)
    &          +2.*alpha_d(4)*alpha_d(4)+2.*alpha_d(5)*alpha_d(5)
    &          +2.*alpha_d(6)*alpha_d(6)))
!----------------------------------update the state variables
     hsv(1)   = (alpha_d(1)) + 1.
     hsv(2)   = (alpha_d(2)) + 1.
     hsv(3)   = (alpha_d(3)) + 1.
     hsv(4)   = (alpha_d(4))
     hsv(5)   = (alpha_d(5))
```

```
      hsv(6)   = (alpha_d(6))
      hsv(7)   = alpha0
      hsv(8)   = pmi
      hsv(9)   = xx
      hsv(10)  = pmd
      hsv(12)  = pb
c    write(*,*) 'dpmi=',dpmi,'  pmi=',pmi
c    write(*,*) 'dxx=',dxx,'  xx=',xx
c    write(*,*) 'alpha0=',alpha0,'  pmd=',pmd
!-----------------------------------------------            set convergence criterion
      converged=.false.
c
        End If     !  if (yield.gt.1d-7) then
        end do
!-----------------------------------------------             to keep eps as input
101    Sig_new  = (-1.0)*Sig
      Return
      End Subroutine GCAM_model
!-------------------------------------------------------------------------------------------
! End the subroutine of GCM
!-------------------------------------------------------------------------------------------

!-------------------------------------------------------------------------------------------
       SUBROUTINE MATRIXDE_ANI(SIG,cm,hsv,CC)
!=================================================================================
!                 Calculation of Anisotropic Elastic Matrix D for SIG=D*EPS
!          Remark: E_ratio = Eh / Ev ( = 1 for isotropic elasticity)
!                 E_para can be dermined from kapa-line ( = 1 for isotropic elasticity)
!=================================================================================
      INCLUDE 'ABA_PARAM.INC'
      Character*80 cmname
      Double Precision SIG(6),cm(*),hsv(*),CC(6,6)
      ! get ny value
        pois     = cm(1)
        dkappa   = cm(2)
        ee       = hsv(11)  ! keep constant e as the
        E_ratio  = cm(17)   ! Anisotropy of elastic modulus
      ! determine shear modulus from Youngs' modulus
        pdash = (Sig(1)+ Sig(2)+ Sig(3))/3.                      ! global mean stress
      if ((pdash.gt.-5.).and.(pdash.le.0.)) pdash=-5.
      if ((pdash.lt.5.).and.(pdash.gt.0.))  pdash= 5.
      dKdash = abs((1.+ee)*pdash/dkappa)                          ! ...compression modulus
!          E      = 3.*(1.-2.*pois)*dKdash*E_para                 ! E (Young's) from kappa
      E = dKdash*(1.-4.*pois+2./E_ratio-2.*pois/sqrt(E_ratio))
       pois = pois*sqrt(E_ratio)
      ! compose liner elastic material stiffness matrix
      CC       = 0.                                               ! initial Matrix
      F1       = E/(1.+pois)/(1.-2.*pois)
      alph     = sqrt(E_ratio)
      CC(1,1)  = F1*(1.-pois)*alph**2.
      CC(2,2)  = F1*(1.-pois)
      CC(3,3)  = F1*(1.-pois)*alph**2.
      CC(4,4)  = F1*(1.-2.*pois)*alph  ! =2G,using the normal strain; tau_xy=G*gamma_xy=2G*epsilon_12
      CC(5,5)  = F1*(1.-2.*pois)*alph
      CC(6,6)  = F1*(1.-2.*pois)*alph**2.
      CC(1,2)  = F1*pois*alph
      CC(1,3)  = F1*pois*alph**2.
      CC(2,3)  = F1*pois*alph
      CC(2,1)  = CC(1,2)
      CC(3,1)  = CC(1,3)
      CC(3,2)  = CC(2,3)

      Return
      End SUBROUTINE MATRIXDE_ANI
!-------------------------------------------------------------------------------------------
```

```
!-----------------------------------------------------------------------------------------
       SUBROUTINE MATRIXDE_ANI_SS(SIG,cm,hsv,CC)
!=================================================================================
!
!                   Calculation of Anisotropic Elastic Matrix D for SIG=D*EPS
!
!        Remark: E_ratio = Eh / Ev ( = 1 for isotropic elasticity)
!                E_para can be dermined from kapa-line ( = 1 for isotropic elasticity)
!=================================================================================
      INCLUDE 'ABA_PARAM.INC'
      Character*80 cmname
      Double Precision SIG(6),cm(*),hsv(*),CC(6,6)

      ! get ny value
        pois      = cm(1)
        dkappa    = cm(2)
        ee        = hsv(11)              ! keep constant e as the definition of compression curve
        E_ratio   = cm(17)               ! Anisotropy of elastic modulus E_ratio=Eh/Ev, <=1
        eps70     = cm(20)               ! parameter of small strain stiffness
 !----------------------------------------------------------------------------------------
      ! determine shear modulus from Youngs' modulus
      pdash = ( Sig(1) + Sig(2) + Sig(3) ) /3.        ! global mean stress
      if ((pdash.gt.-1.).and.(pdash.le.0.)) pdash=-1.    ! only for Pa=========
      if ((pdash.lt.1.).and.(pdash.gt.0.))  pdash=1.
c-------For small strain stiffness (pref=101.325 kPa)
        D_reverse = hsv(39)
        eps_d_m   = hsv(40)


      if (eps_d_m .le. 1.d-3) then
          dLL = 1.
      if(D_reverse .gt. 0.) dLL = 2.

        dGref0 = (1.+ee)*3.*(1.-2.*pois)/2./(1.+pois)*(1.+3./7./dLL*
     &          abs(0.001/eps70))**2*101.325/dkappa
        dkappa_s = (1.+ee)*3.*(1.-2.*pois)/2./(1.+pois)*(1.+3./7./dLL*
     &          abs(eps_d_m/eps70))**2*101.325/dGref0
      if (dkappa_s.gt.dkappa) dkappa_s = dkappa
            dkappa = dkappa_s
      end if

      if (dkappa .lt. 0.1*cm(2)) dkappa = 0.1*cm(2)     ! a reasonable value
      if (eps70 .gt. 0.001) dkappa = cm(2)               ! the condition to reduce back to no sss
! we still need to introduce the evolution of G with loading angle cos(theta)
c==================================end if small strain stiffness
      dKdash= abs((1.+ee)*pdash/dkappa)                    ! ...compression modulus
      Ev = dKdash*(1.-4.*pois+2./E_ratio-2.*pois/sqrt(E_ratio))
!      Ev = 3.*(1.-2.*pois)*dKdash*E_para
      pois = pois*sqrt(E_ratio)                            ! to calculate nu_vh

      ! compose liner elastic material stiffness matrix
      CC = 0.                                             ! initial Matrix
      F1 = Ev/(1.+pois)/(1.-2.*pois)
      alph = sqrt(E_ratio)
      CC(1,1) = F1*(1.-pois)*alph**2.
      CC(2,2) = F1*(1.-pois)
      CC(3,3) = F1*(1.-pois)*alph**2.
      CC(4,4) = F1*(1.-2.*pois)*alph
      CC(5,5) = F1*(1.-2.*pois)*alph
      CC(6,6) = F1*(1.-2.*pois)*alph**2.
      CC(1,2) = F1*pois*alph
      CC(1,3) = F1*pois*alph**2.
      CC(2,3) = F1*pois*alph
      CC(2,1) = CC(1,2)
      CC(3,1) = CC(1,3)
      CC(3,2) = CC(2,3)

      Return
      End  SUBROUTINE MATRIXDE_ANI_SS
!-----------------------------------------------------------------------------------------
```

```
!-------------------------------------------------------------------------------------
        Subroutine MatVec(xMat,IM,Vec,N,VecR)
!=======================================================================
!
!     Calculate VecR = xMat*Vec
!
! I   xMat  : (Square) Matrix (IM,*)
! I   Vec   : Vector
! I   N     : Number of rows/colums
! O   VecR  : Resulting vector
!
!=======================================================================
      INCLUDE 'ABA_PARAM.INC'
      Character*80 cmname
      Double Precision xMat(IM,*),Vec(*),VecR(*)

      Do I=1,N
        X=0
        Do J=1,N
          X=X+xMat(I,J)*Vec(J)
        End Do
        VecR(I)=X
      End Do
      Return
      End Subroutine MatVec
!-------------------------------------------------------------------------------------


!-------------------------------------------------------------------------------------
      SUBROUTINE Lode_sheng(Sig,cm,dMratio)
!=======================================================================
!                    calculate the crrection of dM by Lode angle
!                   [Sheng D., Sloan S.W. and Yu H.S. 2000]
!     |- Aspects of finite element implementation of critical state models
!=======================================================================
      INCLUDE 'ABA_PARAM.INC'
      Character*80 cmname
      Double Precision SIG(6),cm(50)

      ! compression positive
       dMc      = cm(5)
       ttt      = 3./(3.+dMc)   ! based on MC or SMP
       dM_ratio = cm(19)     ! dM_ratio=Me/Mc only for the case of Sheng
       if (dM_ratio .ge. 1.)  dM_ratio = ttt
       if (dM_ratio .lt. ttt) dM_ratio = ttt
c------------------------------------------------------------------
       dJ2 = ((Sig(1)-Sig(2))**2+(Sig(1)-Sig(3))**2
     &        +(Sig(2)-Sig(3))**2)/6.+Sig(4)**2+Sig(5)**2+Sig(6)**2
        pdash = (Sig(1)+Sig(2)+Sig(3))/3.
        s1    = Sig(1) - pdash
        s2    = Sig(2) - pdash
        s3    = Sig(3) - pdash
        s4    = Sig(4)
        s5    = Sig(5)
        s6    = Sig(6)
        dJ3   = s1*s2*s3+2.*s4*s5*s6-s1*s6**2-s2*s5**2-s3*s4**2

       if (dJ2 .lt. 1.d-6) then     ! dJ2 .lt. 1.d-6
          g_theta = 1. !(2.*dM_ratio**4/(1.+dM_ratio**4))**0.25
      else
          sin3fi  = -3.*sqrt(3.)*dJ3/2./sqrt(abs(dJ2)**3)
      if (sin3fi .gt. 1.)  sin3fi = 1.
      if (sin3fi .lt. -1.) sin3fi = -1.
      g_theta=2.*dM_ratio**4/
     &        (1.+dM_ratio**4+(1.-dM_ratio**4)*sin3fi)
          g_theta= sqrt(g_theta)    ! for ^0.25
      end if
      dMratio = g_theta
c--------------------------------------------------------------
      if (dMratio .gt. 1.)  dMratio = 1.
      if (dMratio .lt. ttt)  dMratio = ttt
```

```
c------------------------------------------------------------
      Return
      end Subroutine Lode_sheng
!-----------------------------------------------------------------------------
!-----------------------------------------------------------------------------

      Subroutine Lode_TS (Sig,cm,dMratio)
!============================================================================
!                   calculate the crrection of dM by  TS of Yao
!                           [Yao, Hou & Zhou (2009)]
!                      - Geotechnique 59, No.5, 451-469
!============================================================================
      INCLUDE 'ABA_PARAM.INC'
      Character*80 cmname
      double precision Sig(6),cm(50)

      dMc = cm(5)
      ttt = 3./(3.+dMc)  ! based on MC or SMP
!
     dI1 = sig(1)+sig(2)+sig(3)
     dI2 = sig(1)*sig(2)+sig(2)*sig(3)+sig(1)*sig(3)
    &       -sig(4)**2-sig(5)**2-sig(6)**2

     dI3 = sig(1)*sig(2)*sig(3)+2.*sig(4)*sig(5)*sig(6)
    &        -sig(1)*sig(4)**2-sig(2)*sig(5)**2-sig(3)*sig(6)**2

     p  = dI1/3.
     s1 = Sig(1) - p
     s2 = Sig(2) - p
     s3 = Sig(3) - p
     s4 = Sig(4)
     s5 = Sig(5)
     s6 = Sig(6)
     dJ2= 0.5*(s1*s1+s2*s2+s3*s3+2.*s4*s4+2.*s5*s5+2.*s6*s6)
     q  = sqrt(3.*dJ2)

     term0 = abs(dI1*dI2-9.*dI3)
     if (term0 .lt. 1.) then
         term1 =0.
     else
        term1 = abs(dI1*dI2-dI3)/term0
     end if
     term2 = abs(3.*sqrt(term1)-1.)
     if (term2 .lt. 1.) then     ! revised on 19/07/2016
        yqc=0.
     else
        yqc=2.*abs(dI1)/term2                   ! SMP criteria
     end if
       If (yqc .lt. 1.)  then
            dMratio=1.
        else
             dMratio = q/yqc
        if (q .lt. 1.) dMratio=1. ! for numerical calculation
     end if
!-------------------modified the SMP criteria for modifying the lode angle---------------
     if (dMratio .ge. 1.)  dMratio = 1.
     if (dMratio .lt. ttt)  dMratio = ttt
!=-------------------------------------------------------------------------------------

!=-----------------------------------------------------------------------------------
     Return
     End Subroutine Lode_TS
!------------------------------------------------------------------------------------
!------------------------------------------------------------------------------------
     Subroutine strain_invariant(eps, eps_v, eps_d)  ! use normal strain, not engineering strain
     INCLUDE 'ABA_PARAM.INC'
     Character*80 cmname
     double precision eps(6)

     eps_v = eps(1) + eps(2) + eps(3)
     e1 = eps(1) - eps_v / 3.
     e2 = eps(2) - eps_v / 3.
     e3 = eps(3) - eps_v / 3.
     e4 = eps(4)
     e5 = eps(5)
     e6 = eps(6)
     eps_d =Sqrt(2./3.*(e1*e1+e2*e2+e3*e3+2.*e4*e4+2.*e5*e5+2.*e6*e6))

     Return
     End Subroutine strain_invariant
!------------------------------------------------------------------------------------
```

```
!-----------------------------------------------------------------------------------------
      Subroutine stress_invariant(Sig,dI1, dI2, dI3,p, q)
    ! calculate p and q from Sig
      INCLUDE 'ABA_PARAM.INC'
      Character*80 cmname
      double precision Sig(6)
c
      dI1 = sig(1)+sig(2)+sig(3)
      dI2 = sig(1)*sig(2)+sig(2)*sig(3)+
     &         sig(1)*sig(3)-sig(4)**2-sig(5)**2-sig(6)**2
        dI3 = sig(1)*sig(2)*sig(3)+2.*sig(4)*sig(5)*sig(6)
     &           -sig(1)*sig(4)**2.-sig(2)*sig(5)**2.-sig(3)*sig(6)**2

      p  = dI1/3.
      s1 = Sig(1) - p
      s2 = Sig(2) - p
      s3 = Sig(3) - p
      s4 = Sig(4)
      s5 = Sig(5)
      s6 = Sig(6)
      dJ2= 0.5*(s1*s1+s2*s2+s3*s3+2.*s4*s4+2.*s5*s5+2.*s6*s6)
      q  = sqrt(3.*dJ2)

      Return
      End Subroutine stress_invariant



          SUBROUTINE smallstrain (hsv,deps)
!c======================================================================
!c
!c    reverse loading judgement based on previous and current dstrain (deps0 and deps)
!c    current version can only be used for strain or displacement control
!c
!c======================================================================
        IMPLICIT DOUBLE PRECISION(A-H,O-Z)
        Double Precision deps(6),hsv(*)
        Double Precision deps0(6),eps0(6),epsR(6),eps(6),dnij0(6),dnij(6)
!c
      ! hsv(20+i) = 0.     ! initial dstrain deps0(i)
      ! hsv(26+i) = 0.     ! initial strain eps0(i)
      ! hsv(32+i) = 0.     ! initial reverse strain point epsR(i)
        ! hsv(39)= 0.      ! ID_reverse = 0
        ! hsv(40)= 0.       ! eps_d_m=0.
      ! hsv(41)  =  costheta

        do 11 i = 1, 6
           deps0(i)=hsv(20+i) ! previous dstrain
           eps0(i) =hsv(26+i) ! previous strain
           epsR(i) =hsv(32+i) ! previous stored reverse strain point
           eps(i)=eps0(i)+deps(i)
11      continue      ! Next i

       deps0_scalar = deps0(1)*deps0(1)+deps0(2)*deps0(2)
     &               +deps0(3)*deps0(3)+deps0(4)*deps0(4)
     &               +deps0(5)*deps0(5)+deps0(6)*deps0(6)
       deps0_scalar = sqrt(deps0_scalar)
      if (deps0_scalar .gt. 1.d-7) then
           dnij0 = deps0/deps0_scalar
        else
           dnij0 = 0.
      end if

       deps_scalar = deps(1)*deps(1)+deps(2)*deps(2)
     &            +deps(3)*deps(3)+deps(4)*deps(4)
     &            +deps(5)*deps(5)+deps(6)*deps(6)
       deps_scalar = sqrt(deps_scalar)
      if (deps_scalar .gt. 1.d-7) then
           dnij = deps/deps_scalar
        else
```

```
          else
             dnij = 0.
         end if

         summm = dnij0(1)*dnij(1)+dnij0(2)*dnij(2)+dnij0(3)*dnij(3)
          &  +dnij0(4)*dnij(4)+dnij0(5)*dnij(5)+dnij0(6)*dnij(6)

         if ((deps0_scalar .gt. 1.d-7) .and. (deps_scalar .gt. 1.d-7)) then
            costheta = summm
          else
            costheta = 1.    ! no reverse
         end if

c        if (abs(summm - 1.) .gt. 1.d-6) hsv(41) = costheta

c        if (abs(costheta-1.) .gt. 1.d-7) then  ! with reverse
          if (costheta .lt. 0.) then  ! with reverse
             hsv(39) = hsv(39) + 1.
             epsR = eps0
          end if

! to calculate shear strain
          eps_m1 = eps(1)-epsR(1)
          eps_m2 = eps(2)-epsR(2)
          eps_m3 = eps(3)-epsR(3)
          eps_m4 = eps(4)-epsR(4)
          eps_m5 = eps(5)-epsR(5)
          eps_m6 = eps(6)-epsR(6)
          eps_m_v = eps_m1 + eps_m2 + eps_m3
          e1 = eps_m1 - eps_m_v / 3.
          e2 = eps_m2 - eps_m_v / 3.
          e3 = eps_m3 - eps_m_v / 3.
          e4 = eps_m4
          e5 = eps_m5
          e6 = eps_m6
          eps_d_m=Sqrt(2./3.*(e1*e1+e2*e2+e3*e3+2.*e4*e4+2.*e5*e5+2.*e6*e6)
         &           +eps_m_v*eps_m_v)  ! total strain, modified by yin 4 oct 2013
          hsv(40) = eps_d_m


      do 12 i = 1, 6
      hsv(20+i)=deps(i)      ! update dstrain for next loading step
      hsv(26+i)=eps(i)       ! update strain for next loading step
      hsv(32+i)=epsR(i) ! update reverse strain point
12    continue     ! Next i
!      test=1.
      END
!-----------------------------------------------------------------------------------------
```

Appendix C: SIMSAND

(Copyright (c) 2017, geoinvention www.geoinvention.com)

```
! ---------------------------------------------------------------
c***********************************************************
      SUBROUTINE UMAT(STRESS,STATEV,DDSDDE,SSE,SPD,SCD,
     1 RPL,DDSDDT,DRPLDE,DRPLDT,
     2 STRAN,DSTRAN,TIME,DTIME,TEMP,DTEMP,PREDEF,DPRED,CMNAME,
     3 NDI,NSHR,NTENS,NSTATV,PROPS,NPROPS,COORDS,DROT,PNEWDT,
     4 CELENT,DFGRD0,DFGRD1,NOEL,NPT,LAYER,KSPT,KSTEP,KINC)
c
      INCLUDE 'ABA_PARAM.INC'
c

      CHARACTER*80 CMNAME
      DIMENSION STRESS(NTENS),STATEV(NSTATV),
     1 DDSDDE(NTENS,NTENS),
     2 DDSDDT(NTENS),DRPLDE(NTENS),
     3 STRAN(NTENS),DSTRAN(NTENS),TIME(2),PREDEF(1),DPRED(1),
     4 PROPS(NPROPS),COORDS(3),DROT(3,3),DFGRD0(3,3),DFGRD1(3,3)
c
      Double Precision   cm(NPROPS),CC(6,6),depsp(6)
      Double Precision   Sig(6),eps(6),dSig(6),dEps(6),Epsp(6)
      Double Precision   hsv(NSTATV)
      Double Precision   StVar(NSTATV),StVar_DDS(NSTATV)
      Double Precision   Sig0(6),Sig_DDS(6),Sig_new(6),Sig_temp(6)
      Double Precision   Stress_star(6),sig_star(6),stress_old(6)
      Double Precision   Eps0(6),dEps_dds(6),deps_temp(6)
      Double Precision   dstran_star_dt(6),deps_ini(6)
      Double Precision   D(6,6),DDS(6,6),DD(6,6),DDE(6,6)
      Double Precision   DDSDD(NTENS,NTENS)
      Double Precision   theta
      integer            ntens,ii,jj,kk,i,j
      integer            CSL,psi_selection,sub_nstep
      Parameter( zero = 0.D0, one = 1.D0, two = 2.D0, three = 3.D0,
     1  third = one/three, half = 0.5D0, twoThirds = two/three,
     2  threeHalfs = 1.5D0 )
c
c***************************************************************************c
c       parameters definition
c***************************************************************************c
       cm     = props
c------- Relating to Elasticity
      e0      = cm(1)    ! Initial void ratio
      dK0     = cm(2)    ! dK=dK0*pat*(2.97-ee)**2/(1.+ee)*((p_module+bond)/pat)**dnn
      dG0     = cm(3)    ! dG=bG0*pat*(2.97-ee)**2/(1.+ee)*((p_module+bond)/pat)**dnn
```


Z. Yin and Y. Jin, *Practice of Optimisation Theory in Geotechnical Engineering*,
https://doi.org/10.1007/978-981-13-3408-5

```
      dnn    = cm(4)    ! defaut value: 0.6667

c------- Relating to Critical State Lines: eec=ecu+(ec0-ecu)*dexp(-1.*(pp0/pat/dlamda)**dksi)
      ec0    = cm(5)    ! Initial critical void ratio for initial grading when p=0
      ecu0   = cm(6)    ! =emin, Initial critical void ratio for initial grading when p=inifinity
      dlamda = cm(7)
      dksi   = cm(8)
      fi     = cm(9)    ! critical friction angle to computer M
      fi     = fi*3.14159265358979323 9/180.  ! change the unity
      pat    = 101.325

c------- Relating to shear mechanism f1=q/(p+bond)-Mp*epspd/(akp+epspd), d=ad*(Mpt-q/(p+bond))
      akp    = cm(10)   ! f=q/(p+bond)-Mp*epspd/(akp+epspd)
      ad     = cm(11)   ! d = ad*(Mpt-q/(p+bond))
      dnp    = cm(12)   ! fi_b=atan(tan(fi)*(ec/e)**dnp) defaut value=1
      dnd    = cm(13)   ! fi_d=atan(tan(fi)*(e/ec)**dnd) defaut value=2


c------- Relating to normal mechanism f2= 1/dn*(eta/dM)^(dn+1)*p+p-pm, associate flow rule
      dn     = cm(14)   ! canbe calculated by K0=1-sin(fi)
      pm0    = cm(15)   !
      dr0    = cm(16)   ! compression coefficient in loge-logp
c------- Relating to grain breakage
      dBr    = cm(17)   ! initial grading index =0.
      dParBr = cm(18)   ! Br=wp/(dParBr+wp)
      ecuf   = cm(19)   ! eminf, fractal critical void ratio for fractal grading when p=inifinity
      Parecu = cm(20)   ! obtained from emin-Br, ecu = ecuf+(ecu0-ecuf)*dexp(-1.*Parecu*dBr)

c------- Relating to bondong and debonding of cemented sand
      bond0  = cm(21)   ! for cemented sand
      if (bond0 .lt. 0.01) bond0=0.01
      dw     = cm(22)   ! degradation: dbond=-dw*bond*dEpsDP
c---------------------------------------------------------------------------------------
      psi_selection=cm(23)   !psi_selection=1, e/ec; psi_selection=2, e-ec
      eps70     = cm(24)
c---------------------------------------------------------------------------------------
c ---------------- select the critical state line ---------------------------------------
      if (dlamda .lt. 1. .AND. dksi .eq. 1) then
          CSL=1       ! eec=ec0-dlamda*(pp0/pat)**dksi
      else if (dlamda .lt. 1. .AND. dksi .lt. 1) then
          CSL=2       ! eec=ec0-dlamda*(pp0/pat)**dksi
      else
          CSL=3       ! eec=ecu0+(ec0-ecu0)*exp(-1.*(pp0/pat/dlamda)**dksi)
      endif
c      write(*,*) 'CSL=',CSL
c*********************************************************************************c
```

```
21    format(6(1x,1p,e12.5))
c          stop
! ----------------------------------------------------------------------------------------
! Read STRESS STRAIN
! ----------------------------------------------------------------------------------------
      Sig0(1:4)    = STRESS(1:4)
      Eps0(1:3)    = STRAN(1:3)
      Eps0(4)      = STRAN(4)*0.5  ! change engineering strain to normal strain
      dEps(1:3)    = DSTRAN(1:3)
      dEps(4)      = DSTRAN(4)*0.5  ! change engineering strain to normal strain
      Sig0(5:6)    = zero
      Eps0(5:6)    = zero
      dEps(5:6)    = zero
      if( NSHR .GT. 1 ) then
          Sig0(5:6)   = STRESS(5:6)
          Eps0(5:6)   = STRAN(5:6)*0.5  ! change engineering strain to normal strain
          dEps(5:6)   = DSTRAN(5:6)*0.5  ! change engineering strain to normal strain
      end if
c change sign due to EVP-MCC formulation (compression positive)
c      write(*,*)  NTENS
c      write(*,21) dEps
! ----------------------------------------------------------------------------------------
! Read state variables
! ----------------------------------------------------------------------------------------
      hsv=STATEV
!     write(*,21) STATEV
! ----------------------------------------------------------------------------------------
! Initialize state variables
! ----------------------------------------------------------------------------------------
      IF (KSTEP == 1 .AND. TIME(2) ==zero) THEN
! ----------------------------------------------------------------------------------------
! Initialize state variables
! ----------------------------------------------------------------------------------------
      ! just for print for triaxial condition
       Sig_temp=( -1.)*Sig0
       call stress_invariant(Sig_temp, dI1, dI2, dI3, pp0, qq0)
c       write(*,*) pp0,qq0
      ! initialization
       evol  = 0.
       Eps   = 0.
       hsv   = 0.
```

```
c***************************************************************************
          select case(CSL)

          case(1)
           if (pp0 .le. 1.d-2) then
              eec= ec0-dlamda*log(0.01/pat)
           else
              eec= ec0-dlamda*log(pp0/pat)
           end if

          case(2)

           if (pp0 .le. 1.d-2) then
              eec=ec0-dlamda*(0.01/pat)**dksi
           else
              eec=ec0-dlamda*(pp0/pat)**dksi
           end if

          case(3)
           if (pp0 .le. 1.d-2) then
              eec=ecu0+(ec0-ecu0)*exp(-1.*(0.01/pat/dlamda)**dksi)
           else
              eec=ecu0+(ec0-ecu0)*exp(-1.*(pp0/pat/dlamda)**dksi)
           end if

         end select
c***************************************************************************
      ! calculate stress invariants and lode angle effect
c        Call Lode_TS (Sig_temp,cm,dMratio)

         dMratio=1.0
        select case(psi_selection)

          case(1)

          psi=eec/e0
          if (psi .lt. 1.d-7) psi=1.d-7
          fi_p=atan(tan(fi)*psi**dnp)
          fi_d=atan(tan(fi)*(1./psi)**dnd)
c
          dMp    = 6.*sin(fi_p)/(3.-sin(fi_p))*dMratio
          dMpt   = 6.*sin(fi_d)/(3.-sin(fi_d))*dMratio


     case (2)

     psi=e0-eec
     dM = 6.*sin(fi)/(3.-sin(fi))*dMratio
     dMp = dM*exp(-dnp*psi)
     dMpt = dM*exp(dnd*psi)

     case default

     psi=eec/e0
     if (psi .lt. 1.d-7) psi=1.d-7
     fi_p=atan(tan(fi)*psi**dnp)
     fi_d=atan(tan(fi)*(1./psi)**dnd)
c
     dMp    = 6.*sin(fi_p)/(3.-sin(fi_p))*dMratio
     dMpt   = 6.*sin(fi_d)/(3.-sin(fi_d))*dMratio

   end select
c***************************************************************************
      epsd_p = qq0*akp/(-qq0+dMp*pp0+dMp*bond0)+0.0001
      evol_p = ad*(dMpt-qq0/(pp0+bond0))*epsd_p
      epsp1 = epsd_p+(1./3.)*evol_p
      epsp3 = (1./3.)*evol_p-(1./2.)*epsd_p
      hsv(2) = epsp3*1.1
      hsv(3) = epsp1*1.1 ! to make sure the first step is elastic, "1" is vertical
      hsv(4) = epsp3*1.1
```

```
      hsv(1) = e0                    ! void ratio
      hsv(8) = pm0
      hsv(9) = bond0                 ! bonding parameters
      hsv(18)= 0.                    ! equivalent plastic strain
      hsv(10)= 0.                    ! plastic work for breakage
      hsv(19)= eec                   ! void ratio in critical state line eec
      hsv(20)= psi                   ! the state parameter
      hsv(21)= pp0                   ! effective mean stress
      END IF
! ------------------------------------------------------------------------------------
! Strain Load
! ------------------------------------------------------------------------------------
      deps_temp  = deps*(-1.)
      Eps0       = -Eps0

     Call CSsand (cm,deps,sig0,hsv,sig)
! ------------------------------------------------------------------------------------
! UPDATE STRESS STRAIN AND STATE VARIABLES
! ------------------------------------------------------------------------------------
      STRESS(1:4) = (-1.0)*SIG(1:4)
      if( NSHR .EQ. 3 ) then
          STRESS(5:6) = (-1.0)*SIG(5:6)
      end if
      STATEV = hsv
c     write(*,21) hsv(1)
! ------------------------------------------------------------------------------------
! Give the DDSDDE BY NUMERICAL DISTURBED METHOD
! ------------------------------------------------------------------------------------
      Call MATRIXDE_SS(sig,cm,hsv,D)
      Do i=1,NTENS
         Do j=1,NTENS
            DDSDDE(i,j)=D(i,j)
         End Do
      End Do
      DDSDDE(4,4)=D(4,4)*0.5     ! change to engineering strain for abaqus-umat
      if( NSHR .GT. 1 ) then
          DDSDDE(5,5)=D(5,5)*0.5
          DDSDDE(6,6)=D(6,6)*0.5
      end if
c-----------------------------------------------------------------------------------
      Return
      END SUBROUTINE UMAT

!-----------------------------------------------------------------------------------
      Subroutine CSsand (cm,deps,sig,hsv,sig_new)
      INCLUDE 'ABA_PARAM.INC'
c     Implicit none
      Character*80 cmname
      Double Precision cm(*),hsv(*),deps(6),deps0(6),epsp(6),depsp(6)
      Double Precision sig0(6),sig(6),CC(6,6),dsig(6),dpdsig(6)
      Double Precision sig_new(6),dqdsig(6),dum(6),Eps0(6)
      Double Precision dF1dsig(6),dG1dsig(6),dF2dsig(6),dG2dsig(6)
      Double Precision dumG1(6),dumG2(6)
      INTEGER   CSL, psi_selection
      Logical   converged            ! convergence for iteration

      Parameter( zero = 0.D0, one = 1.D0, two = 2.D0, three = 3.D0,
     1  third = one/three, half = 0.5D0, twoThirds = two/three,
     2  threeHalfs = 1.5D0 )

21    format(6(1x,1p,e12.5))
      !parameters
c------- Relating to Elasticity
      e0     = cm(1)   ! Initial void ratio
      dK0    = cm(2)   ! dK=dK0*pat*(2.97-ee)**2/(1.+ee)*((p_module+bond)/pat)**dnn
      dG0    = cm(3)   ! dG=bG0*pat*(2.97-ee)**2/(1.+ee)*((p_module+bond)/pat)**dnn
      dnn    = cm(4)   ! defaut value: 0.6667
```

```
c------- Relating to Critical State Lines: eec=ecu+(ec0-ecu)*dexp(-1.*(pp0/pat/dlamda)**dksi)
      ec0    = cm(5)   ! Initial critical void ratio for initial grading when p=0
      ecu0   = cm(6)   ! =emin, Initial critical void ratio for initial grading when p=inifinity
      dlamda = cm(7)
      dksi   = cm(8)
      fi     = cm(9)  ! critical friction angle to computer M
      fi     = fi*3.14159265358979323 9/180.  ! change the unity
      pat    = 101.325
c------- Relating to shear mechanism f1=q/(p+bond)-Mp*epspd/(akp+epspd), d=ad*(Mpt-q/(p+bond))
      akp    = cm(10)  ! f=q/(p+bond)-Mp*epspd/(akp+epspd)
      ad     = cm(11)  ! d = ad*(Mpt-q/(p+bond))
      dnp    = cm(12)  ! fi_b=atan(tan(fi)*(ec/e)**dnp) defaut value=1
      dnd    = cm(13)  ! fi_d=atan(tan(fi)*(e/ec)**dnd) defaut value=2
c------- Relating to normal mechanism f2= 1/dn*(eta/dM)^(dn+1)*p+p-pm, associate flow rule
      dn     = cm(14)  ! can be calculated by K0=1-sin(fi)
      pm0  = cm(15)  !
      dr0    = cm(16)  ! compression coefficient in loge-logp
c------- Relating to grain breakage
     dBr     = cm(17)  ! initial grading index =0.
     dParBr  = cm(18)  ! Br=wp/(dParBr+wp)
      ecuf   = cm(19)  ! eminf, fractal critical void ratio for fractal grading when p=inifinity
     Parecu  = cm(20)  ! obtained from emin-Br, ecu = ecuf+(ecu0-ecuf)*dexp(-1.*Parecu*dBr)
c------- Relating to bondong and debonding of cemented sand
      bond0  = cm(21)  ! for cemented sand
      dw     = cm(22)  ! degradation: dbond=-dw*bond*dEpsDP
c-----------------------------------------------------------------------------------------
      psi_selection=cm(23)  !psi_selection=1, e/ec; psi_selection=2, e-ec
      eps70     = cm(24)
c-----------------------------------------------------------------------------------------
c        WRITE(*,21) cm(22)
c --------------- select the critical state line ---------------------------------------
      if (dlamda .lt. 1. .AND. dksi .eq. 1) then
          CSL=1      ! eec=ec0-dlamda*(pp0/pat)**dksi
      else if (dlamda .lt. 1. .AND. dksi .lt. 1) then
          CSL=2      ! eec=ec0-dlamda*(pp0/pat)**dksi
      else
          CSL=3      ! eec=ecu0+(ec0-ecu0)*exp(-1.*(pp0/pat/dlamda)**dksi)
      end if
! ---------------------------------------------------------------------------------------
! ---------------------------------------------------------------------------------------
! Change sign due to SCLAY formulation (compression positive)
! ---------------------------------------------------------------------------------------
      Sig    = -1.*Sig
```

```
      dEps   = -1.*dEps

c          dMratio= 1.0
c          WRITE(*,21) Sig
c************************small strain ****************************************************
        Call MATRIXDE_SS(Sig,cm,hsv,CC)
c*******************************************************************************************
c          stop
c--------initialization-----------------
      depspv = 0.
      depspd = 0.
      depsp  = 0.
      deps0  = deps ! to regive the eps back in the end
      sig0   = sig ! to regive the eps back in the end for the case p<0.1
c
      ee          = hsv(1)
      epsp(1:6) = hsv(2:7)
      pm          = hsv(8)
      bond        = hsv(9)
      workp       = hsv(10)
      dBr         = hsv(11)
c
      call stress_invariant(Sig, dI1, dI2, dI3, p, q)
      Call strain_invariant(deps,depsv,depsd)
      Call strain_invariant(epsp,epspv,epspd)
c
      ee=ee-depsv*(1.+e0)
      if (ee .lt. 1.d-7) ee=1.d-7
          hsv(1) = ee
      if (dParBr .lt. 1.d-7) then
         dBr =0.                      ! not use breakage
      else
         dBr=workp/(dParBr+workp) ! Br=wp/(par+wp)
      end if
      hsv(11) = dBr
c
c        calculate critical void ratio due to breakage
      ecu  = ecuf+(ec0-ecuf)*exp(-1.*Parecu*dBr)
c
      select case(CSL)

      case(1)

          if (p .le. 1.d-2) then
             ecrit= ecu-dlamda*log(0.01/pat)
          else
             ecrit= ecu-dlamda*log(p/pat)
          end if

          case(2)

          if (p .le. 1.d-2) then
             ecrit=ecu-dlamda*(0.01/pat)**dksi
          else
             ecrit=ecu-dlamda*(p/pat)**dksi
          end if

          case(3)

          ecu  = ecuf+(ecu0-ecuf)*exp(-1.*Parecu*dBr)
          if (p .le. 1.d-2) then
             ecrit=ecu+(ec0-ecu0)*exp(-1.*(0.01/pat/dlamda)**dksi)
          else
             ecrit=ecu+(ec0-ecu0)*exp(-1.*(p/pat/dlamda)**dksi)
          end if

          end select
    c*********************************************************************************
```

```
    ! calculate stress invariants and lode angle effect
      Call Lode_TS (Sig,cm,dMratio)
c*******************************************************************************
c     save the critical void ration into state varibales Hsv
      hsv(19) = ecrit
      if (ecrit .lt. 1.d-7) ecrit=1.d-7

      select case(psi_selection)

      case(1)

      psi=ecrit/ee
      if (psi .lt. 1.d-7) psi=1.d-7
      fi_p = atan(tan(fi)*psi**dnp)
      fi_d = atan(tan(fi)*(1./psi)**dnd)
      dM   = 6.*sin(fi)/(3.-sin(fi))*dMratio
      dMp  = 6.*sin(fi_p)/(3.-sin(fi_p))*dMratio
      dMpt = 6.*sin(fi_d)/(3.-sin(fi_d))*dMratio

      case(2)

      psi=ee-ecrit
      dM = 6.*sin(fi)/(3.-sin(fi))*dMratio
      dMp = dM*exp(-dnp*psi)
      dMpt = dM*exp(dnd*psi)

      case default

      psi=ecrit/ee
      if (psi .lt. 1.d-7) psi=1.d-7
      fi_p = atan(tan(fi)*psi**dnp)
      fi_d = atan(tan(fi)*(1./psi)**dnd)
      dM   = 6.*sin(fi)/(3.-sin(fi))*dMratio
      dMp  = 6.*sin(fi_p)/(3.-sin(fi_p))*dMratio
      dMpt = 6.*sin(fi_d)/(3.-sin(fi_d))*dMratio

      end select
c
      hsv(20)=psi

      converged =.false.        ! do at least one iteration for each increment including
      inewt     = 0

  !-----------------------------------------------------------------------------------
  !     New ton-Raphson iteration for plasticity start
        ! loop until convergence is reached
        Do While (.not.converged)
        inewt = inewt + 1
        converged=.true. ! gets false again during iteration when plasticity occurs

        ! solve plasticity
        epsp(1:6) = hsv(2:7)
        pm        = hsv(8)
        bond      = hsv(9)
        Call strain_invariant(epsp,epspv,epspd)

        ! set stress strain increment
         deps = deps - depsp
c        write(*,21) depsp
c        write(*,21) deps
         Call MatVec( CC, 6, deps, 6, dsig)
              sig = sig0+dsig
         call stress_invariant(Sig, dI1, dI2, dI3, p, q)
c       if (abs(p+bond) .le. 1.d-7) bond=bond+0.001
        if ((p+bond) .le. 1.d-7) bond=bond+0.001
        f1= q/(p+bond)-dMp*epspd/(akp+epspd)
        f2= 1./dn*(q/(p+bond)/dM)**(dn+1.)*(p+bond)+p-pm
c       write(*,*) f1
c       write(*,*) f2
  !-----------------------------------------------------------------------------------
```

```
!---------------------------------------------------------------------------------
      If (f1 .gt. 1.d-7) Then   !======== for the case of shear, compression is included

      df1dp      = -1.*q/(p+bond)**2
      df1dq      = 1./(p+bond)
      df1dbond   = -1.*q/(p+bond)**2
      df1depspd  = -1.*dMp/(akp+epspd)+dMp*epspd/(akp+epspd)**2
      dbonddepspd=-dw*bond
      depspdddepspd=1.

      dG1dp = ad*(dMpt-q/(p+bond))
      dG1dq = 1.

      call dpqdsig(Sig, dpdsig, dqdsig)
      dF1dsig = df1dp*dpdsig + df1dq*dqdsig
      dG1dsig = dG1dp*dpdsig + dG1dq*dqdsig

      Call MatVec( CC, 6, dG1dsig, 6, dum)
      dd = -dF1dsig(1)*dum(1)-dF1dsig(2)*dum(2)-dF1dsig(3)*dum(3)
     &  -2.*(dF1dsig(4)*dum(4)+dF1dsig(5)*dum(5)+dF1dsig(6)*dum(6))
      hh = df1dbond*dbonddepspd*dG1dq + df1depspd*depspdddepspd*dG1dq

      dF1dLam = dd + hh
      XLam1   = -f1 / dF1dLam
c*****************************************************************************
       if(isnan(XLam1))then
         write(*,*) 'XLam1',XLam1
         XLam1=0.
       end if
c*****************************************************************************c
      depsp   = XLam1 * dG1dsig
c
       if(isnan(depsp(1)))then
         depsp=0.
       end if
c
      epsp    = epsp + depsp
      Call strain_invariant(depsp,depspv,depspd)

      hsv(2:7) = epsp(1:6)
      dkappa=(1.+e0)*(p+bond)/(dK0*pat*(2.97-ee)**2/(1.+ee)
     &       *(abs(p+bond)/pat)**dnn)
c*****************************************************************************

      if (depspv .lt. 0.) depspv=0.
         dr1=dr0*ee
      if (dr1 .le. dkappa) dr1 = 1.01*dkappa
      hsv(8)    = pm + (1.+e0)*pm*depspv/(dr1-dkappa)
      hsv(9)    = bond - dw*bond*depspd

        !set convergence criterion
        converged = .False.

      End If                             !======== end for the case of shear

!---------------------------------------------------------------------------------
      If ((f1 .lt. 1.d-7) .and. (f2 .gt. 1.d-7)) Then  !======== for the case of normal compression
!   f2= 1./dn*(q/(p+bond)/dM)**(dn+1.)*(p+bond)+p-pm
c
       df2dp = 1.-(q/((p+bond)*dM))**(dn+1.)
       df2dq =  (q/((p+bond)*dM))**dn*(dn+1.)/dn/dM
       df2dbond = df2dp-1.
       df2dpm = -1.
       dbonddepspd=-dw*bond
       dkappa=(1.+e0)*(p+bond)/(dK0*pat*(2.97-ee)**2/(1.+ee)
     &       *(abs(p+bond)/pat)**dnn)
       dr1=dr0*ee
      if (dr1 .le. dkappa) dr1 = 1.01*dkappa
      dpmdepspv = (1.+e0)*pm/(dr1-dkappa)
```

```
      ! associated flow rule
      Call dpqdsig(Sig, dpdsig, dqdsig)
      dF2dsig = df2dp*dpdsig + df2dq*dqdsig
      dG2dsig = dF2dsig

      Call MatVec( CC, 6, dG2dsig, 6, dum)
       dd = -dF2dsig(1)*dum(1) - dF2dsig(2)*dum(2) - dF2dsig(3)*dum(3)
     &  -2.*(dF2dsig(4)*dum(4) + dF2dsig(5)*dum(5) + dF2dsig(6)*dum(6))
       hh = df2dbond*dbonddepspd*df2dq + df2dpm*dpmdepspv*df2dp
       dF2dLam = dd + hh
       XLam2 = -f2 / dF2dLam
       depsp = XLam2 * dG2dsig
c
       if(isnan(depsp(1)))then
         depsp=0.
       end if
c
      epsp  = epsp + depsp
      Call strain_invariant(depsp,depspv,depspd)

      hsv(2:7) = epsp(1:6)
      hsv(8)    = pm + (1.+e0)*pm*depspv/(dr1-dkappa)
      hsv(9) = bond-dw*bond*depspd

       !set convergence criterion
       converged = .False.
      End If                                  !======== end for the case of normal compression
!------------------------------------------------------------------------------------

      End Do ! While not converged end of plasticity

!================================================================== end of plastic iteration

101   depsp = deps0-deps
         if (inewt .gt. 1000) then
            write(*,*) 'goto101'
            write(*,21) depsp
         end if

      Call strain_invariant(depsp,depspv,depspd)
      dpeeq = depspd
      hsv(18)= hsv(18)+dpeeq

           hsv(12:17) = depsp(1:6)
           dworkp=sig(1)*depsp(1)+sig(2)*depsp(2)+sig(3)*depsp(3)
          &    +2.*(sig(4)*depsp(4)+sig(5)*depsp(5)+sig(6)*depsp(6))
           hsv(10) = workp + dworkp
           p=(sig(1)+sig(2)+sig(3))/3.0
      c*************************************************************************
            if (p .lt. -1.*bond) then
                sij11=sig(1)-p
                sij22=sig(2)-p
                sij33=sig(3)-p
                p = -1.*bond+0.001
                sig(1)=sij11+p
                sig(2)=sij22+p
                sig(3)=sij33+p
           end if
      c
           hsv(21)=p
      c-------------------------------------------------------------------------
           sig_new=sig

           Return
           End
```

```
!========================================================================
      Subroutine stress_invariant(Sig, dI1, dI2, dI3, p, q)
      INCLUDE 'ABA_PARAM.INC'
c     Implicit none
      character*80 cmname
      Double Precision Sig(6)

      dI1 = sig(1)+sig(2)+sig(3)
      dI2 = sig(1)*sig(2)+sig(2)*sig(3)+sig(1)*sig(3)
     &    -sig(4)**2-sig(5)**2-sig(6)**2
      dI3 = sig(1)*sig(2)*sig(3)+2.*sig(4)*sig(5)*sig(6)
     &      -sig(1)*sig(4)**2-sig(2)*sig(5)**2-sig(3)*sig(6)**2

      p  = dI1/3.
      s1 = Sig(1) - p
      s2 = Sig(2) - p
      s3 = Sig(3) - p
      s4 = Sig(4)
      s5 = Sig(5)
      s6 = Sig(6)
      dJ2= 0.5*(s1*s1+s2*s2+s3*s3+2.*s4*s4+2.*s5*s5+2.*s6*s6)
      q  = sqrt(3.*dJ2)

      Return
      End
!-------------------------------------------------------------------------------------
      Subroutine strain_invariant(eps, eps_v, eps_d)
      INCLUDE 'ABA_PARAM.INC'
c     Implicit none
      character*80 cmname
      Double Precision eps(6)

      eps_v = eps(1) + eps(2) + eps(3)
      e1 = eps(1) - eps_v / 3.
      e2 = eps(2) - eps_v / 3.
      e3 = eps(3) - eps_v / 3.
      e4 = eps(4)
      e5 = eps(5)
      e6 = eps(6)
      eps_d =Sqrt(2./3.*(e1*e1+e2*e2+e3*e3+2.*e4*e4+2.*e5*e5+2.*e6*e6))

      Return
      End

!-------------------------------------------------------------------------
      Subroutine multip_vec(C, m, n, A, B)
      INCLUDE 'ABA_PARAM.INC'
      Character*80 cmname
      Double precision A(m,n),B(n),C(m)
c
      do 10 i = 1, m
         C(i) = 0.
         do 20 j = 1, n
          C(i) = C(i) + A(i, j) * B(j)
20       continue    !  Next j
10     continue     !  Next i

       Return
      End
!-------------------------------------------------------------------------------------

C***********************************************************************
      Subroutine MatVec(xMat,IM,Vec,N,VecR)
C***********************************************************************
C
C     Calculate VecR = xMat*Vec
C
C I   xMat  : (Square) Matrix (IM,*)
C I   Vec   : Vector
C I   N     : Number of rows/colums
C O   VecR  : Resulting vector
C
C***********************************************************************
```

```
C**********************************************************************
      INCLUDE 'ABA_PARAM.INC'
c     Implicit none
      Character*80 cmname
      Double Precision xMat(IM,*),Vec(*),VecR(*)
C**********************************************************************
      Do I=1,N
        X=0
        Do J=1,N
          X=X+xMat(I,J)*Vec(J)
        End Do
        VecR(I)=X
      End Do
      Return
      End
!-----------------------------------------------------------------------
      Subroutine ZeroM(m, n, A)
      ! make a matrix zero
       INCLUDE 'ABA_PARAM.INC'
       Character*80 cmname
       Double Precision A(m,n)

      do 10 i = 1, m
      do 20 j = 1, n
        A(i, j) = 0.
20     continue    !  Next j
10    continue     !  Next i

      Return
      End
!-----------------------------------------------------------------------
      Subroutine dpqdsig(Sig, dpdsig, dqdsig)
      INCLUDE 'ABA_PARAM.INC'
c     Implicit none
      Character*80 cmname
      Double Precision Sig(6), dpdsig(6), dqdsig(6)

      dpdsig(1) = 1. / 3.
      dpdsig(2) = 1. / 3.
      dpdsig(3) = 1. / 3.
      dpdsig(4) = 0.
      dpdsig(5) = 0.
      dpdsig(6) = 0.


     p = (Sig(1) + Sig(2) + Sig(3)) / 3.
     s1 = Sig(1) - p
     s2 = Sig(2) - p
     s3 = Sig(3) - p
     s4 = Sig(4)
     s5 = Sig(5)
     s6 = Sig(6)
     dJ2 = 0.5*(s1*s1+s2*s2+s3*s3+2.*s4*s4+2.*s5*s5+2.*s6*s6)
     q = Sqrt(3. * dJ2)

     If (q .gt. 1.d-7) Then
       dqdsig(1) = 3. * s1 / (2. * q)
       dqdsig(2) = 3. * s2 / (2. * q)
       dqdsig(3) = 3. * s3 / (2. * q)
       dqdsig(4) = 3. * s4 / (2. * q)
       dqdsig(5) = 3. * s5 / (2. * q)
       dqdsig(6) = 3. * s6 / (2. * q)
      Else
       dqdsig = 1.5
     End If

     Return
     End
```

```
c-----------------------------------------------------------------------------
|       SUBROUTINE MATRIXDE(Sig,cm,hsv,CC)
C=======================================================================
c                    Calculation Elastic Matrix D for SIG=D*EPS
C=======================================================================
      IMPLICIT DOUBLE PRECISION(A-H,O-Z)
      DIMENSION SIG(6),cm(*),Hsv(*),CC(6,6)

        dK0    = cm(2)    ! dK=dK0*pat*(2.97-ee)**2/(1.+ee)*((p_module+bond)/pat)**dnn
        dG0    = cm(3)    ! dG=bG0*pat*(2.97-ee)**2/(1.+ee)*((p_module+bond)/pat)**dnn
        dnn    = cm(4)    ! defaut value: 0.6667
        p_module = (Sig(1) + Sig(2) + Sig(3)) / 3.
        If (p_module .lt. 0.1)  p_module = 0.1
        pat = 101.325
        ee  = hsv(1)
        bond= hsv(9)
!         calculate bulk and shear modulus
        if (ee .lt. 1.d-7) ee=1.d-7
        gg=dG0*pat*(2.97-ee)**2/(1.+ee)*(abs(p_module+bond)/pat)**dnn
        bkk=dK0*pat*(2.97-ee)**2/(1.+ee)*(abs(p_module+bond)/pat)**dnn
        emodule = 9.*bkk*gg/(3.*bkk+gg)    ! need to be modified for anisotropic elasticity
        pois = (3.*bkk-2.*gg)/(6.*bkk+2.*gg)
        if (pois .gt. 0.49) pois = 0.49
        Call ZeroM(6, 6, CC)          ! initial Matrix

        F1 = emodule/(1.+pois)/(1.-2.*pois)
        CC(1,1) = F1*(1.-pois)
        CC(2,2) = F1*(1.-pois)
        CC(3,3) = F1*(1.-pois)
        CC(4,4) = F1*(1.-2.*pois)
        CC(5,5) = F1*(1.-2.*pois)
        CC(6,6) = F1*(1.-2.*pois)
        CC(1,2) = F1*pois   ! Engineering shear strain is used, if use normal shear strain keep constant
        CC(1,3) = F1*pois
        CC(2,3) = F1*pois
        CC(2,1) = CC(1,2)
        CC(3,1) = CC(1,3)
        CC(3,2) = CC(2,3)

         Return
         end
!----------------------------------------------------------------------------------------------


c---------------------------------------------------------------------------------
        SUBROUTINE MATRIXDE_SS(Sig,cm,hsv,CC)
C=======================================================================
c
c                    Calculation Elastic Matrix D for SIG=D*EPS
c
C=======================================================================
      IMPLICIT DOUBLE PRECISION(A-H,O-Z)
      DIMENSION SIG(6),cm(*),Hsv(*),CC(6,6)

        dK0    = cm(2)    ! dK=dK0*pat*(2.97-ee)**2/(1.+ee)*((p_module+bond)/pat)**dnn
        dG0    = cm(3)    ! dG=bG0*pat*(2.97-ee)**2/(1.+ee)*((p_module+bond)/pat)**dnn
        dnn    = cm(4)    ! defaut value: 0.6667
        eps70  = cm(24)   ! parameter of small strain stiffness
        if(eps70 .eq. 0.) eps70=0.1

        p_module = (Sig(1) + Sig(2) + Sig(3)) / 3.
        If (p_module .lt. 0.1)  p_module = 0.1
        pat = 101.325
        ee  = hsv(1)
        bond= hsv(9)

c-------For small strain stiffness
         D_reverse = hsv(49)
         eps_d_m   = hsv(22)
      if (eps_d_m .le. 1.d-3) then
          dLL = 1.
```

```
      if(D_reverse .gt. 0.) dLL = 2.
         dRatio_SS = dLL*((1.+0.003/eps70/7.)/(1.+eps_d_m*3./eps70/7.))**2
      else
         dRatio_SS = 1.
      end if
      dG0_SS = dRatio_SS*dG0
      dK0_SS = dRatio_SS*dK0
c===================================end if small strain stiffness
c        calculate bulk and shear modulus
        if (ee .lt. 1.d-7) ee=1.d-7
       gg=dG0_SS*pat*(2.97-ee)**2/(1.+ee)*(abs(p_module+bond)/pat)**dnn
       bkk=dK0_SS*pat*(2.97-ee)**2/(1.+ee)*(abs(p_module+bond)/pat)**dnn
c
        emodule = 9.*bkk*gg/(3.*bkk+gg)   ! need to be modified for anisotropic elasticity
        pois = (3.*bkk-2.*gg)/(6.*bkk+2.*gg)
        if (pois .gt. 0.49) pois = 0.49
        Call ZeroM(6, 6, CC)         ! initial Matrix

        F1 = emodule/(1.+pois)/(1.-2.*pois)
        CC(1,1) = F1*(1.-pois)
        CC(2,2) = F1*(1.-pois)
        CC(3,3) = F1*(1.-pois)
        CC(4,4) = F1*(1.-2.*pois)
        CC(5,5) = F1*(1.-2.*pois)
        CC(6,6) = F1*(1.-2.*pois)
        CC(1,2) = F1*pois   ! Engineering shear strain is used, if use normal shear strain keep constant
        CC(1,3) = F1*pois
        CC(2,3) = F1*pois
        CC(2,1) = CC(1,2)
        CC(3,1) = CC(1,3)
        CC(3,2) = CC(2,3)

         Return
         end
!-------------------------------------------------------------------------------------------

!-------------------------------------------------------------------------------------------
      SUBROUTINE smallstrain_sim (hsv,Eps0)
!c=========================================================================
!c
!c    reverse loading judgement based on previous and current dstrain (deps0 and deps)
!c    current version can only be used for strain or displacement control
!c
!c=========================================================================
      INCLUDE 'ABA_PARAM.INC'
      Character*80 cmname
      Double Precision hsv(*)
      Double Precision Eps0(6),Eps_temp(6)

      Eps_temp=-1.*Eps0
      call strain_invariant(Eps_temp, eps_v, eps_d)
      eps_d_m = sqrt(eps_d*eps_d+eps_v*eps_v)
      hsv(22) = eps_d_m
      hsv(49) = 0.

      Return
      END  SUBROUTINE smallstrain_sim
!-------------------------------------------------------------------------------------------


         Subroutine Lode_TS (Sig,cm,dMratio)
!=========================================================================
!                  calculate the crrection of dM by  TS of Yao
!                        [Yao, Hou & Zhou (2009)]
!                  - Geotechnique 59, No.5, 451-469
!=========================================================================
         INCLUDE 'ABA_PARAM.INC'
         Character*80 cmname
         double precision Sig(6),cm(50)

          fi0  = cm(9)
          fi   = fi0*3.1415926535897932 39/180.  ! change the unity
          dMc = 6.*sin(fi)/(3.-sin(fi))
          ttt = 3./(3.+dMc)   ! based on MC or SMP
         dI1 = sig(1)+sig(2)+sig(3)

         dI2 = sig(1)*sig(2)+sig(2)*sig(3)+sig(1)*sig(3)
        &       -sig(4)**2-sig(5)**2-sig(6)**2
```

```
      dI2 = sig(1)*sig(2)+sig(2)*sig(3)+sig(1)*sig(3)
     &       -sig(4)**2-sig(5)**2-sig(6)**2

      dI3 = sig(1)*sig(2)*sig(3)+2.*sig(4)*sig(5)*sig(6)
     &        -sig(1)*sig(4)**2-sig(2)*sig(5)**2-sig(3)*sig(6)**2

      p  = dI1/3.
      s1 = Sig(1) - p
      s2 = Sig(2) - p
      s3 = Sig(3) - p
      s4 = Sig(4)
      s5 = Sig(5)
      s6 = Sig(6)
      dJ2= 0.5*(s1*s1+s2*s2+s3*s3+2.*s4*s4+2.*s5*s5+2.*s6*s6)
      q  = sqrt(3.*dJ2)

      term0 = abs(dI1*dI2-9.*dI3)
      if (term0 .lt. 1.) then
          term1 =0.
      else
         term1 = abs(dI1*dI2-dI3)/term0
      end if
      term2 = abs(3.*sqrt(term1)-1.)
      if (term2 .lt. 1.) then     ! revised on 19/07/2016
         yqc=0.
      else
         yqc=2.*abs(dI1)/term2                 ! SMP criteria
      end if
        If (yqc .lt. 1.)  then
              dMratio=1.
         else
               dMratio = q/yqc
         if (q .lt. 1.) dMratio=1. ! for numerical calculation
      end if
!--------------------modified the SMP criteria for modifying the lode angle---------------------------
      if (dMratio .ge. 1.)  dMratio = 1.
      if (dMratio .lt. ttt)  dMratio = ttt
!=--------------------------------------------------------------------------------------------
      Return
      End Subroutine Lode_TS
!====================== end of subroutines for the soil model Iso sand ===============
```

Appendix D: Some Selected Optimization Algorithms in This Book

1. **Genetic algorithm (GA) (Copyright (c) 2015, Yarpiz** www.yarpiz.com)

```
clc;
clear;
close all;

%% Problem Definition

CostFunction=@(x) Sphere(x);      % Cost Function

nVar=10;                % Number of Decision Variables

VarSize=[1 nVar];   % Decision Variables Matrix Size

VarMin=-10;         % Lower Bound of Variables
VarMax= 10;         % Upper Bound of Variables


%% GA Parameters

MaxIt=1000;      % Maximum Number of Iterations

nPop=100;        % Population Size

pc=0.7;                 % Crossover Percentage
nc=2*round(pc*nPop/2);  % Number of Offsprings (also Parnets)
gamma=0.4;              % Extra Range Factor for Crossover

pm=0.3;                 % Mutation Percentage
nm=round(pm*nPop);      % Number of Mutants
mu=0.1;         % Mutation Rate

ANSWER=questdlg('Choose selection method:','Genetic Algorith',...
    'Roulette Wheel','Tournament','Random','Roulette Wheel');

UseRouletteWheelSelection=strcmp(ANSWER,'Roulette Wheel');
UseTournamentSelection=strcmp(ANSWER,'Tournament');
UseRandomSelection=strcmp(ANSWER,'Random');

if UseRouletteWheelSelection
    beta=8; % Selection Pressure
end

if UseTournamentSelection
    TournamentSize=3;   % Tournamnet Size
end

pause(0.01); % Due to a bug in older versions of MATLAB
```


Z. Yin and Y. Jin, *Practice of Optimisation Theory in Geotechnical Engineering*,
https://doi.org/10.1007/978-981-13-3408-5

```
%% Initialization

empty_individual.Position=[];
empty_individual.Cost=[];

pop=repmat(empty_individual,nPop,1);

for i=1:nPop

    % Initialize Position
    pop(i).Position=unifrnd(VarMin,VarMax,VarSize);

    % Evaluation
    pop(i).Cost=CostFunction(pop(i).Position);

end

% Sort Population
Costs=[pop.Cost];
[Costs, SortOrder]=sort(Costs);
pop=pop(SortOrder);
 BestSol=pop(1);

 % Array to Hold Best Cost Values
 BestCost=zeros(MaxIt,1);

 % Store Cost
 WorstCost=pop(end).Cost;

 %% Main Loop

 for it=1:MaxIt

     % Calculate Selection Probabilities
     if UseRouletteWheelSelection
         P=exp(-beta*Costs/WorstCost);
         P=P/sum(P);
     end

     % Crossover
     popc=repmat(empty_individual,nc/2,2);
     for k=1:nc/2

         % Select Parents Indices
         if UseRouletteWheelSelection
             i1=RouletteWheelSelection(P);
             i2=RouletteWheelSelection(P);
         end
         if UseTournamentSelection
             i1=TournamentSelection(pop,TournamentSize);
             i2=TournamentSelection(pop,TournamentSize);
         end
         if UseRandomSelection
             i1=randi([1 nPop]);
             i2=randi([1 nPop]);
         end
```

```
        % Select Parents
        p1=pop(i1);
        p2=pop(i2);

        % Apply Crossover
        [popc(k,1).Position,
        popc(k,2).Position]=Crossover(p1.Position,p2.Position,gamma,VarMin,VarMax);

        % Evaluate Offsprings
        popc(k,1).Cost=CostFunction(popc(k,1).Position);
        popc(k,2).Cost=CostFunction(popc(k,2).Position);

    end
    popc=popc(:);


    % Mutation
    popm=repmat(empty_individual,nm,1);
    for k=1:nm

        % Select Parent
        i=randi([1 nPop]);
        p=pop(i);

        % Apply Mutation
        popm(k).Position=Mutate(p.Position,mu,VarMin,VarMax);

        % Evaluate Mutant
        popm(k).Cost=CostFunction(popm(k).Position);

    end

    % Create Merged Population
    pop=[pop
         popc

                  popm]; %#ok

            % Sort Population
            Costs=[pop.Cost];
            [Costs, SortOrder]=sort(Costs);
            pop=pop(SortOrder);

            % Update Worst Cost
            WorstCost=max(WorstCost,pop(end).Cost);

            % Truncation
            pop=pop(1:nPop);
            Costs=Costs(1:nPop);

            % Store Best Solution Ever Found
            BestSol=pop(1);

            % Store Best Cost Ever Found
            BestCost(it)=BestSol.Cost;

            % Show Iteration Information
            disp(['Iteration ' num2str(it) ': Best Cost = ' num2str(BestCost(it))]);

        end

        %% Results

        figure;
        semilogy(BestCost,'LineWidth',2);
        % plot(BestCost,'LineWidth',2);
        xlabel('Iteration');
        ylabel('Cost');
        grid on;
```

```
function [y1, y2]=Crossover(x1,x2,gamma,VarMin,VarMax)

    alpha=unifrnd(-gamma,1+gamma,size(x1));

    y1=alpha.*x1+(1-alpha).*x2;
    y2=alpha.*x2+(1-alpha).*x1;

    y1=max(y1,VarMin);
    y1=min(y1,VarMax);

    y2=max(y2,VarMin);
    y2=min(y2,VarMax);

end

function y=Mutate(x,mu,VarMin,VarMax)

    nVar=numel(x);

    nmu=ceil(mu*nVar);

    j=randsample(nVar,nmu);

    sigma=0.1*(VarMax-VarMin);

    y=x;
    y(j)=x(j)+sigma*randn(size(j));

    y=max(y,VarMin);
    y=min(y,VarMax);

end

function i=RouletteWheelSelection(P)

    r=rand;

    c=cumsum(P);

    i=find(r<=c,1,'first');

end

function i=TournamentSelection(pop,m)

    nPop=numel(pop);

    S=randsample(nPop,m);

    spop=pop(S);

    scosts=[spop.Cost];

    [~, j]=min(scosts);

    i=S(j);

end

function z=Sphere(x)

    z=sum(x.^2);

end
```

Note that the presented cost function “z = Sphere(x)” is a mathematical benchmark test for showing how to use the optimization algorithms, such as GA, DE, PSO, and SA.

For identifying the parameters of MC based on geotechnical testing, a new cost function should be defined using the error function presented in previous chapters, as shown:

```
function error=Cost_function(x)
% x is the parameters of MC
%% experimental data
file='Artifical_data.xlsx';      % the datafile containing the experimental data
test_results=xlsread(file,'PMT1');
eps_exp=test_results(:,1);
pre_exp=test_results(:,2);
%% change the parameters which in the abaqus input file
E=x(1);
nu=0.33;
phi=x(3);
psi=x(4);
c=x(2);
temp=0.0;
%%
% a=1 means the first pressuremeter test;
%%%%%%%%%%%%%%%%%
name_o='PMT1_origin.inp';
fid_o = fopen(name_o);
tline = fgets(fid_o);        % Read the first line and continue in the while loops
%%%%%%%%%%%%%%%%%
%  file_to_open='SH_ALE_new.inp';
file_to_open=['PMT','.inp'];
fid_temp= fopen(file_to_open,'w+');
%%%%%%%%%%%%%%%%%
m=1;
while ischar(tline)
    %%%%%%%%%%%%
    %disp(tline)
    %%%%%%%%%%%%
    % first layer
    if m==2185
        tline=[num2str(E),',  ',num2str(nu),'\n'];
    end
    %
    if m==2187
        tline=[num2str(phi),',  ',num2str(psi),'\n'];
    end


    if m==2189
        tline=[num2str(c),',  ',num2str(temp),'\n'];
    end
    fprintf(fid_temp,tline,'\n\r');
    %%%%%%%%%%%%%%%%%%%%%
    tline = fgets(fid_o);           % read the following 1 line
    m=m+1;
end
fclose(fid_temp);
fclose(fid_o);
%%
%% call the abaqus to calculate
dos('abq6111 job=PMT cpus=1 int');   % PMT is the name of abaqus input file
%% post
dos('abq6111 cae noGUI=Post.py');    % Post.py is the postdata file
%%
load U1.dat;
load 'S11.dat';
%
eps_num=U1;
sig_num=S11;
%% process the data
eps_exp=eps_exp;
eps_temp=eps_exp;
%  interpolation
P_num=csaps(eps_num,sig_num,1,eps_temp);
P_exp=csaps(eps_exp,pre_exp,1,eps_temp);
%% error
error=sqrt((mean(((P_num-P_exp)/max(P_exp)).^2)))*100
```

For other optimization, operators used in the enhanced RCGA are shown as follows:

```
function [child1,child2] = Crossover_SBX(parent1,parent2,Num_v,upbound,lowbound,~,etaC)
% Function performs crossover of two individuals-SBX crossover
EPS = 1e-14;
% Now do if loop
for qq = 1:Num_v,
    if (rand <= 0.5),
        if abs(parent1(qq)-parent2(qq)) > EPS,
            if parent1(qq) < parent2(qq),
                y1 = parent1(qq); y2 = parent2(qq);
            else
                y1 = parent2(qq); y2 = parent1(qq);
            end
            yl = lowbound(qq); yu = upbound(qq);
            rnd = rand;
            beta = 1.0 + (2.0*(y1-yl)/(y2-y1));
            alpha = 2.0 - beta^(-(etaC+1.0));
            %
            if (rnd <= (1.0/alpha)),
                betaq = (rnd*alpha)^(1.0/(etaC+1.0));
            else
                betaq = (1.0/(2.0 - rnd*alpha))^(1.0/(etaC+1.0));
            end;
            %
            c1 = 0.5*((y1+y2)-betaq*(y2-y1));
            beta = 1.0 + (2.0*(yu-y2)/(y2-y1));
            alpha = 2.0 - beta^(-(etaC+1.0));
```

```
            if (rnd <= (1.0/alpha)),
                betaq = (rnd*alpha)^(1.0/(etaC+1.0));
            else
                betaq = (1.0/(2.0 - rnd*alpha))^(1.0/(etaC+1.0));
            end;
            %
            c2 = 0.5*((y1+y2)+betaq*(y2-y1));
            if (c1<yl), c1=yl; end;
            if (c2<yl), c2=yl; end;
            if (c1>yu), c1=yu; end;
            if (c2>yu), c2=yu; end;
            %
            if (rand<=0.5),
                child1(qq) = c2; child2(qq) = c1;
            else
                child1(qq) = c1; child2(qq) = c2;
            end;
        else
            child1(qq) = parent1(qq); child2(qq) = parent2(qq);
        end;
    else
        child1(qq) = parent1(qq); child2(qq) = parent2(qq);
    end;
end;
```

```
function [child1,child2] = Crossover_LX(parent1,parent2,Num_v,upbound,lowbound,a,b,~)
% Function performs crossover of two individuals,
% Using the Laplace crossover from Deep et al. 2007
%%
EPS = 1e-14;
% Now do if loop
for qq = 1:Num_v
    if abs(parent1(qq)-parent2(qq)) > EPS,
        if parent1(qq) < parent2(qq),
            y1 = parent1(qq); y2 = parent2(qq);
        else
            y1 = parent2(qq); y2 = parent1(qq);
        end
        yu = upbound(qq); yl = lowbound(qq);
        %
        u=rand;
        if (u<=0.5)
            beta = a-b*log(u);
        else
            beta = a+b*log(u);
        end
        %
        c1 = y1+beta*abs(y1-y2);
        c2 = y2+beta*abs(y1-y2);
        %
        if (c1<yl), c1=yl; end;
        if (c2<yl), c2=yl; end;
        if (c1>yu), c1=yu; end;
        if (c2>yu), c2=yu; end;
        %
        child1(qq) = c1; child2(qq) = c2;
    else
        child1(qq) = parent1(qq); child2(qq) = parent2(qq);
    end
end
```

```
function [child1,child2] = Crossover_BEX(parent1,parent2,Num_v,upbound,lowbound,~)
% Function performs crossover of two individuals,
for qq = 1:Num_v
    if abs(parent1(qq)-parent2(qq)) > EPS,
        if parent1(qq) < parent2(qq),
            y1 = parent1(qq); y2 = parent2(qq);
        else
            y1 = parent2(qq); y2 = parent1(qq);
        end
        yu = upbound(qq); yl = lowbound(qq);
        %
        ui=rand();
        lambda=0.25;
        aa1=exp((yl-y1)/(lambda*(y2-y1)));
        aa2=exp((yl-y2)/(lambda*(y2-y1)));
        bb1=exp((y1-yu)/(lambda*(y2-y1)));
        bb2=exp((y2-yu)/(lambda*(y2-y1)));
        if (rand<=0.5)
            beta1=lambda*log(aa1+ui*(1-aa1));
            beta2=lambda*log(aa2+ui*(1-aa2));
        else
            beta1=-lambda*(1-ui*(1-bb1));
            beta2=-lambda*(1-ui*(1-bb2));
        end
        c1=y1+beta1*(y2-y1);
        c2=y2+beta2*(y2-y1);
        %
        if (c1<yl), c1=yl; end;
        if (c2<yl), c2=yl; end;
        if (c1>yu), c1=yu; end;
        if (c2>yu), c2=yu; end;
        %
        child1(qq) = c1; child2(qq) = c2;
    else
        child1(qq) = parent1(qq); child2(qq) = parent2(qq);
    end
end
```

```
function [ChildGen] = Mutate_DRM(ChildGen,upbound,lowbound,pM,gen,Num_gen)
% This function performs Mutation of Children
% Dynamic random mutation proposed by Yao-Chen Chuang et al. 2015
% First determine the dimension of the Children
[nmbOfIndivs,Num_v] = size(ChildGen);
% Now loop over individuals
for qq = 1:nmbOfIndivs
    % Loop over individual elements of each individual
    for jj = 1:Num_v
        % Check whether this number if larger than mutation factpr
        y = ChildGen(qq,jj);
        if (rand <= pM)
            % Define boundaries
            yu = upbound(jj); yl = lowbound(jj);
            %
            pertur_factor=0.5;
            m=4;
            %                          b=20;
            %                          a=0.5;
            sm=(1-gen/Num_gen)^m;
            %                          sm=1/3*(atan(b*(1-2*(gen/Num_gen).^a)))+0.5;
            rad=(rand()*2-1)*pertur_factor;
            y=y+sm*rad*(yu-yl);
            % Check found bounds
            if (y<yl), y = yl; end;
            if (y>yu), y = yu; end;
        end
        % Now update ChildGen
        ChildGen(qq,jj) = y;
    end
end
```

```
function [ChildGen] = Mutate_NSGA(ChildGen,upbound,lowbound,pM,etaM)
% This function performs Mutation of Children
% First determine the dimension of the Children
[nmbOfIndivs,Num_v] = size(ChildGen);
% Now loop over individuals
for qq = 1:nmbOfIndivs
    % Loop over individual elements of each individual
    for jj = 1:Num_v
        % Check whether this number if larger than mutation factor
        if (rand <= pM)
            y = ChildGen(qq,jj);
            % Define boundaries
            yl = lowbound(jj); yu = upbound(jj);
            % Normalize range
            delta1 = (y-yl)/(yu-yl); delta2 = (yu-y)/(yu-yl);
            % Again draw random number
            rnd = rand;
            % Compute mutation power
            mut_pow = 1/(etaM+1);
            if rnd <= 0.5
                xy = 1-delta1;
                val = 2.0*rnd+(1.0-2.0*rnd)*(xy^(etaM+1.0));
                deltaq = val^mut_pow - 1;
            else
                xy = 1.0-delta2;
                val = 2.0*(1.0-rnd)+2.0*(rnd-0.5)*(xy^(etaM+1.0));
                deltaq = 1.0 - val^mut_pow;
            end;
            % Check found bounds
            y = y + deltaq*(yu-yl);
            if (y<yl), y = yl; end;
            if (y>yu), y = yu; end;
            % Now update ChildGen
            ChildGen(qq,jj) = y;
        end;
    end;
end;
```

2. **Particle swarm optimization (PSO) (Copyright (c) 2015, Yarpiz** www.yarpiz.com)

```
clc;
clear;
close all;

%% Problem Definition

CostFunction=@(x) Sphere(x);        % Cost Function

nVar=10;             % Number of Decision Variables

VarSize=[1 nVar];    % Size of Decision Variables Matrix

VarMin=-10;          % Lower Bound of Variables
VarMax= 10;          % Upper Bound of Variables


%% PSO Parameters

MaxIt=1000;      % Maximum Number of Iterations

nPop=100;        % Population Size (Swarm Size)

% PSO Parameters
w=1;            % Inertia Weight
wdamp=0.99;     % Inertia Weight Damping Ratio
c1=1.5;         % Personal Learning Coefficient
c2=2.0;         % Global Learning Coefficient

% If you would like to use Constriction Coefficients for PSO,
% uncomment the following block and comment the above set of parameters.

% % Constriction Coefficients
% phi1=2.05;
% phi2=2.05;
% phi=phi1+phi2;
% chi=2/(phi-2+sqrt(phi^2-4*phi));
% w=chi;          % Inertia Weight
% wdamp=1;        % Inertia Weight Damping Ratio
% c1=chi*phi1;    % Personal Learning Coefficient
% c2=chi*phi2;    % Global Learning Coefficient

% Velocity Limits
VelMax=0.1*(VarMax-VarMin);
VelMin=-VelMax;

%% Initialization

empty_particle.Position=[];
empty_particle.Cost=[];
empty_particle.Velocity=[];
empty_particle.Best.Position=[];
empty_particle.Best.Cost=[];

particle=repmat(empty_particle,nPop,1);

GlobalBest.Cost=inf;

for i=1:nPop

    % Initialize Position
    particle(i).Position=unifrnd(VarMin,VarMax,VarSize);

    % Initialize Velocity
    particle(i).Velocity=zeros(VarSize);

    % Evaluation
    particle(i).Cost=CostFunction(particle(i).Position);

    % Update Personal Best
    particle(i).Best.Position=particle(i).Position;
```

```
        particle(i).Best.Cost=particle(i).Cost;

        % Update Global Best
        if particle(i).Best.Cost<GlobalBest.Cost

            GlobalBest=particle(i).Best;

        end

    end

    BestCost=zeros(MaxIt,1);

    %% PSO Main Loop

    for it=1:MaxIt

        for i=1:nPop

            % Update Velocity
            particle(i).Velocity = w*particle(i).Velocity ...
                +c1*rand(VarSize).*(particle(i).Best.Position-particle(i).Position) ...
                +c2*rand(VarSize).*(GlobalBest.Position-particle(i).Position);

            % Apply Velocity Limits
            particle(i).Velocity = max(particle(i).Velocity,VelMin);
            particle(i).Velocity = min(particle(i).Velocity,VelMax);

            % Update Position
            particle(i).Position = particle(i).Position + particle(i).Velocity;

            % Velocity Mirror Effect
            IsOutside=(particle(i).Position<VarMin | particle(i).Position>VarMax);
            particle(i).Velocity(IsOutside)=-particle(i).Velocity(IsOutside);

            % Apply Position Limits
            particle(i).Position = max(particle(i).Position,VarMin);
            particle(i).Position = min(particle(i).Position,VarMax);

            % Evaluation
            particle(i).Cost = CostFunction(particle(i).Position);

            % Update Personal Best
            if particle(i).Cost<particle(i).Best.Cost

                particle(i).Best.Position=particle(i).Position;
                particle(i).Best.Cost=particle(i).Cost;

                % Update Global Best
                if particle(i).Best.Cost<GlobalBest.Cost

                    GlobalBest=particle(i).Best;

                end

            end

        end

        BestCost(it)=GlobalBest.Cost;

        disp(['Iteration ' num2str(it) ': Best Cost = ' num2str(BestCost(it))]);

        w=w*wdamp;

    end

    BestSol = GlobalBest;

    %% Results
```

3. **Differential evolution (DE) (Copyright (c) 2015, Yarpiz** www.yarpiz.com)

```
clc;
clear;
close all;

%% Problem Definition

CostFunction=@(x) Sphere(x);     % Cost Function

nVar=10;              % Number of Decision Variables

VarSize=[1 nVar];   % Decision Variables Matrix Size

VarMin=-10;          % Lower Bound of Variables
VarMax= 10;          % Upper Bound of Variables


%% GA Parameters

MaxIt=1000;      % Maximum Number of Iterations

nPop=100;        % Population Size

pc=0.7;                  % Crossover Percentage
nc=2*round(pc*nPop/2);   % Number of Offsprings (also Parnets)
gamma=0.4;               % Extra Range Factor for Crossover

pm=0.3;                  % Mutation Percentage
nm=round(pm*nPop);       % Number of Mutants
mu=0.1;          % Mutation Rate

ANSWER=questdlg('Choose selection method:','Genetic Algorith',...
    'Roulette Wheel','Tournament','Random','Roulette Wheel');

UseRouletteWheelSelection=strcmp(ANSWER,'Roulette Wheel');
UseTournamentSelection=strcmp(ANSWER,'Tournament');
UseRandomSelection=strcmp(ANSWER,'Random');

if UseRouletteWheelSelection
    beta=8; % Selection Pressure
end

if UseTournamentSelection
    TournamentSize=3;   % Tournamnet Size
end

pause(0.01); % Due to a bug in older versions of MATLAB

%% Initialization

empty_individual.Position=[];
empty_individual.Cost=[];

pop=repmat(empty_individual,nPop,1);

for i=1:nPop

    % Initialize Position
    pop(i).Position=unifrnd(VarMin,VarMax,VarSize);

    % Evaluation
    pop(i).Cost=CostFunction(pop(i).Position);

end

% Sort Population
Costs=[pop.Cost];
[Costs, SortOrder]=sort(Costs);
pop=pop(SortOrder);

% Store Best Solution
```

```
BestSol=pop(1);

% Array to Hold Best Cost Values
BestCost=zeros(MaxIt,1);

% Store Cost
WorstCost=pop(end).Cost;

%% Main Loop

for it=1:MaxIt

    % Calculate Selection Probabilities
    if UseRouletteWheelSelection
        P=exp(-beta*Costs/WorstCost);
        P=P/sum(P);
    end

    % Crossover
    popc=repmat(empty_individual,nc/2,2);
    for k=1:nc/2

        % Select Parents Indices
        if UseRouletteWheelSelection
            i1=RouletteWheelSelection(P);
            i2=RouletteWheelSelection(P);
        end
        if UseTournamentSelection
            i1=TournamentSelection(pop,TournamentSize);
            i2=TournamentSelection(pop,TournamentSize);
        end
        if UseRandomSelection
            i1=randi([1 nPop]);
            i2=randi([1 nPop]);
        end

        % Select Parents
        p1=pop(i1);
        p2=pop(i2);

        % Apply Crossover
        [popc(k,1).Position,
        popc(k,2).Position]=Crossover(p1.Position,p2.Position,gamma,VarMin,VarMax);

        % Evaluate Offsprings
        popc(k,1).Cost=CostFunction(popc(k,1).Position);
        popc(k,2).Cost=CostFunction(popc(k,2).Position);

    end
    popc=popc(:);


    % Mutation
    popm=repmat(empty_individual,nm,1);
    for k=1:nm

        % Select Parent
        i=randi([1 nPop]);
        p=pop(i);

        % Apply Mutation
        popm(k).Position=Mutate(p.Position,mu,VarMin,VarMax);

        % Evaluate Mutant
        popm(k).Cost=CostFunction(popm(k).Position);

    end

    % Create Merged Population
    pop=[pop
         popc
```

```
                popm]; %#ok

            % Sort Population
            Costs=[pop.Cost];
            [Costs, SortOrder]=sort(Costs);
            pop=pop(SortOrder);

            % Update Worst Cost
            WorstCost=max(WorstCost,pop(end).Cost);

            % Truncation
            pop=pop(1:nPop);
            Costs=Costs(1:nPop);

            % Store Best Solution Ever Found
            BestSol=pop(1);

            % Store Best Cost Ever Found
            BestCost(it)=BestSol.Cost;

            % Show Iteration Information
            disp(['Iteration ' num2str(it) ': Best Cost = ' num2str(BestCost(it))]);

        end

        %% Results

        figure;
        semilogy(BestCost,'LineWidth',2);
        % plot(BestCost,'LineWidth',2);
        xlabel('Iteration');
        ylabel('Cost');
        grid on;

function i=RouletteWheelSelection(P)

    r=rand;

    c=cumsum(P);

    i=find(r<=c,1,'first');

end
```

Some operators for enhancing the original DE are shown as follows:

```
function [ParGen,ObjVals]=ANMS(myfunction,ParGen,ObjVals,tol,max_feval,upbound,lowbound,step_size)
%
% Adaptive Nelder-Mead Simplex Algorithm for
% solving the unconstrained optimization problem:   min f(x).
%
% Inputs:
%  xinit--initial guess
%  tol--tolerance for termination (Recommended value: 10^-4)
%  max_feval--maximum number of function evaluations
%  myfunction--objective function. (In myfunction(x), x is a ROW vector)
%
% Outputs:
%  xmin--approximate optimal solution at termination. It is a row vector.
%  fmin--minimum function value at termination
%  ct--number of function evaluations at termination
%
%%  for save results
x_total=[];
obj_total=[];
count=0;
%% Saves the elite best positions
Num_p=length(ParGen(:,1));
elite_ind=length(ParGen(1,:))+1;
[~,posi] = sort(ObjVals,'ascend');  % sort(a,'descend')  9 8 7
elites = posi(1:elite_ind);
xinit=ParGen(elites,:);
FX=ObjVals(elites,:);
%%  initialization
x0=xinit;  % x0 is a row vector.
myfunction = fcnchk(myfunction);
dim=max(size(x0(1,:))); % dimension of the problem
% disp(dim)
%% set up adaptive parameters
alpha=1; beta=1+2/dim; gamma=0.75-0.5/dim; delta=1-1/dim;
%
% FX=feval(myfunction,x0,order);
[FX,I]=sort(FX);     % up
x_total=[x_total; xinit];
obj_total=[obj_total;FX];

    ct=dim+1;
    X=x0(I,:);
    %% Main iteration
    while max(max(abs(X(2:dim+1,:)-X(1:dim,:))))>=tol
        if ct>=max_feval
            break;
        end
        M=mean(X(1:dim,:));  % Centroid of the dim best vertices
        %%  reflection
        xref=(1+alpha)*M- alpha*X(dim+1,:);
        % check the bounds
        xref=check_bounds(xref,upbound,lowbound,step_size);
        Fref=feval(myfunction,xref);
        if Fref<FX(1)
            %% expansion
            xexp=(1+alpha*beta)*M-alpha*beta*X(dim+1,:);
            xexp=check_bounds(xexp,upbound,lowbound,step_size);
            Fexp=feval(myfunction,xexp);
            ct=ct+1;
            if Fexp < Fref
                X(dim+1,:)=xexp;
                FX(dim+1)=Fexp;
            else
                X(dim+1,:)=xref;
                FX(dim+1)=Fref;
            end;
            x_total=[x_total; X(dim+1,:)];
            obj_total=[obj_total; FX(dim+1)];
        else
            if Fref<FX(dim)
                %% accept reflection point
                ct=ct+1;
                X(dim+1,:)=xref;
                FX(dim+1)=Fref;
                x_total=[x_total; X(dim+1,:)];
                obj_total=[obj_total; FX(dim+1)];
            else
```

```
                if Fref<FX(dim+1)
                    %% Outside contraction
                    xoc=(1+alpha*gamma)*M-alpha*gamma*X(dim+1,:);
                    xoc=check_bounds(xoc,upbound,lowbound,step_size);
                    Foc=feval(myfunction,xoc);
                    if Foc<=Fref
                        ct=ct+1;
                        X(dim+1,:)=xoc;
                        FX(dim+1)=Foc;
                        x_total=[x_total; X(dim+1,:)];
                        obj_total=[obj_total; FX(dim+1)];
                    else
                        %% shrink
                        ct=ct+dim;
                        for i=2:dim+1
                            X(i,:)=X(1,:)+ delta*(X(i,:)-X(1,:));
                            X(i,:)=check_bounds(X(i,:),upbound,lowbound,step_size);
                            FX(i)=feval(myfunction,X(i,:));
                            x_total=[x_total;X(i,:)];
                            obj_total=[obj_total;  FX(i)];
                        end;
                    end;
                else
                    %% inside contraction
                    xic=(1-gamma)*M+gamma*X(dim+1,:);
                    xic=check_bounds(xic,upbound,lowbound,step_size);
                    Fic=feval(myfunction,xic);
                    if Fic<FX(dim+1)
                        ct=ct+1;
                        X(dim+1,:)=xic;
                        FX(dim+1)=Fic;
                        x_total=[x_total; X(dim+1,:)];
                        obj_total=[obj_total; FX(dim+1)];
                    else


                        %% shrink
                        ct=ct+dim;
                        for i=2:dim+1
                            X(i,:)=X(1,:)+ delta*(X(i,:)-X(1,:));
                            X(i,:)=check_bounds(X(i,:),upbound,lowbound,step_size);
                            FX(i)=feval(myfunction,X(i,:));
                            x_total=[x_total;X(i,:)];
                            obj_total=[obj_total; FX(i)];
                        end;
                    end;
                end;
            end;
        end
        [FX,I]=sort(FX);
        X=X(I,:);
    end
    xmin=X(1,:);
    fmin=FX(1);
```

```
aa=1;
switch aa
    case 1
        ParGen(posi(1),:)=xmin;
        ObjVals(posi(1),:)=fmin;
    case 2
        ParGen(posi(end),:)=xmin;
        ObjVals(posi(end),:)=fmin;
    case 3
        for jj=1:elite_ind
            if FX(jj)<ObjVals(posi(jj),:)
                ParGen(posi(jj),:)=X(jj,:);
                ObjVals(posi(jj),:)=FX(jj);
            end
        end
end
%% check bounds
function [inputs]=check_bounds(inputs,upbound,lowbound,step_size)
numv=length(inputs(1,:));
nump=length(inputs(:,1));
%%
for i=1:nump
    for j=1:numv
        inputs(i,j)=round(inputs(i,j)/step_size(j))*step_size(j);
        if inputs(i,j)<lowbound(j)
            inputs(i,j)=lowbound(j);
        end;
        if inputs(i,j)>upbound(j)
            inputs(i,j)=upbound(j);
        end;
    end
end
```

4. **Artificial bee colony (ABC) (Copyright (c) 2015, Yarpiz** www.yarpiz.com)

```
clc;
clear;
close all;

%% Problem Definition

CostFunction=@(x) Sphere(x);        % Cost Function

nVar=5;             % Number of Decision Variables

VarSize=[1 nVar];   % Decision Variables Matrix Size

VarMin=-10;         % Decision Variables Lower Bound
VarMax= 10;         % Decision Variables Upper Bound

%% ABC Settings

MaxIt=200;              % Maximum Number of Iterations

nPop=100;               % Population Size (Colony Size)

nOnlooker=nPop;         % Number of Onlooker Bees

L=round(0.6*nVar*nPop); % Abandonment Limit Parameter (Trial Limit)

a=1;                    % Acceleration Coefficient Upper Bound

%% Initialization

% Empty Bee Structure
empty_bee.Position=[];
empty_bee.Cost=[];

% Initialize Population Array
pop=repmat(empty_bee,nPop,1);

% Initialize Best Solution Ever Found
BestSol.Cost=inf;

% Create Initial Population
for i=1:nPop
    pop(i).Position=unifrnd(VarMin,VarMax,VarSize);
    pop(i).Cost=CostFunction(pop(i).Position);
    if pop(i).Cost<=BestSol.Cost
        BestSol=pop(i);
    end
end

% Abandonment Counter
C=zeros(nPop,1);

% Array to Hold Best Cost Values
BestCost=zeros(MaxIt,1);

%% ABC Main Loop

for it=1:MaxIt

    % Recruited Bees
    for i=1:nPop

        % Choose k randomly, not equal to i
        K=[1:i-1 i+1:nPop];
        k=K(randi([1 numel(K)]));

        % Define Acceleration Coeff.
        phi=a*unifrnd(-1,+1,VarSize);
```

```
            % New Bee Position
            newbee.Position=pop(i).Position+phi.*(pop(i).Position-pop(k).Position);

            % Evaluation
            newbee.Cost=CostFunction(newbee.Position);

            % Comparision
            if newbee.Cost<=pop(i).Cost
                pop(i)=newbee;
            else
                C(i)=C(i)+1;
            end

        end

        % Calculate Fitness Values and Selection Probabilities
        F=zeros(nPop,1);
        MeanCost = mean([pop.Cost]);
        for i=1:nPop
            F(i) = exp(-pop(i).Cost/MeanCost); % Convert Cost to Fitness
        end
        P=F/sum(F);

        % Onlooker Bees
        for m=1:nOnlooker

            % Select Source Site
            i=RouletteWheelSelection(P);

            % Choose k randomly, not equal to i
            K=[1:i-1 i+1:nPop];
            k=K(randi([1 numel(K)]));

            % Define Acceleration Coeff.
            phi=a*unifrnd(-1,+1,VarSize);

            % New Bee Position
            newbee.Position=pop(i).Position+phi.*(pop(i).Position-pop(k).Position);

            % Evaluation
            newbee.Cost=CostFunction(newbee.Position);

            % Comparision
            if newbee.Cost<=pop(i).Cost
                pop(i)=newbee;
            else
                C(i)=C(i)+1;
            end

        end

        % Scout Bees
        for i=1:nPop
            if C(i)>=L
                pop(i).Position=unifrnd(VarMin,VarMax,VarSize);
                pop(i).Cost=CostFunction(pop(i).Position);
                C(i)=0;
            end
        end

        % Update Best Solution Ever Found
        for i=1:nPop
            if pop(i).Cost<=BestSol.Cost
                BestSol=pop(i);
            end
        end

        % Store Best Cost Ever Found
        BestCost(it)=BestSol.Cost;

    % Display Iteration Information
    disp(['Iteration ' num2str(it) ': Best Cost = ' num2str(BestCost(it))]);

end

%% Results

figure;
%plot(BestCost,'LineWidth',2);
semilogy(BestCost,'LineWidth',2);
xlabel('Iteration');
ylabel('Best Cost');
grid on;
```

5. Simulated annealing (SA) (Copyright (c) 2015, Yarpiz www.yarpiz.com)

```
clc;
clear;
close all;

%% Problem Definition

CostFunction=@(x) Sphere(x);     % Cost Function

nVar = 5;               % Number of Decision (Unknwon) Variables

VarSize = [1 nVar];      % Decision Variables Matrix Size

VarMin = -10;            % Lower Bound of Decision Variables
VarMax =  10;            % Upper Bound of Decision Variables


%% SA Parameters

MaxIt=1000;      % Maximum Number of Iterations

MaxSubIt=20;     % Maximum Number of Sub-iterations

T0=0.1;        % Initial Temp.

alpha=0.99;      % Temp. Reduction Rate

nPop=10;         % Population Size

nMove=5;         % Number of Neighbors per Individual

mu = 0.5;        % Mutation Rate

sigma = 0.1*(VarMax-VarMin);     % Mutation Range (Standard Deviation)

%% Initialization

% Create Empty Structure for Individuals
empty_individual.Position=[];
empty_individual.Cost=[];

% Create Population Array
pop=repmat(empty_individual,nPop,1);

% Initialize Best Solution
BestSol.Cost=inf;

% Initialize Population
for i=1:nPop

    % Initialize Position
    pop(i).Position=unifrnd(VarMin, VarMax, VarSize);

    % Evaluation
    pop(i).Cost=CostFunction(pop(i).Position);

    % Update Best Solution
    if pop(i).Cost<=BestSol.Cost
        BestSol=pop(i);
    end

end

% Array to Hold Best Cost Values
BestCost=zeros(MaxIt,1);

% Intialize Temp.
T=T0;
```

```
for it=1:MaxIt

    for subit=1:MaxSubIt

        % Create and Evaluate New Solutions
        newpop=repmat(empty_individual,nPop,nMove);
        for i=1:nPop
            for j=1:nMove

                % Create Neighbor
                newpop(i,j).Position=Mutate(pop(i).Position,mu,sigma,VarMin,VarMax);

                % Evaluation
                newpop(i,j).Cost=CostFunction(newpop(i,j).Position);

            end
        end
        newpop=newpop(:);

        % Sort Neighbors
        [~, SortOrder]=sort([newpop.Cost]);
        newpop=newpop(SortOrder);

        for i=1:nPop

            if newpop(i).Cost<=pop(i).Cost
                pop(i)=newpop(i);

            else
                DELTA=(newpop(i).Cost-pop(i).Cost)/pop(i).Cost;
                P=exp(-DELTA/T);
                if rand<=P
                    pop(i)=newpop(i);
                end
            end

            % Update Best Solution Ever Found
            if pop(i).Cost<=BestSol.Cost
                BestSol=pop(i);
            end

        end

    end

    % Store Best Cost Ever Found
    BestCost(it)=BestSol.Cost;

    % Display Iteration Information
    disp(['Iteration ' num2str(it) ': Best Cost = ' num2str(BestCost(it))]);

    % Update Temp.
    T=alpha*T;

    sigma = 0.98*sigma;

end

%% Results

figure;
%plot(BestCost,'LineWidth',2);
semilogy(BestCost,'LineWidth',2);
xlabel('Iteration');
ylabel('Best Cost');
grid on;
```

Printed by Printforce, the Netherlands